Urban Aquaculture

Urban Aquaculture

Editor:

Dan Piestun, Ph.D.

Delve Publishing
www.delvepublishing.com

Urban Aquaculture

Editor: Dan Piestun, Ph.D.

708 3rd Avenue, 6th Floor
New York, NY 10017
United States of America

www.delvepublishing.com

ISBN: 978-1-68095-843-0

Library of Congress Control Number: 2016951596

About the Editor

Dan Piestun, Ph.D.

Dan Piestun (PhD) is currently a startup entrepreneur in Israel working on the interface of Agriculture and Biomedical Sciences and was formerly president-CEO of the National Institute of Agricultural Research (INIA) in Uruguay. Dan is a widely published scientist who has received many honours during his career including being a two-time recipient of the Amit Golda Meir Prize from the Hebrew University of Jerusalem, his areas of expertise includes stem cell molecular biology, plant and animal genetics and bioinformatics. Dan's passion for applied science and technological solutions did not stop him from pursuing a deep connection to the farmer, his family and nature. Among some of his interest and practices counts enjoying working as a beekeeper and onboard fishing.

Contents

Section III: Reproduction and Physiology

Section IV: Fish Health

Section VI: Fish postharvest

List of Contributors

GUI JianFang
State Key Laboratory of Freshwater Ecology and Biotechnology, Institute of Hydrobiology, Chinese Academy of Sciences, Wuhan 430072, China

ZHU ZuoYan
State Key Laboratory of Freshwater Ecology and Biotechnology, Institute of Hydrobiology, Chinese Academy of Sciences, Wuhan 430072, China

Kyoung-Duck Kim
Inland Aquaculture Research Center, National Fisheries Research & Development Institute, Jinhae 645-806, Korea

Sang Gu Lim
Inland Aquaculture Research Center, National Fisheries Research & Development Institute, Jinhae 645-806, Korea

Yong Jin Kang
Inland Aquaculture Research Center, National Fisheries Research & Development Institute, Jinhae 645-806, Korea

Kang-Woong Kim
Inland Aquaculture Research Center, National Fisheries Research & Development Institute, Jinhae 645-806, Korea

Maeng Hyun Son
Inland Aquaculture Research Center, National Fisheries Research & Development Institute, Jinhae 645-806, Korea

Meng-Yao Chen
Inland Aquaculture Research Center, National Fisheries Research & Development Institute, Jinhae 645-806, Korea

Ji-Dan Ye
Inland Aquaculture Research Center, National Fisheries Research & Development Institute, Jinhae 645-806, Korea

Wei Yang
Inland Aquaculture Research Center, National Fisheries Research & Development Institute, Jinhae 645-806, Korea

Kun Wang
Xiamen Key Laboratory for Feed Quality Testing and Safety Evaluation, Jimei University, Xiamen 361021, China

Till S. Harter
Aquaculture and Fisheries Group, Wageningen University, Wageningen, The Netherlands

Johan A. J. Verreth
Aquaculture and Fisheries Group, Wageningen University, Wageningen, The Netherlands

Leon T. N. Heinsbroek
Aquaculture and Fisheries Group, Wageningen University, Wageningen, The Netherlands

Johan W. Schrama
Aquaculture and Fisheries Group, Wageningen University, Wageningen, The Netherlands

Chao Ran
Key Laboratory for Feed Biotechnology of the Ministry of Agriculture, Feed Research Institute, Chinese Academy of Agricultural Sciences, Beijing, People's Republic of China

Lu Huang
Key Laboratory for Feed Biotechnology of the Ministry of Agriculture, Feed Research Institute, Chinese Academy of Agricultural Sciences, Beijing, People's Republic of China

Zhi Liu
Key Laboratory for Feed Biotechnology of the Ministry of Agriculture, Feed Research Institute, Chinese Academy of Agricultural Sciences, Beijing, People's Republic of China

Li Xu
Key Laboratory for Feed Biotechnology of the Ministry of Agriculture, Feed Research Institute, Chinese Academy of Agricultural Sciences, Beijing, People's Republic of China

Yalin Yang
Key Laboratory for Feed Biotechnology of the Ministry of Agriculture, Feed Research Institute, Chinese Academy of Agricultural Sciences, Beijing, People's Republic of China

Philippe Tacon
Société Industrielle Lesaffre, Phileo Lesaffre Animal Care, Marcq-en-Baroeul, France

Eric Auclair
Société Industrielle Lesaffre, Phileo Lesaffre Animal Care, Marcq-en-Baroeul, France

Zhigang Zhou
Key Laboratory for Feed Biotechnology of the Ministry of Agriculture, Feed Research Institute, Chinese Academy of Agricultural Sciences, Beijing, People's Republic of China

Tae-Hyun Yoon
College of Animal Life Sciences, Kangwon National University, Chuncheon 200-701, Korea

Dong-Hoon Lee
Gyenoggi Province Maritime and Fisheries Research Institute, Yangpyeong 476-841, Korea.
College of Animal Life Sciences, Kangwon National University, Chuncheon 200-701, Korea

Seung-Gun Won
College of Animal Life Sciences, Kangwon National University, Chuncheon 200-701, Korea

Chang-Six Ra
College of Animal Life Sciences, Kangwon National University, Chuncheon 200-701, Korea

Jeong-Dae Kim
College of Animal Life Sciences, Kangwon National University, Chuncheon 200-701, Korea

Ruchi Singh
National Bureau of Fish Genetic Resources, Canal Ring Road, Lucknow-226002, Uttar Pradesh, India

A. K. Singh
National Bureau of Fish Genetic Resources, Canal Ring Road, Lucknow-226002, Uttar Pradesh, India

Madhu Tripathi
Department of Zoology, University of Lucknow, Lucknow-226010, Uttar Pradesh, India.
National Bureau of Fish Genetic Resources, Canal Ring Road, Lucknow-226002, Uttar Pradesh, India

Paulino Martínez
Departamento de Genética, Facultad de Veterinaria, Universidad de Santiago de Compostela, Lugo, Spain

Ana M. Viñas
Departamento de Genética, Facultad de Biología, Universidad de Santiago de Compostela, Santiago de Compostela, Spain

Laura Sánchez
Departamento de Genética, Facultad de Veterinaria, Universidad de Santiago de Compostela, Lugo, Spain

Noelia Díaz
Institut de Ciències del Mar, Consejo Superior de Investigaciones Científicas, Barcelona, Spain

Laia Ribas
Institut de Ciències del Mar, Consejo Superior de Investigaciones Científicas, Barcelona, Spain

Francesc Piferrer
Institut de Ciències del Mar, Consejo Superior de Investigaciones Científicas, Barcelona, Spain

Lisa Liu
Department of Chemistry and Chemical Biology, College of Arts and Sciences, Cornell University, Ithaca, New York, United States of America

Chao Li
Department of Fisheries and Allied Aquacultures, Auburn University, Auburn, Alabama, United States of America

Baofeng Su
Department of Fisheries and Allied Aquacultures, Auburn University, Auburn, Alabama, United States of America

Benjamin H. Beck
United States Department of Agriculture, Agricultural Research Service, Stuttgart National Aquaculture Research Center, Stuttgart, Arkansas, United States of America

Eric Peatman
Department of Fisheries and Allied Aquacultures, Auburn University, Auburn, Alabama, United States of America

ZHANG QiYa
State Key Laboratory of Freshwater Ecology and Biotechnology, Institute of Hydrobiology, Chinese Academy of Sciences, University of the Chinese Academy of Sciences, Wuhan 430072, China

Wiboonluk Pungrasmi
Department of Environmental Engineering, Faculty of Engineering, Chulalongkorn University, Bangkok 10330, Thailand

Ratchadaporn Chaisri
Department of Environmental Engineering, Faculty of Engineering, Chulalongkorn University, Bangkok 10330, Thailand

Ekachai Malaphol
Center of Excellence for Marine Biotechnology, Department of Marine Science, Faculty of Science, Chulalongkorn University, Bangkok 10330, Thailand

Sorawit Powtongsook
Center of Excellence for Marine Biotechnology, Department of Marine Science, Faculty of Science, Chulalongkorn University, Bangkok 10330, Thailand
National Center for Genetic Engineering and Biotechnology, Pathum Thani 12120, Thailand

Joshua A. Haslun
Michigan State University - 288 Farm Lane, East Lansing, MI 48824 USA
Annis Water Resources Institute, Grand Valley State University - 131 Lake Michigan Center, 740 W. Shoreline Drive, Muskegon, MI 49441 USA

Eudes Correia
AgriLife Research Mariculture Laboratory - 4301 Waldron Rd., Corpus Christi, TX 78418 USA

Kevin Strychar
Annis Water Resources Institute, Grand Valley State University - 131 Lake Michigan Center, 740 W. Shoreline Drive, Muskegon, MI 49441 USA

Timothy Morris
AgriLife Research Mariculture Laboratory - 4301 Waldron Rd., Corpus Christi, TX 78418 USA

Tzachi Samocha
AgriLife Research Mariculture Laboratory - 4301 Waldron Rd., Corpus Christi, TX 78418 USA

Patrik J. G. Henriksson
Department of Industrial Ecology, Leiden University, 2333 CC Leiden, the Netherlands

Jeroen B. Guinée
Department of Industrial Ecology, Leiden University, 2333 CC Leiden, the Netherlands

René Kleijn
Department of Industrial Ecology, Leiden University, 2333 CC Leiden, the Netherlands

Geert R. de Snoo
Department of Industrial Ecology, Leiden University, 2333 CC Leiden, the Netherlands

Werner Kloas
Department of Ecophysiology and Aquaculture, Leibniz-Institute of Freshwater Ecology and Inland Fisheries, Müggelseedamm 310, 12587 Berlin, Germany
Department of Endocrinology, Institute of Biology, Humboldt University, Unter den Linden 6, 10099 Berlin, Germany
Albrecht Daniel Thaer-Institute of Agricultural and Horticultural Sciences, Department of Crop and Animal Science, Division of Biosystem Engineering, Humboldt University, Invalidenstraße 42, 10115 Berlin, Germany

Roman Groß
Department of Ecophysiology and Aquaculture, Leibniz-Institute of Freshwater Ecology and Inland Fisheries, Müggelseedamm 310, 12587 Berlin, Germany

Daniela Baganz
Department of Biology and Ecology of Fishes, Leibniz-Institute of Freshwater Ecology and Inland Fisheries, Müggelseedamm 310, 12587 Berlin, Germany

Johannes Graupner
Department of Ecophysiology and Aquaculture, Leibniz-Institute of Freshwater Ecology and Inland Fisheries, Müggelseedamm 310, 12587 Berlin, Germany

Henrik Monsees
Department of Ecophysiology and Aquaculture, Leibniz-Institute of Freshwater Ecology and Inland Fisheries, Müggelseedamm 310, 12587 Berlin, Germany

Uwe Schmidt
Albrecht Daniel Thaer-Institute of Agricultural and Horticultural Sciences, Department of Crop and Animal Science, Division of Biosystem Engineering, Humboldt University, Invalidenstraße 42, 10115 Berlin, Germany

Georg Staaks
Department of Biology and Ecology of Fishes, Leibniz-Institute of Freshwater Ecology and Inland Fisheries, Müggelseedamm 310, 12587 Berlin, Germany

Johanna Suhl
Department of Biology and Ecology of Fishes, Leibniz-Institute of Freshwater Ecology and Inland Fisheries, Müggelseedamm 310, 12587 Berlin, Germany
Albrecht Daniel Thaer-Institute of Agricultural and Horticultural Sciences, Department of Crop and Animal Science, Division of Biosystem Engineering, Humboldt University, Invalidenstraße 42, 10115 Berlin, Germany

Martin Tschirner
Department of Ecophysiology and Aquaculture, Leibniz-Institute of Freshwater Ecology and Inland Fisheries, Müggelseedamm 310, 12587 Berlin, Germany
Albrecht Daniel Thaer-Institute of Agricultural and Horticultural Sciences, Department of Crop and Animal Science, Division of Biosystem Engineering, Humboldt University, Invalidenstraße 42, 10115 Berlin, Germany

Bernd Wittstock
Department of Ecophysiology and Aquaculture, Leibniz-Institute of Freshwater Ecology and Inland Fisheries, Müggelseedamm 310, 12587 Berlin, Germany

Sven Wuertz
Department of Ecophysiology and Aquaculture, Leibniz-Institute of Freshwater Ecology and Inland Fisheries, Müggelseedamm 310, 12587 Berlin, Germany

Andrea Zikova
Department of Ecophysiology and Aquaculture, Leibniz-Institute of Freshwater Ecology and Inland Fisheries, Müggelseedamm 310, 12587 Berlin, Germany

Bernhard Rennert
Department of Ecophysiology and Aquaculture, Leibniz-Institute of Freshwater Ecology and Inland Fisheries, Müggelseedamm 310, 12587 Berlin, Germany

Peggy Schrobback
School of Economics and Finance, Queensland University of Technology, Brisbane, QLD, Australia

Sean Pascoe
CSIRO Oceans and Atmospheric Flagship, EcoSciences Precinct, Brisbane, QLD, Australia

Louisa Coglan
School of Economics and Finance, Queensland University of Technology, Brisbane, QLD, Australia

Chen-Hsing Yu
Department of Food Science, National Pingtung University of Science and Technology, 1 Shuefu Road, Neipu, Pingtung 91201, Taiwan

Tzou-Chi Huang
Department of Biological Science and Technology, National Pingtung University of Science and Technology, 1 Shuefu Road, Neipu, Pingtung 91201, Taiwan

Chao-Chin Chung
Department of Food Science, National Pingtung University of Science and Technology, 1 Shuefu Road, Neipu, Pingtung 91201, Taiwan

Hao-Hsun Huang
Department of Food Science, National Pingtung University of Science and Technology, 1 Shuefu Road, Neipu, Pingtung 91201, Taiwan

Ho-Hsien Chen
Department of Food Science, National Pingtung University of Science and Technology, 1 Shuefu Road, Neipu, Pingtung 91201, Taiwan

Odoli Cyprian
Kenya Marine and Fisheries Research Institute, Mombasa, Kenya
Faculty of Food Science and Nutrition, University of Iceland, Sæmundargotu 2, Reykjavík, Iceland

Hélène L. Lauzon
Matís ohf./Icelandic Food and Biotech R&D Vínlandsleið 12, Reykjavik, Iceland

Ragnar Jóhannsson
Matís ohf./Icelandic Food and Biotech R&D Vínlandsleið 12, Reykjavik, Iceland

Kolbrún Sveinsdóttir
Matís ohf./Icelandic Food and Biotech R&D Vínlandsleið 12, Reykjavik, Iceland

Sigurjón Arason
Faculty of Food Science and Nutrition, University of Iceland, Sæmundargotu 2, Reykjavík, Iceland
Matís ohf./Icelandic Food and Biotech R&D Vínlandsleið 12, Reykjavik, Iceland

Emilía Martinsdóttir
Matís ohf./Icelandic Food and Biotech R&D Vínlandsleið 12, Reykjavik, Iceland

Maria Lúcia Guerra Monteiro
Programa de Pós-Graduação em Higiene Veterinária e Processamento Tecnológico de Produtos de Origem Animal, Faculdade de Veterinária – UFF, Rua Vital Brazil Filho, n 64, 24.230-340, Santa Rosa, Niterói, Rio de Janeiro, Brazil

Eliane Teixeira Mársico
Departamento de Tecnologia de Alimentos, Faculdade de Veterinária – UFF, Rua Vital Brazil Filho, n 64, 24.230-340, Santa Rosa, Niterói, Rio de Janeiro, Brazil

Sérgio Borges Mano
Departamento de Tecnologia de Alimentos, Faculdade de Veterinária – UFF, Rua Vital Brazil Filho, n 64, 24.230-340, Santa Rosa, Niterói, Rio de Janeiro, Brazil

Claudia Emília Teixeira
Programa de Pós-Graduação em Higiene Veterinária e Processamento Tecnológico de Produtos de Origem Animal, Faculdade de Veterinária – UFF, Rua Vital Brazil Filho, n 64, 24.230-340, Santa Rosa, Niterói, Rio de Janeiro, Brazil

Anna Carolina Vilhena da Cruz Silva Canto
Programa de Pós-Graduação em Higiene Veterinária e Processamento Tecnológico de Produtos de Origem Animal, Faculdade de Veterinária – UFF, Rua Vital Brazil Filho, n 64, 24.230-340, Santa Rosa, Niterói, Rio de Janeiro, Brazil

Helio de Carvalho Vital
Centro Tecnológico do Exército – CTEx, Avenida das Américas, n 28.705, 23020-470, Guaratiba, Rio de Janeiro, Brazil

Carlos Adam Conte-Júnior
Departamento de Tecnologia de Alimentos, Faculdade de Veterinária – UFF, Rua Vital Brazil Filho, n 64, 24.230-340, Santa Rosa, Niterói, Rio de Janeiro, Brazil

Preface

Our goal is to provide a broad treatment of Intensive Aquaculture topics with a simple resource addressing from genetics, through reproductive behavior, fish health, up to water ecosystem dynamics as affected by human activity.

Audience

The comprehensive theme of the book is aimed at veterinary and agronomy professionals most likely involved in aquaculture and aquaponics. As technological advances in sustainable and specialized or accessory fish-raising farms evolve, there is an increasing need for permanent education for professionals working in the field and advanced university and technical students.

The following readers may benefit from this resource:

Students in professional veterinary training programs

Students in agronomy education programs

Product development and commercialization teams who are working towards aquaculture.

Organization

Meeting present and future protein needs. Urban Aquaculture is organized into six sections:

- "Fish Genetics." Presents a relevant genetics framework. Molecular basis and genetic improvement of economically important traits in aquaculture animals: The authors present candidate trait-related genes in the context of the recent advances in genomics and molecular cell biology. Further, the study explores into some breeding applications for coping with biotic and abiotic stresses.
- "Nutrition". This section presents an overview of some aspects of energy and nutrient economy of fish species. Effects of Dietary Protein and Lipid Levels on Growth and Body Composition of Juvenile Far Eastern Catfish (Silurus asotus): In order to tackle feed balance and efficiency, this research group studies the influence of dietary protein and lipid content on the growth and body composition of fish juveniles. Growth, Feed Utilization, and Blood Metabolic Responses to Different Amylose-amylopectin Ratio Fed Diets in Tilapia (Oreochromis niloticus): Studied whether growth performance and nutrient digestibility are affected by high-dietary amylose-amylopectin ratio.

  Isoenergetic Replacement of Fat by Starch in Diets for African Catfish (Clarias gariepinus): Effect on Water Fluxes in the Gastro-Intestinal Tract: This study focus on detecting a causal relationship, between dietary characteristics, the resulting development of chyme features in the gastro-intestinal tract and processes of water balance. A Comparison of the Beneficial Effects of Live and Heat-Inactivated Baker's Yeast on Nile Tilapia: Suggestions on the Role and Function of the Secretory Metabolites Released from the Yeast: Studied the influence of live and heat-inactivated yeast on weight gain, gut morphology, immune system, intestinal microbiota, and disease resistance of Nile tilapia. Effects of Dietary Supplementation of Magnesium Hydrogen Phosphate ($MgHPO4$) as an Alternative Phosphorus Source on Growth and Feed Utilization of Juvenile Far Eastern Catfish (Silurus asotus): Studied how phosphorus availability and supplementation affects eastern catfish growth and feed utilization.
- "Reproduction and Physiology." Presents a compilation of issues related to conventional and other innovative approaches to fish reproductive behavior and management. Melatonin Induced Changes in Specific Growth Rate, Gonadal Maturity, Lipid and Protein Production in Nile Tilapia Oreochromis niloticus: Studied the effect of melatonin on fish growth, the maturity of the gonad, total gonadal protein and lipid in Nile tilapia. Genetic architecture of sex determination in fish: applications to sex ratio control in aquaculture: reviews the information available on the genetic basis of sex determination in fish and in which fashion is sex ratio controlled in commercial aquaculture. The authors explore how sex determination varies across fish species and model organisms, as well as environmental clues affecting it.

  "Fish Health." This section is a compilation of articles stressing the importance of some critical concerns in pathogens and ways to contra arrest their potential effects. The research studies are: Short-Term Feed Deprivation Alters Immune Status of Surface Mucosa in Channel Catfish (Ictalurus punctatus): Analyzes how food deprivation strategies employed as part of a seasonal response to overproduction, or as a response to disease outbreak affects the immune system at the molecular level. Virus genomes and virus-host interactions in aquaculture animals: The authors review advances in the characterization of some important pathogenic viruses, like iridoviruses, herpesviruses, reoviruses, and rhabdoviruses, and virus–host interactions in aquaculture animals.

  Solid digestion-denitrification column in suspended solid and nitrate removal from recirculating aquaculture system: Brings an example of an alternative approach to removing the suspended solids from indoor recirculating aquaculture system by simple gravity sedimentation process. The system complements by degradation of the precipitated organic matter through anaerobic decomposition and further recycling of carbon to the fish pond.

- "Aquaculture Systems and Fish Management." Characterization of Bioflocs in a No Water Exchange Super-intensive System for the Production of Food Size Pacific White Shrimp Litopenaeus vannamei: This study evaluates the efficacy of a nursery trial to act as a natural probiotic community to grow L. vannamei juveniles to food-size in greenhouse-enclosed super-intensive recirculating aquaculture system under no water exchange. Life cycle assessment of aquaculture systems—a review of methodologies: Reviewed literature focusing on life cycle assessment of aquaculture systems in Asia and intended to explain the methodological alternatives taken, recognize possible data gaps, and contribute recommendations for future development within this field of research. A new concept for aquaponic systems to improve sustainability, increase productivity, and reduce environmental impacts: The aim of this new notion for aquaponics is to promote sustainability and productivity concomitant with dropping environmental emissions. The aquaponic method for emission-free tomato and fish production in greenhouses, consisting of two detached recirculating units: an aquaculture operation (RAS) for fish and a hydroponic unit for plants. Shape Up or Ship Out: Can We Enhance Productivity in Coastal Aquaculture to Compete with Other Uses?: In this research, the productivity of the Moreton Bay SRO oyster industry was assessed through measures such as technical efficiency, scale efficiency, allocative efficiency and capacity utilization.
- "Fish Postharvest." Comprises studies on managing hygiene and freshness of fish produce. The section includes the following subjects. Application of Highly Purified Electrolyzed Chlorine Dioxide for Tilapia Fillet Disinfection: The authors explored how to develop a novel electrolysis equipment to produce highly purified, low-cost ClO_2 to disinfect with water, while at the same time avoiding trihalomethane (THM) residuals on tilapia fillets.

Shelf life of air and modified atmosphere-packaged fresh tilapia (Oreochromis niloticus) fillets stored under chilled and superchilled conditions: Studied the joint effect of modified atmosphere packaging (50% CO_2/50% N_2) and superchilling on the sensory, microbiological, and chemical modifications during storage of fresh tilapia fillets. Influence of good manufacturing practices on the shelf life of refrigerated fillets of tilapia (Oreochromis niloticus) packed in modified atmosphere and gamma-irradiated: The objective of this study was to assess the effects of GMP (good manufacturing practices) on the shelf life of refrigerated fillets of Nile tilapia exposed to irradiation and preserved in a modified atmosphere. It is our wish that readers not only learn or deepen on principles of Urban Aquaculture but moreover renew and actualize their awareness in broad concepts and new findings on fish science and water ecosystems.

Editor
Dan Piestun, Ph.D.

INTRODUCTION

Urban aquaculture is the "farming" of aquatic organisms, including fish, molluscs, crustaceans and aquatic plants within the urban environment (rivers, ponds, lakes, canals). Farming implies some form of intervention in the rearing process to enhance production, such as the regular stocking, feeding, protection from predators, etc. Farming also implies individual or corporate ownership of the stock being cultivated. Urban aquaculture systems can de differentiated according to location, species, environment concerned, and the intensity of production. Urban aquaculture holds various advantages of being located close to urban markets and being able to access urban waste resources. But there are potential constraints associated with undertaking aquaculture in urban and peri-urban areas: producers are often unable to exert control over the prevailing hydrology and regulation of all inflows and discharges which may allow contaminants, predators and diseases to enter and nutrients, food resources and stock to escape. The physical openness of many extensive and semi-intensive aquaculture systems also means risks from airborne pollution and theft by poachers. For an extensive introduction to this subject please go to State of the Art. This paper assesses the status of urban aquaculture and the most important literature and knowledge sources are discussed providing a comprehensive overview that highlights challenges facing decision-makers, planners and stakeholders in developing policies, programmers and management strategies that facilitate sustainable, equitable and safe urban aquaculture. The prevailing characteristics of existing urban aquaculture activities are described and the associated benefits are discussed. The recognized constraints and emerging threats to urban aquaculture are then presented. Following this assessment important knowledge gaps and challenges facing planners, managers and other stakeholders are identified and potential approaches to deal with the issues raised are proposed.

Urban agriculture can be defined as the growing of plants and the raising of animals for food and other uses within and around cities and towns, and related activities such as the production and delivery of inputs, and the processing and marketing of products. Urban Agriculture is located within or on the fringe of a city and comprises of a variety of production systems, ranging from subsistence production and processing at household level to fully commercialized agriculture. Urban agriculture is generally characterized by closeness to markets, high competition for land, limited space, use of urban resources such as organic solid wastes and wastewater, low degree of farmer organization, mainly perishable products, high degree of specialization, to name a few. By supplying perishable products such as vegetables, fresh milk and poultry products, urban agriculture to a large extent complements rural agriculture and increases the efficiency of national food systems. Having reviewed the related literature, Mougeot (2000) concludes that the most important distinguishing character of urban agriculture is not so much its location - or any other of aforementioned criteria - but the fact that it is an integral part of the urban economic, social and ecological system: urban agriculture uses urban resources (land, labor, urban organic wastes, water), produces for urban citizens, is strongly influenced by urban conditions (policies, competition for land, urban markets and prices) and impacts the urban system (effects on urban food security and poverty, ecological and health impacts). Growing urban poverty, hunger and lack of formal employment, as well as the special opportunities that a city provides for farmers (including the growing urban demand for food, herbs and plants, proximity to markets and availability of cheap resources such as urban organic wastes and wastewater) have stimulated the development of a diversity of agricultural production systems in and around cities, often specialized in perishable products, such as green leafy vegetables, milk, eggs and meat, taking advantage of vacant open spaces in and around cities. Although some forms of urban and peri-urban agriculture are based on temporary use of vacant lands, urban agriculture as such is a permanent feature of many cities in developing as well as developed countries.

Potentials and Risks of Urban Agriculture Urban agriculture, as with other urban activities, has both positive and negative social, environmental, and economic impacts and externalities. Externalities are transformations of the physical or social environment caused, in this case, by urban farming beyond the limits of its productive system. The risks of urban agriculture relate to human health and the environment, if certain associated risks are not taken into account and consequently proper preventive and guiding measures are not implemented. The main potentials and risks of urban agriculture can be summarized as follows:

Urban food security and nutrition

The contribution of urban agriculture to food security and healthy nutrition is probably its most important asset. Food production in the city is in many cases a response of the urban poor to inadequate, unreliable and irregular access to food, and the lack of purchasing power. In urban settings, lack of income translates more directly into lack of food than in rural settings (cash is needed to buy food).The costs of supplying and distributing food from rural areas to urban areas, or importing food for the cities, is rising continuously, and its distribution within the cities is uneven. As a consequence, urban food insecurity will continue to increase (). In addition to enhanced food security and nutrition of the urban producers themselves, urban agriculture produces large amounts of food for other categories of the population. It was estimated that 200 million urban residents produce food for the urban market providing 15 to 20percent of the world's food.

Health

Improved access to fresh food as described above, directly relates to improved health. In most discussions, though, the health risks of urban agriculture are referred to in a negative context. A review of literature indicates that, although insight into the potential health risks of urban agriculture is growing, detailed information on the actual health impacts of urban agriculture is still scanty (which pretty much is the case today). However, the health risks associated to urban agriculture should be taken seriously and health impacts assessments and adequate regulatory and preventive measures should be put in place. But the fear of contaminated food and other health risks should not be exaggerated and need to be compared to those of rural agriculture.

The main health risks associated with urban agriculture can be grouped into the following categories:

- Contamination of crops with pathogenic organisms as a result of irrigation with water from polluted streams and insufficiently treated wastewater or the unhygienic handling of the products during transport, processing and marketing of fresh products;
- Spread of certain human diseases by mosquitoes and scavenging animals attracted by agricultural activities;
- Contamination of crops due to prolonged intensive use of agrochemicals;
- Contamination of soils and products with heavy metals due to traffic emissions and industrial effluents;
- Certain diseases transmitted to humans by keeping livestock in close proximity without proper precautions being taken.

Local economic development

Urban agriculture is an important source of income for a substantial number of urban households. In addition to income from sales of surpluses, farming households save on household expenditure by growing their own food. Since poor people generally spend a substantial part of their income (60 – 80 percent, Mougeot, 2005) on food, the savings can be substantial. Urban agriculture also stimulates the development of micro-enterprises for the production of necessary agricultural inputs (e.g. fodder, compost, and earthworms), the processing, packaging and marketing of products and the rendering of other services (e.g. animal health services, bookkeeping, transportation).

Social Inclusion and Gender

Urban agriculture may function as an important strategy for poverty alleviation and social integration of disadvantaged groups (such as immigrants, HIV-AIDS affected households, disabled people, female-headed households with children, elderly people without pension, youngsters without a job) by integrating them more strongly into the urban network, providing them with a decent livelihood and preventing social problems (Gonzalez Novo and Murphy 2000). Urban and peri-urban farms may also take on an important role in providing recreational and educational functions to urban citizens or play a role in landscape and biodiversity management. A large majority of the world's urban farmers are women (around 65percent). Promotion of urban agriculture without due attention to gender aspects may lead to a (further) increase in women's burden of work.

Urban environmental management

The disposal of waste has become a serious problem in many cities. Urban agriculture can contribute to solving this problem by turning urban wastes into a productive resource through compost production, permaculture, and irrigation with wastewater. Urban agriculture and forestry can also positively impact upon the greening of the city, the improvement of the urban micro-climate (wind breaks, dust reduction, shade) and the maintenance of biodiversity as well as the reduction of the ecological foot print of the city by producing fresh foods close to the consumers and thereby reducing energy use for transport, packaging, cooling, etc. Research in the Netherlands has shown that greenery around homes has a positive effect on people's health.

However, urban agriculture could contaminate local water sources if overly high inputs of chemical fertilizers and pesticides are used. The wastewater discharge from intensive poultry farms for instance can be high in micro-organisms and could contaminate drinking water supplies. Inappropriate farming practices may – under certain situations - lead to reduction of vegetation or siltation of water bodies. Because of the under-valuation of urban agriculture and the stiff competition for land, urban agriculture is often pushed back to the marginal areas within a city such as wetlands and hill-slopes, where it may harm the fragile ecosystems if not properly guided.

Central research issues

Today's centralized infrastructures for the provision of water or energy are prevailingly nonfunctional in the Western world. Their design and performance depend on fossil fuel and relate predominantly to linear and long-distance resource flows. Two key problems with this are in a nutshell, from a natural perspective, (1) a lack of landscape ecosystem service support, and, from a cultural perspective, (2) a lack of everyday perception and participative interaction.

Problems of Western Water Infrastructures and Urban Resource Management

1. Lacking landscape ecosystem service support for common livability: Although contemporary urban water infrastructures make use of biological-natural processes (micro-organismic water purification via mineralization, nitrification or mechanization, etc.), they do not actively regenerate landscape ecosystem services in watersheds. In fact, the majority of current water production and treatment technologies more or less undermine landscape processes as their basic ecosystem service providers. Collecting water from lakes, and wetlands often causes "landscape dehydration," and the release of sewage treatment plant outflow into natural water bodies causes eutrophication. The current standards of drinking water and wastewater management are primarily based on fossil fuel. The operation of either the centralized pump network or treatment process designs aiming to quickly break down organic water compounds (e.g. converting ammonia (NH_3) into nitrate (NO_3) via nitrification) requires high amounts of energy. On the other hand, nitrate – the final product of artificial fertilizer production – is technically produced via the energy-intensive Haber-Bosch process. Guterstam stresses it as follows: "The production of ammonia-nitrogen by the Haber-Bosch technique uses 1.5 liters of oil for each kilogram of extracted nitrogen from the atmosphere.
2. Lacking everyday perception and participative intervention (centralized vs. decentralized): The prevailing centralized water infrastructure technologies, comprising vast networks of pipes, pumps, transport lines, and treatment plants, integrate and rely on socio-cultural processes, such as proper maintenance and management. Nevertheless, they exclude experiences on a more human scale. This issue is linked to the current gap of the sensorial perception of water along with the missing relational experiences and interaction in open spaces and urbanized landscapes. The sociologist, Detlev Ipsen, stresses the phenomena of "hidden infrastructures" and "the invisible city." He refers to the ideal of the "hygienic city" arising in the 19th century when "water disappears from the senses of urbanites and responsibility of the citizens." Beyond improved public health, however, the increasing lack of the everyday tangibility to water in urban space leads to the loss of aesthetic, emotional and recreational values. Hence, it limits the human experience of the natural water processes in the city.

By facing the urban challenges of climate-responsive strategies mentioned, the primary focus of this research is on water-sensitive design-management approaches in the city and landscape. Referring to the close interrelation between water and climatic extremes, as mentioned in the introduction, it thereby opens a door to a broader climate change debate. This integrates issues of sustainable water and resource management, and the design of related infrastructures as well as stressing obvious interrelations, e.g. regarding place-based water-centric climate changes and climate chances. With this background, the next section introduces the central research approach of the dissertation to tackle the problems of prevailing Western water infrastructures and urban resource management.

Central research questions, objectives, limitations

The dissertation investigates so far underexposed facets of aquaculture – as a complementary part of agriculture – in the context of the city embracing an urban aquaculture. Accordingly, it explores water in its hybrid role as a natural landscape element and cultural human infrastructure from a global-local perspective. By widening the classic understanding of water farming, it integrates other urban aqua cultural practices of living and wellbeing such as those related to a city's shipping or swimming culture aquaculture typologies ranging from swimming gardens and ponds-and-pools to types of water farm greenhouse are the central subject of research. They are newly recognized and interpreted as specific types of blue-green infrastructure facing postindustrial urban challenges, such as fresh water and food provision, multifunctional infrastructure design at a human-scale or bathing quality in urban rivers.

In this regard, the three central questions within the normative framework of sustainability are:

1. What are the characteristic facets of urban aquaculture forming the water-based identities, morphologies and relationships of cities on a human scale?
2. What cross-cultural aquaculture types are there and what multifunctional (bluegreen) design and service potentials can they fulfill as urban building-blocks and specific blue-green infrastructures?
3. How can similar multifunctional infrastructures catalyze a water-sensitive transformation of cityscapes and contribute to a sustainable urban aquaculture in the 21st century?

Aquacultural typologies – Aquaponics – Hydroponics – Water-Farms – River Pools

Aquaculture typologies, such as swimming gardens, ponds-and-pools, and types of water-farm greenhouses are newly interpreted and investigated as specific blue-green infrastructures aquaculture blue-green infrastructures (in short, aquaculture infrastructures or aquaculture typologies) describe aquatic ecosystem technologies for natural-cultural life support and the renewal of livelihoods. The term water-farm addresses water-based farming technologies and practices, such as hydroponics (soil-less plant cultivation), aquaculture (fish and aquatic animal cultivation, e.g. crayfish, mussels) or aquaponics (combination of aquaculture and hydroponics). Similar to the meaning of urban aquaculture, the adjective aquaculture intends to address more than solely water-farm productive contents. Corresponding aquaculture practices comprise everyday activities from shipping, fishing, drinking, washing and bathing, to swimming. Affiliated aquaculture infrastructures that are investigated and recognized in the past and contemporary urban landscape context comprised, for example, bathing ships or ecologically cleaned urban river pools. Thus, aquaculture blue-green infrastructures were recognized. Aquaculture services additionally correlate with water-cultural rituals, and individual or societal water use in place-specific contexts. Their close intertwining water-natural landscape ecosystem processes reflect the hybrid natural-cultural infrastructure character.

Blue-green infrastructures – Everyday life infrastructures – Regenerative infrastructures

The overarching term blue-green infrastructures stands for water ecosystem-based technologies for regenerative hybrid (natural-cultural) life-support, including integrated water, food, energy, and other quality of life services. The perception of food, thereby, is linked to the "Cradle-to-Cradle concept." Food consists of various green materials, including living biomass and biological waste, which can be food or soil substrate for other living beings. The attribute blue stands for daily fresh water flows, and green for everyday vegetative ecosystem services (e.g. biomass production, evaporative cooling). Thus, they specify key landscape elements and process drivers and are similarly used in the concept of "blue and green water." They, also refer to Anglo-Saxon spatial planning terminology, such as green infrastructures (vegetative spaces), blue infrastructures (water spaces) or grey infrastructures (roads, sealed surfaces, etc.). Green infrastructure is interpreted as follows: "all natural, semi-natural, and artificial networks of multifunctional ecological systems within, around, and between urban areas, at all spatial scales." The combined terminology blue-green is used to express the intertwined process character of serving natural-cultural life-support in a hybrid manner. The notion of blue-green infrastructure in Germany is used, for example, within the regional landscape scale context of the Regionale 2010 in the Cologne region.

Nature-Culture

The interpretation of nature in this research context primarily addresses the landscape ecosystem dimension by referring to the life-supporting role of ecosystem services. The understanding of culture, first of all, stresses the everyday human dimension. It is affiliated with the interpretation of water culture by Ipsen et al., which is analogously used as culinary culture or living culture. Thus, the focus is on the everyday cultural dimension. It addresses various socio-cultural meanings of water as a natural element, technical infrastructure or aesthetic medium. Furthermore, it includes the design of using and handling water alongside the human-intentive production of meanings.

Urban water histories

Water as the cradle of cities: water-based identities, morphologies and aquacultural practices. Water as the cradle of the city has always played a substantial role. Most famous large cities are most commonly those that originated by the water: beside rivers, lakes or seashores. Water has shaped urban prosperity and morphologies, particularly along the waterfronts, by enabling urban trade and enterprise.

Waterways and natural water cycles have nourished the metabolism of the city. Spirn refers to the life-supporting role in the urban context by saying: "Water is the city's life blood: it drives industry, heats and cools homes, nurtures food, quenches thirst, and carries waste."111 She further points out the role of urban water creating a specific urban geo-hydrological pattern: "Taken together, urban activities, the density of urban form and the impervious materials of which it is built, the pattern of settlement and its relation to the natural drainage network, and the design of the drainage and flood control system produce a characteristic urban water regime."112

Furthermore, water sustains and facilitates the everyday infrastructure services of food and resource provision and, consequently, has shaped each city's culinary culture closely intertwined to its water culture. Fish and food caught in local rivers and the sea served as typical local dishes, and urban aquaculture practices of fishing or shipping have also created place based waterscapes for many centuries. In the face of 21st century urban resource challenges, famous old cities can, thus, be urban mirrors which can be reflected upon and learnt from. In addition to unveiling a city's cultural water history, the urban form reflects on the natural history of the water. Often similar in experience, it tells a common hi-story of natural-cultural changes in spaces and landscapes, of growth and shrinkage, of rise and fall, and of urban quality of life over time. Urban history's most often link to everyday stories of basic provision with common goods and services telling about people's food and water culture. Consequently, they reflect on characteristic facets of urban aquaculture, as well as a city's and its citizens' individual water based identity in the sense of a waterscape biography.

Policy Making and Action Planning on Urban Agriculture

When local authorities come to understand the role urban farmers can play in various urban policy areas rather than just prohibit or (temporarily) tolerate urban agriculture, they will seek to formulate policies that facilitate and regulate urban agriculture so as to maximize the benefits of urban agriculture whilst preventing or reducing the associated risks. Urban agriculture has a significant share in the food supply of many cities in the world, supplying especially perishable products such as vegetables, fresh milk and poultry products, but important differences are found between urban and rural agriculture. This has important consequences for the design of policies and support programmers.

Policy dimensions of Urban Agriculture

Dabbling describe three main policy dimensions of urban agriculture that may help to focus and differentiate policies regarding urban agriculture. A first, the social policy dimension refers mainly (but not exclusively) to subsistence oriented types of urban agriculture that form part of the livelihood strategies of (especially) the urban poor and that are mainly focused on producing food and medicinal plants for home consumption. In addition, the expenses of the family on food and medicines are reduced and minor cash income is generated from sales of surpluses. These households need additional income from other sources than agriculture to survive. Examples include home gardening, community gardening, institutional gardens at schools and hospitals, and open field farming at micro scale with low levels of investment. These systems show little direct profitability but have important social impacts such as social inclusion, poverty alleviation, community development, HIV-AIDS mitigation etc.

The economic policy dimension is more related to market oriented types of urban agriculture. Activities are undertaken (mainly) by small-scale family-based enterprises and (some) larger scale entrepreneurial farms run by private investors or producer associations. The activities not only refer to food production (e.g. irrigated vegetable production, stall-fed dairy production) but also to non-food products (medicinal and aromatic herbs, flowers, and ornamental plants). These commercial farms are embedded in a chain of small-scale and larger enterprises involved in inputs delivery (e.g. compost, fodder), processing and marketing.

These types of urban agriculture have more economic impact and higher profitability, but their externalities for the city and urban populations, especially those of the intensive larger scale enterprises, tend to be higher (e.g. risk of contamination of soils and water due to intensive use of agro chemicals, health risks due to use of contaminated water for irrigation and risks of zoonosis).

The ecological policy dimension refers to types of urban agriculture that have a multifunctional character: besides provision of food and generating income they play a role in environmental management and provide other services demanded by urban citizens: decentralized composting and reuse of organic wastes and wastewater (including nutrients), urban greening and improvement of the urban climate (shade, O_2, dust reduction), landscape management (parks, buffer zones, flood or earthquake prone or ecologically valuable zones that should be kept free from construction), provision of opportunities for leisure and recreational activities, water storage etc. In order to allow such a combination of functions, multi-functional agriculture will have to adopt agro-ecological production methods, link up with eco-sanitation and decentralized sustainable waste management, as well as with parks, nature and recreation planning and management. The analysis of the actual urban farming systems and the discussion of alternative policy measures could be directly linked with one or more of the above mentioned policy areas, depending upon the actual situation in the city and the existing policy priorities. A local government concerned about the growing food insecurity or the exclusion of certain categories of the population will probably focus on the "social" dimension of urban agriculture. Another city - mainly interested in local economic development- will focus on the "economic" dimension of urban agriculture or seek to stimulate subsistence farmers to move into the market sector. Local authorities mainly concerned about the poor urban living climate, the growing waste management problems, or the negative environmental and health effects of market-oriented urban agriculture, may concentrate on the 'environmental' dimension of urban agriculture, or promote a shift from high-input commercial agricultural production towards sustainable and multi-functional agriculture.

Strategies for the Development of Safe and Sustainable Urban Agriculture Although urban agriculture takes place under varying socio-political conditions and policy regimes urban policy makers and support institutions can substantially contribute to the development of safe and sustainable urban agriculture by:

- Creating a conducive policy environment and formal acceptance of urban agriculture as an urban land use;
- Enhancing access to vacant open urban spaces and the security of agricultural land use;
- Enhancing the productivity and economic viability of urban agriculture by improving access of urban farmers to training, technical advice, and credit;

SECTION I:
FISH GENETICS

CHAPTER 1 MOLECULAR BASIS AND GENETIC IMPROVEMENT OF AQUACULTURE ANIMALS

GUI JianFang and ZHU ZuoYan

Keywords: Aquaculture genome technology, somatic cell nuclear transfer, stem cell technology, reproduction-related gene, sex determination gene, growth-related gene, disease resistance-related genes, cold tolerance-related gene, hypoxia tolerance-related gene, genetic improvement.

Aquaculture has a history of about 4000 years, and China is the earliest country for raising fish in fishponds. However, it is only in the recent 50 years that aquaculture has rapidly developed and become a true production industry because artificial propagation technique has been successfully established since the early 1960s for the "four important domestic fish" including silver carp, bighead carp, grass carp and black carp [1] and aquaculture biotechnology for genetic breeding and disease control has been quickly developed since the 1980s [2]. In China, the annual aquaculture production capacity has increased to 38.3 million tons in 2010 from less than 1 million tons in 1960s, and possesses the fastest growing rate among agriculture and food industry. The aquaculture proportion has risen to more than 71.3% of the food fish, and reaches about 70% of the world's aquaculture production. Aquaculture industry has become a major power to promote sustainable, rapid and stable development of China's fishery, and played an important role in guaranteeing supply, stabilizing the market, ensuring national food security and promoting trade development. As an optimal method for exchanging animal proteins by minimum plant proteins, aquaculture has been believed as the most efficient technology by some authoritative economists, which is a major Chinese contribution to the world.

Just entering the 21st century, sustainable fisheries have become a common theme for the world, and the effect and contribution of aquaculture on world fish supplies have been widely recognized by many scientists in the developed countries[3,4]. Moreover, fish and other aquatic products as safer food sources have been well identified with most of the people [5]. At the same time, China's aquaculture is in a crucial change stage from traditional farming to modern industry, in which the severely lagged basic research has become the outstanding issue for restricting aquaculture development. Especially, the improved varieties were very few, and major aquaculture fish and shellfish were still de- pendent on the wild species. The frequent incidences of infectious diseases led to serious economic losses of about 15 billion RMB annually, and the unreasonable responses and non-standard medication to the diseases further resulted in drug residues, which affected quality and food safety of aquatic products, and in turn restricted the sustainable development of aquaculture [6]. In order to solve the above problems, the National Key Basic Research

Citation: Gui, J. & Zhu, Z., "Molecular basis and genetic improvement of aquaculture animals,". doi:10.1007/s11434-012-5213-0.

Program of China (973 Program) has successively supported 7 projects since 1999, such as "Fundamental research on the disease occurrence and disease resistance of the commercially important organisms in mariculture, "Studies on genetic and developmental biological basis for improvement of important aquaculture fishes", "Studies on the disease outbreak and immunological control of commercially important organisms in mariculture", "Investigation of crucial scientific issues in intensive aquaculture of freshwater ponds in China", "Basic studies on functional genomics and molecular design breeding in important aquaculture fishes", "The molecular basis for the important economic traits of the mollusks and the study on the molecular design breeding", and "Basic studies on outbreak mechanism and immunological control of main virus diseases in marine aquaculture animals", which have promoted significant advances in basic studies on genetic improvement and disease control of aquaculture animals. Especially in basic studies on genetic improvement in the aquaculture performance, the molecular basis of several economically important traits including reproduction, sex, growth, disease resistance, cold tolerance, and hypoxia tolerance have begun to elucidate, although these traits are highly complex and problematic. Here, we will attempt to retrace and review some significant breakthrough progress in the field for the recent 10 years, especially the major achievements obtained by Chinese scientists, and attempt to forecast developmental prospects for the future.

GENETIC IMPROVEMENT TECHNOLOGIES OF AQUACULTURE ANIMALS

Aquaculture Animals in Which Complete Genomes Have Been Sequenced

As complete genomes have been sequenced and released from human, model animals and plants, as well as several model fish including zebrafish (Danio rerio), fugu (Fugu rubripes), Tetraodon (Tetraodon nigroviridis), medaka (Oryzias latipes), and three-spined stickleback (Gasterosteus aculeatus) since this century, the genomes of some important aquaculture fish, such as channel catfish (Ictalurus punctatus), Nile tilapia (Oreochromis niloticus), Atlantic salmon (Salmo salar), rainbow trout (Oncorhynchus mykiss), gilthead seabream (Sparus aurata), European seabass (Dicentrarchus labrax), and Atlantic cod (Gadus morhua), have been also sequenced [7,8]. From 2010, the complete genomes of some main aquaculture animals including tongue sole (Cynoglossus semilaevis), Pacific oyster (Crassostrea gigas), large yellow croaker (Pseudosciaena crocea), orange-spotted grouper (Epinephelus coioides), common carp (Cyprinus carpio) and Japanese flounder (Paralichthys olivaceu) have been announced to be sequenced by Chinese fisheries-related institutes (Table 1). In 2011, Institute of Hydrobiology, Chinese Academy of Sciences and the collaborators have also started to sequence the complete genomes of grass carp (Ctenopharyngodon idellus), silver carp (Hypophthalmichthys molitrix) and bighead carp (Aristichthys nobilis) of the "four important domestic fish", and small RNA transcriptomes of bighead and silver carp have been characterized [9]. Therefore, we are convinced that the complete genome bioinformatics and the detailed molecule anatomy will provide significantly referent and even directive knowledge for their trait improvement and disease control of these major aquaculture animals.

Genome Technologies of Aquaculture Animals

In 2007, Prof. Zhanjiang Liu at Auburn University of USA, together with more than 40 aquaculture genomics experts in the world, edited a highly technical and comprehensive book Aquaculture Genome Technologies[10]. According to the book, the major aquaculture genome technologies include DNA marker technologies, such as amplified fragment length polymorphism (AFLP), microsatellites and single nucleotide polymorphism, novel DNA sequencing technologies, such as the SOLiD sequencing platform, the Solexa sequencing platform and the 454 sequencing platform, gene discovery technologies of expressed sequence tags and de novo sequencing of whole transcriptomes, genome mapping technologies, such as genetic linkage mapping, QTL mapping, radiation hybrid mapping and BACbased physical mapping, and genome expression analysis technologies, such as microarray technology and sequence tag-based technology.

Recently, numerous polymorphic DNA markers for aquaculture animals have been developed and significant advances have been made in China. In polyploid gibel carp (Carassius auratus gibelio), several significant molecular markers including transferrin alleles [11], SCAR [12], microsatellites [13], AFLP [14] and mitochondria DNA sequences [15,16] were identified, and dual modes of unisexual gynogenesis and sexual reproduction were revealed by these markers[17,18]. Significantly, the unique dual reproduction mode, as commented by Avise (2008), was the first documented instance in vertebrates [19], and the discovery firstly explained reproduction mechanism of high genetic diversity and an evolutionarily long-lived timescale of unisexual species [18]. In common carp (Cyprinus carpio), along with advances of polymorphic markers and genome sequencing, various density linkage maps were constructed [20–23], and used to locate cold tolerance [20] and muscle fiber-related [22] QTLs. A physical map was constructed from a BAC library containing a total of 92160 BAC clones with an average insert size of 141 kb [24,25]. In addition, a consensus linkage map was also constructed in grass carp [26]. As a required approach for positional gene cloning, physical mapping, and genome sequencing, several BAC libraries have been developed and characterized in other important aquaculture animals of China, such as gibel carp [27], zhikong scallop (Chlamys farreri)[28], pacific white shrimp (Litopenaeus vannamei) [29], grass carp [30], and halfsmooth tongue sole (Cynoglossus semilaevis) [31].

Table 1: Aquaculture animals that have been announced to be completely sequenced in China

Species	Genome size	Predicted gene number	Mainly announced institute	Announced time
Tongue sole (*Cynoglossus semilaevis*)	520 Mb	More than 20000	Yellow Sea Fisheries Research Institute, Chinese Academy of Fishery Sciences	August 2010
Pacific oyster (*Crassostrea gigas*)	8 Gb	More than 20000	Institute of Oceanology, Chinese Academy of Sciences	August 2010
Large yellow croaker (*Pseudosciaena crocea*)	750 Mb	Data not given	Zhejiang Ocean University	January 2011
Orange-spotted grouper (*Epinephelus coioides*)	1.1 Gb	More than 20000	Sun Yat-Sen University	March 2011
Common carp (*Cyprinus carpio*)	1.7 Gb	Data not given	The Centre for Applied Aquatic Genomics, Chinese Academy of Fishery Sciences	May 2011
Japanese flounder (*Paralichthys olivaceu*)	620 Mb	About 23000	Yellow Sea Fisheries Research Institute, Chinese Academy of Fishery Sciences	December 2011

Somatic Cell Nuclear Transfer and Stem Cell Technologies in Fish

Nuclear transfer has been successfully applied to fish genetic breeding for about half a century in China [32], and their developmental potentiality and reprogramming have been observed from the transplanted nuclei[33,34]. During this century, the cloned zebrafish were obtained by nuclear transfer from long-term-cultured cells [35], and the cytoplasmic impact was analyzed in cross-genus cloned fish derived from transgenic common carp (Cyprinus carpio) nuclei and goldfish (Carassius auratus) enucleated eggs [36]. Moreover, some differentially expressed genes were identified from the cloned zebrafish embryos at the dome stage [37], and the critical developmental stages for high integration efficiency were determined in zebrafish [38,39]. Stem cell culture and application have been rapidly developed in model fishes of medaka and zebrafish [40,41], and some stem cell technologies, such as gene targeting, germ cell transplantation and semi-cloning by nuclear transfer, have been successfully formed because of specific potentials of fish germ stem cells and embryonic stem cells [42,43]. Especially, Yi et al. [44] used the semi-cloning technology to create the semi-cloned female medaka which shows normal fertility and germline transmission over three generations. This approach represents a significant development for germline transmission of cultured cells, and will be a potential reproduction technology to combat infertility in vertebrates including human. Additionally, primordial germ cells (PGCs) and spermatogonial stem cells could be transplanted between two different salmonid species, and resulted in production of donor-derived offspring [45–47]. In combination with natural and artificial polyploid fishes with special reproduction modes [18,48], the surrogate approach might have significant values for producing polyploidy stocks in aquaculture or for restoring an endangered species in wild life conservation.

MOLECULAR BASES FOR ECONOMIC TRAITS OF AQUACULTURE ANIMALS

Reproduction Trait and Candidate Reproduction Related Genes

In comparison with mammals, fish reproduction modes and strategies exhibit abundant diversities and extreme particularities[49,50]. Owing to the diversities and particularities, much attention has begun to be paid to some important fish with special reproduction modes. Gibel carp, because of the polyploid background and dual reproduction modes of gynogenesis and sexuality, has become a promising fish for screening and isolating reproduction-related genes that are involved in primordial germ cell formation, oocyte maturation and egg fertilization. Using suppression subtractive hybridization and protein purification [51,52], some important genes, such as germ plasm markers vasa [53] and Dazl [54], oocyte maturation-related factors cyclin A2 [55], C-type lectin [52], C1q-like [56–58], histone H2A variant h2af1o [59], and maternal-effect factor spindlin [60], have been identified and functionally characterized from gibel carp. For example, Sun et al. [61] have observed oocytespecific expression pattern and dynamic distribution of Spindlin during oocyte maturation and egg fertilization, demonstrated its association and interaction with β-tubulin and spindle, and found that its elimination destroys spindle assembly, and thereby disturbs the oocyte-to-embryo transition and first cleavage, which provided the first direct evidence for the critical oocyte-to-embryo transition function of Spindlin in vertebrates.

It is well known that fish reproduction is regulated and controlled by a cascade of hormones along the hypothalamus-pituitary-gonad (HPG) axis. In this cascade, gonadotropin-releasing hormones (GnRHs) produced by hypothalamus are found to be the primary neuropeptides, and gonadotropic hormones (GtHs) synthesized and released by pituitary are found to lead ovaries or testicles to produce mature eggs or sperms, whereas dopamine (DA) produced by pituitary has been found to exhibit an inhibiting effect on the functions of GnRHs[62–64]. Multiple GtHs have been reported by Chinese scientists in groupers[65] and Chinese sturgeon (Acipenser sinensis) [66]. In common carp, two GnRH variants have been isolated [67]. Significantly, when common carp sGnRH expression was inhibited by sGnRH antisense, the plasma GtH-II level was significantly reduced, and the gonad development was severely diminished or hampered [68]. Moreover, Xu et al. [69] used the gonad hampered transgenic fish as a model for defining global gene expression changes of the hypothalamic-pituitarygonadal axis, and revealed 9, 28, and 212 differentially expressed genes respectively in hypothalamus, pituitary, and ovary, of which 87 genes were novel. Obviously, this study not only indicated the progressive scaling-up effect of hypothalamic sGnRH antisense acting on the pituitary and ovary, but also provided new insights with respect to the identification of reproduction-related genes and the understanding of the molecular mechanisms and regulative pathways in fish reproduction system. Recently, Zhou et al. (unpublished data) have also conducted a detailed exploration and functional annotation of hypothalamus and pituitary transcriptomes of mature gibel carp by Roche 454 sequencing. By this way, 878430 high quality sequences (average sequence length of (480±50) bp) of hypothalamus and pituitary were assembled into 51111 contigs and 113046 singletons, resulting in 23012 unigenes. Moreover, RNA-Seq approach was applied to analyze the expression level difference of the 23012 unigenes between mature female and male hypothalamus and pituitary. The detailed report will be soon coming.

Sex Determination and Candidate Sex Determination Genes

Fishes are an attractive group of organisms for studies on vertebrate sex determination, because they include various types of sexuality from hermaphroditism to gonochorism and from environmental to genetic sex determination. Fishes can provide important insight into the plasticity of the sex determination process in vertebrates, and can be used to explore the cascade regulatory pathways and their evolutionary conservation of upstream or downstream function of sex determination genes [70,71]. In 2002, the DMY gene, a recent duplicate of DMRT1 (Drosophila doublesex and Caenorhabditis mab-3-related transcription factor 1) in the Japanese medaka, was identified as the first sex-determining gene among non-mammalian vertebrates [72], but it was quickly demonstrated that DMY is not the universal primary sex determining gene in fish [73]. Significantly, DMRT1, the first conserved zinc finger-like DNA binding domain (DM domain) gene identified in vertebrates, has also been shown to be intimately linked to male gonadogenesis and further differentiation among almost all studied fishes including gonochoristic and hermaphroditic species [74]. As a group of protogynous hermaphrodite fishes that the gonad development undergoes transition from ovary to intersexual gonad and then to testis, and cells at different stages during oogenesis and spermatogenesis can be synchronously observed in the transitional gonads, groupers have been used as a good model for mining sex determination or sex differentiation-related genes. Xia et al. [75] have found that grouper DMRT1 is specifically expressed in spermatogenic cells during female-to-male sex reversal, and the spermatogenic cell-specific and dynamic expression pattern was revealed to be basically similar in several sex reversal fishes including protogynous wrasse (Halichoeres tenuispinis)[76] and rice field eel (Monopterus albus) [77] as well as protandrous black porgy (Acanthopagrus schlegeli)[78,79] and gilthead seabream (Sparus aurata)[80].

SOX3 has been suggested to play significant roles in sex determination and differentiation in vertebrates, but the evidence is insufficient and controversial [81]. In grouper gonads, some special cysts synchronously contain primordial germ cells, oogonia and spermatogonia, and SOX3 is expressed very early and can be detected in the nuclei of some primordial germ cells along the germinal epithelium, which provide an excellent chance to examine the functional roles of SOX3. The detailed observation and analysis have indicated that when the SOX3-positive primordial germ cells develop toward oogonia and then oocytes, SOX3 expression continues to increase greatly, whereas, as the SOX3-positive primordial germ cells develop toward spermatogonia, SOX3 expression is stopped and expelled from the nuclei of spermatogonia. The data confirm that SOX3 is involved in oogonium differentiation and oocyte development [82].

In hermaphrodite red-spotted grouper and rice field eel [83,84], the gonad-specific aromatase (cyp19a1a) had been also demonstrated to be expressed by follicular cells of follicular layer around oocytes. The promoter activity analysis showed a number of potential binding sites for some transcriptional factors, such as SOX5, GATA gene family, CREB, AP1, FOXL1, C/EBP, ARE and SF-1, and most of the motifs are conserved among the studied fish species. Recently, a hypothesis proposes that the gonadal aromatase upregulation is necessary not only for triggering but also for maintaining ovarian differentiation, and the down-regulation is the only necessary step for inducing a testicular differentiation pathway [85].

Actually, several important genes, such as Fushi Tarazu factor-1 (FTZ-F1), Sry-related HMG-box gene 9 (SOX9), Wilms tumor gene 1 (WT1), anti-mullerian hormone (AMH), DMRT1 and the zinc finger transcription factor GATA binding proteins 4 (GATA4), have been supposed as candidate sex determination genes, and a hypothetical regulation hierarchy for sex determination and differentiation has been suggested in the hermaphroditic groupers[86]. In mammals, primary sex determination genes have been divided into 4 classes, in which class I comprises Y genes that have a male-specific effect required for spermatogenesis, class II are X genes that are expressed at a higher level in females than in males by virtue of the 2:1 ratio, class III are X genes that receive a parental imprint, and class IV are sex-specific regions of heterochromatin that exert effects on the epigenetic status of the rest of the genome [87]. Of course, sex determination mechanisms are also intricate in fishes, because fishes, like other aquatic animals of amphibians and reptiles, include male or female heterogametic genetic sex determination with or without strongly differentiated sex chromosomes [88,89]. A recent report indicates that sex-chromosome homomorphy in European tree frogs does not result from a recent turnover but is maintained over evolutionary timescales by occasional X-Y recombination, implicating that seemingly ever-young sex chromosomes may thus carry old-established sex-determining genes [90]. Moreover, B chromosomes in addition to standard chromosomes are also found to have a functional effect on sex determination in one cichlid species (Lithochromis rubripinnis) from Lake Victoria [91], which implies that a portion of sex chromosomes has been derived from B chromosomes. Along with the successive revelation of complete genome sequences in some main aquaculture animals and the quick development of next-generation sequencing techniques, more sex determination genes and the underlying mechanisms will be identified and clarified as performed previously in human and chimpanzee [92].

Growth Trait and Candidate Growth-Related Genes

Growth trait is one of the most valuable traits of interest in genetic improvement of aquaculture animals, because growth rate increase can directly reduce aquaculture costs and lead to the benefit increase. Since growth hormone (GH) gene was identified, many important growth-related genes have been demonstrated to play critical roles in growth regulation of vertebrates. Numerous candidate genes are correlated with growth, including main growth-related genes within the somatotropic axis, such as GH, growth hormone receptor (GHR), insulin-like growth factors (IGF-I and -II), growth hormone-releasing hormone (GHRH), Leptin, and growth hormone inhibiting hormone (GHIH or somatostatin), important transforming growth factor genes expressed in muscle tissues, such as myostatin (MSTN) and myogenic regulatory factors (MRFs), and other possible candidate genes have been recommended in fish through reviewing the livestock and fish literatures [93], and the regulative mechanisms of muscle growth have been explored by introducing different growth models including compensatory growth, genetic manipulation and in vitro tissue culture [94].

Recently, several important growth-related genes have been investigated by Chinese fish endocrinologists. For example, multiple somatostatin and somatostatin receptor genes were characterized in the Chinese sturgeon (Acipenser sinensis) [95] and in groupers [96–99]. Orangespotted grouper orexin was found to be involved in feeding and orexin-A was a stimulator of NPY mRNA expression in vivo and in vitro [100]. Zebrafish ghrelin was demonstrated to play an essential role in GH expression during adenohypophysis development [101]. Follistatin 1 (Fst1) overexpression was revealed to promote muscle growth in the transgenic zebrafish [102].

Disease Resistance Trait and Candidate Disease Resistance-Related Genes

Along with rapidly growing and highly intensifying in the past 30 years, many aquaculture species have been threatened by some serious diseases caused by viruses, bacteria, fungi, parasites or other undiagnosed and emerging pathogens. Especially in infectious virus diseases, a lot of pathogenic viruses were identified from freshwater and marine culture animals[103], and the complete genomes of several viruses, such as grass carp hemorrhage virus (GCHV)[104], shrimp white spot syndrome virus (WSSV) [105], mandarin fish infectious spleen and kidney necrosis iridovirus (ISKNV) [106], and flounder lymphocystis disease virus isolated in China (LCDV-C) [107], were sequenced in China. Moreover, their structure proteins were characterized [108–111], and some of them were functionally analyzed [112,113].

In order to identify innate antiviral genes that protect fish from some virus infections, suppressive subtractive hybridization technology has been used to isolate virally induced genes from UV-inactivated GCHV-infected CAB cells [114]. Through this model system, a series of interferon (IFN) induced genes have been cloned and characterized, and great progress has been made in identifying fish antidisease genes involved in IFN antiviral response [115–120]. Significantly, some novel interferon-inducible genes have been found [121], and their functional roles and signaling pathways have been well understood in fish [122–124]. Additionally, several candidate genes, such as PKR [125], PKR-like (or called PKZ) [126], β-Defensin [127], TRBP [128], C-type lectin [129], Enzyme E2 [130], MDA5 and LGP2 [131], have been revealed to play crucial roles in the antiviral defense response of fish and shrimp.

Cold Tolerance Trait and Candidate Cold Tolerance Related Genes

Cold tolerance is an especially attractive trait in fish, and a recent experiment evidence suggests that some fishes might carry sufficient genetic variation to adapt to temperature and the cold tolerance should be under strong selection [132]. To identify candidate cold tolerance-related and cold change-related genes, a large-scale EST sequencing of multi-tissues including brain, liver, head kidney, and ovary had been performed in Antarctic notothenioid (Dissostichus mawsoni), and some important cold-specific expressed genes had been identified by comparative transcriptome analyses between D. mawsoni and non-notothenioid warm-water teleost fishes [133,134]. Moreover, the molecular origination of antifreeze protein III (AFP III) was demonstrated to come from a duplicate copy of sialic acid synthase (SAS) gene, supporting the evolutionary mechanism by neofunctionalization under escape from adaptive conflict [135].

Hypoxia Tolerance Trait and Candidate Hypoxia Tolerance-Related Genes

Variable hypoxia tolerance has been observed in aquatic animals including reptiles, amphibians, and fishes [136,137], but its molecular mechanisms remain largely unknown. Since Carassius species have been found to survive many days or even several months of anoxia at low temperature, Carassius auratus blastulae embryonic cells have been used to screen gene expression changes responding to hypoxia stress by suppression subtractive hybridization (SSH). Through this approach, a total of 211 differentially expressed genes are identified, in which most of them are reported for the first time from hypoxia-treated teleosts[138].

Moreover, several potential hypoxia tolerance-related genes have been characterized and functionally analyzed in fish. For example, heme oxygenase-1 (HO1) has been demonstrated to have protective role in fish under hypoxic stress [139]. Grass carp insulin-like growth factor-binding protein 1 (IGFBP1) has been revealed to play a key role in mediating the hypoxia-induced embryonic growth retardation and developmental delay [140]. A novel function of p53 that directly suppresses BNIP3 expression to protect against hypoxia-induced cell death has been found in model fish zebrafish [141]. Also in zebrafish, a stable transgenic fish has been created by expressing the Vitreoscilla hemoglobin (vhb) gene driven by common carp beta-actin promoter, and the expressed VHb can partially alleviate the hypoxia stress response to improve the survival rate has been confirmed by comparing the expression profiles of the vhb transgenic zebrafish [142].

APPLICATION OF GENETIC IMPROVEMENT IN AQUACULTURE ANIMALS

Genome technologies, especially molecular markers, have been designed to apply to breeding program and genetic improvement of main aquaculture animals including rainbow trout, Atlantic salmon, tilapia, channel catfish and ornamental fishes in the world, and their potential values have been evaluated by aquaculture geneticists[49,143–145].

In China, several nuclear genome markers, such as microsatellite patterns, AFLP profiles and transferrin alleles, and one cytoplasmic marker, the whole mitochondrial genome sequence, have been used to identify genetic organization of a novel clone A+ created by sexual mating between clone D female and clone A male in polyploid gibel carp [18], which have demonstrated that the novel clone is a nucleo-cytoplasmic hybrid between the known clones A and D, because it contains an entire nuclear genome from the paternal clone A and a mtDNA genome (cytoplasm) from the maternal clone D [146]. Therefore, it is suggested to be formed via androgenesis of clone A sperm in clone D ooplasm. Significantly, the selected nucleo-cytoplasmic hybrid female still maintains its gynogenetic ability, and thereby produces a novel clone variety "CAS III" that has been approved by National Certification Committee for Aquatic Varieties. Owing to the significant growth superiority, the novel clone variety has been propagated by gynogenesis in more than 10 hatcheries, and their offspring have been cultured throughout China [146]. Moreover, the finding will be of great significance for exploiting the genetic breeding approaches in gibel carp [18], because the nucleo-cytoplasmic hybrid clone escapes the genetic and developmental destruction caused by drastic treatments of irradiation and physical shocks in induced androgenesis and nuclear transplantation [146].

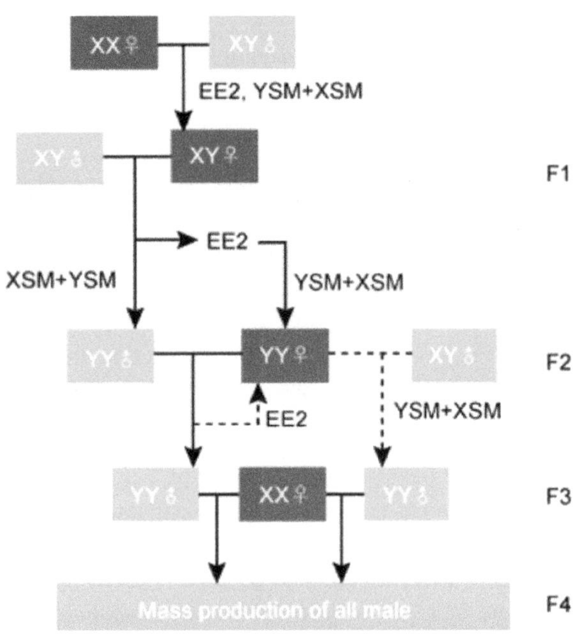

Figure 1: A schematic diagram of mass production of all-male populations through an integration sex control technique of female hormone 17α-ethinyloestradiol (EE2) treatment and genetic identification of Y chromosome-specific marker (YSM) and X chromosome-specific marker (XSM) in aquaculture fish

Additionally, sex-specific molecule markers have been developed and applied in China. Based on artificial propagation families, and gynogenetic XX, XY and YY individuals obtained previously [147], two pairs of Y chromosome- and X chromosome-specific markers were isolated from yellow catfish (Pelteobagrus fulvidraco) by AFLP screening, which thereby develops a Y- and X-specific allele marker-assisted sex control technique for mass production of all-male populations [148]. Using this technique, all-male yellow catfish has been approved as a novel variety "yellow catfish all-male No. 1" by National Certification Committee for Aquatic Varieties. Because yellow catfish males grow faster than females, and result in an ultimate size difference of about two to threefold, the all-male yellow catfish has been widely used for commercial production. Recently, all-female flounder has been also recognized as a novel variety for aquaculture in China. According to the developments, we have proposed an integration sex control technique for mass production of all-male populations in aquaculture fish [148]. As shown in Figure 1, the XY physiological females are selected by the Y chromosome-specific marker (YSM) and X chromosome-specific marker (XSM) from sex reversal progeny through 17α-ethinyloestradiol (EE2) treatment, and about 25% YY super-males can be obtained from the mating progeny of the reversed XY female and normal XY male by YSM and XSM identification. Then, the YY physiological females can be induced from the EE2 treated YY fingerlings, and YY super-males can be produced continually from the mating of the YY physiological females and YY super-males. Subsequently, the YY super-males can be used to mate with XX females for commercial mass production of all-XY males.

CONCLUSION AND PERSPECTIVE

Along with complete genome sequence anatomy of some main aquaculture animals, ongoing development of genome technologies, somatic cell nuclear transfer and stem cell technologies, as well as identification and function analysis of candidate genes relative to economically important traits including reproduction, sex, growth, disease resistance, cold tolerance and hypoxia tolerance, genetic improvement research in aquaculture animals has begun to enter a new era for molecular design breeding. The future issues are mainly aimed at understanding gene regulation and control networks that are related to reproduction, growth, disease resistance, cold tolerance and hypoxia tolerance, at identifying the relative functional genes and molecular markers, and at creating novel breeding materials and technologies, which can be used to develop the theoretical strategies and feasible approaches for molecular design breeding in aquaculture animals. It will establish the theoretical and technical foundation for breeding new varieties with high yield and quality, and will make greater contribution for sustainable fisheries and biotechnological innovation.

REFERENCES

1. Wu H W, Zhong L. Progress and achievement on artificial proliferation of grass carp, black carp, silver carp and big head carp in China. Chin Sci Bull, 1964, 9: 900–907
2. Wu C J, Gui J F. Fish Genetics and Breeding Engineering (in Chinese). Shanghai: Shanghai Scientific and Technical Publishers, 1999
3. Naylor R L, Goldburg R J, Primavera J H, et al. Effect of aquaculture on world fish supplies. Nature, 2000, 405: 1017–1024
4. Pauly D, Christensen V, Guénette S, et al. Towards sustainability in world fisheries. Nature, 2002, 418: 689–695
5. James H T, Geoff L A. Fishes as food: aquaculture's contribution. EMBO Rep, 2001, 21: 958–963
6. Gui J F. The present and future of studies on genetic and developmental biological basis for improvement of important aquaculture fishes. Life Sci, 2005, 17: 112–118
7. Sarropoulou E, Nousdili D, Magoulas A, et al. Linking the genomes of nonmodel teleosts through comparative genomics. Mar Biotechnol, 10: 227–233
8. Star B, Nederbragt A J, Jentoft S, et al. The genome

sequence of Atlantic cod reveals a unique immune system. Nature, 2011, 477: 207–210

9. Chi W, Tong C, Gan X, et al. Characterization and comparative profiling of MiRNA transcriptomes in bighead carp and silver carp. PLoS ONE, 2011, 6: e23549
10. Liu Z J. Aquaculture Genome Technologies. Oxford, UK: Blackwell Publishing, Ames, IA, 2007
11. Yang L, Gui J F. Positive selection on multiple antique allelic lineages of transferrin in the polyploid *Carassius auratus*. Mol Biol Evol, 2004, 21: 1264–1277
12. Zhou L, Wang Y, Gui J F. Molecular analysis of silver crucian carp (*Carassius auratus gibelio* Bloch) clones by SCAR markers. Aquaculture, 2001, 201: 219–228
13. Guo W, Gui J F. Microsatellite marker isolation and cultured strain identification in*Carassius auratus gibelio*. Aquaculture Int, 2008, 16: 497–510
14. Wang D, Mao H L, Peng J X, et al. Discovery of a male-biased mutant family and identification of a male-specific SCAR marker in gynogenetic gibel carp *Carassius auratus gibelio*. Prog Nat Sci, 2009, 19: 1537–1544
15. Li F B, Gui J F. Clonal diversity and genealogical relationships of gibel carp in four hatcheries. Anim Genet, 2008, 39: 28–33
16. Jakovlic I, Gui J F. Recent invasion and low level of divergence between diploid and triploid forms of *Carassius auratus* complex in Croatia. Genetica, 2011, 139: 789–804
17. Zhou L, Wang Y, Gui J F. Genetic evidence for gonochoristic reproduction in gynogenetic silver crucian carp (*Carassius auratus gibelio* Bloch) as revealed by RAPD assays. J Mol Evol, 2000, 51: 498–506
18. Gui J F, Zhou L. Genetic basis and breeding application on clonal diversity and dual reproduction modes in polyploid *Carassius auratus gibelio*. Sci China Life Sci, 2010, 53: 409–415
19. Avise J C. Clonality: The Genetics, Ecology, and Evolution of Sexual Abstinence in Vertebrates. USA: Oxford University Press, 2008
20. Sun X W, Liang L Q. A genetic linkage map of common carp (*Cyprinus carpio* L.) and mapping of a locus associated with cold tolerance. Aquaculture, 2004, 238: 165–172
21. Cheng L, Liu L, Yu X, et al. A linkage map of common carp (*Cyprinus carpio*) based on AFLP and microsatellite markers. Anim Genet, 2009, 41: 191–198
22. Zhang Y, Xu P, Lu C, et al. Genetic linkage mapping and analysis of muscle fiber-related QTLs in common carp (*Cyprinus carpio* L.). Mar Biotechnol, 2010, 13: 376–392
23. Zheng X, Kuang Y, Zhang X, et al. A genetic linkage map and comparative genome analysis of common carp (*Cyprinus carpio* L.) using microsatellites and SNPs. Mol Genet Genomics, 2011, 286: 261–277
24. Li Y, Xu P, Zhao Z X, et al. Construction and characterization of the BAC library for common carp *Cyprinus Carpio* L. and establishment of microsynteny with Zebrafish *Danio Rerio*. Mar Biotechnol, 2010, 13: 706–712
25. Xu P, Li J, Li Y, et al. Genomic insight into the common carp (*Cyprinus carpio*) genome by sequencing analysis of BAC-end sequences. BMC Genomics, 2011, 12: 188
26. Xia J H, Liu F, Zhu Z Y, et al, A consensus linkage map of the grass carp(*Ctenopharyngodon idella*) based on microsatellites and SNPs. BMC Genomics, 2010, 11: 135
27. Geng F S, Zhou L, Gui J F. Construction and characterization of a BAC library for*Carassius auratus gibelio*, a gynogenetic polyploid fish. Anim Genet, 2005, 36: 535–536
28. Zhang Y, Zhang X, Scheuring C F, et al. Construction and characterization of two bacterial artificial chromosome libraries of Zhikong scallop, *Chlamys farreri* Jones et Preston, and identification of BAC clones containing the genes involved in its innate immune system. Mar Biotechnol, 2008, 10: 358–365
29. Zhang X, Zhang Y, Scheuring C, et al. Construction and characterization of a bacterial artificial chromosome (BAC) library of Pacific white shrimp, *Litopenaeus vannamei*. Mar Biotechnol, 2010, 12: 141–149
30. Jang S H, Liu H, Su J G, et al. Construction and characterization of two bacterial artificial chromosome libraries of grass carp. Mar Biotechnol, 2010, 12: 261–266
31. Shao C W, Chen S L, Scheuring C F, et al. Construction of two BAC libraries from half-smooth tongue sole *Cynoglossus semilaevis* and identification of clones containing candidate sex-determination genes. Mar Biotechnol, 2010, 12: 558–568
32. Yan S Y, Lu D Y, Du M, et al. Nuclear transplantation in teleosts. Hybrid fish from the nucleus of crucian and the cytoplasm of carp. Sci Sin B. 1984, 27: 1029–1034
33. Chen H, Yi Y, Chen M, et al. Studies on the developmental potentiality of cultured cell nuclei of fish. Int J Biol Sci, 2010, 6: 192–198
34. Deng C, Liu H. An unknown piece of early work of nuclear reprogramming in fish eggs. Int J Biol Sci, 2010, 6: 190–191
35. Lee K Y, Huang H, Ju B, et al. Cloned zebrafish by nuclear transfer from long-term-cultured cells. Nat Biotechnol, 2002, 20: 795–799
36. Sun Y H, Chen S P, Wang Y P, et al. Cytoplasmic impact on cross-genus cloned fish derived from transgenic common carp (*Cyprinus carpio*) nuclei and goldfish (*Carassius auratus*) enucleated eggs. Biol Reprod, 2005, 72: 510–515
37. Luo D, Hu W, Chen S, et al. Identification of differentially expressed genes between cloned and zygote-developing zebrafish (*Danio rerio*) embryos at the dome stage using suppression subtractive hybridization. Biol Reprod, 2009, 80: 674–684
38. Hu W, Zhu Z Y. Integration mechanisms of transgenes and population fitness of GH transgenic fish. Sci China Life Sci, 2010, 53: 401–408

39. Luo D J, Hu W, Chen S P, et al. Critical developmental stages for the efficiency of somatic cell nuclear transfer in zebrafish. Int J Biol Sci, 2011, 7: 476–486

40. Hong Y, Liu T, Zhao H, et al. Establishment of a normal medakafish spermatogonial cell line capable of sperm production *in vitro*. Proc Natl Acad Sci USA, 2004, 101: 8011–8016

41. Hong N, Li Z, Hong Y. Fish Stem Cell Cultures. Int J Biol Sci, 2011, 7: 392–402

42. Xu H, Li M, Gui J, et al. Fish germ cells. Sci China Life Sci, 2010, 53: 435–446

43. Yi M, Hong N, Li Z, et al. Medaka fish stem cells and their applications. Sci China Life Sci, 2010, 53: 426–434

44. Yi M, Hong N, Hong Y. Generation of medaka fish haploid embryonic stem cells. Science, 2009, 326: 430–433

45. Takeuchi Y, Yoshizaki G, Takeuchi T. Biotechnology: Surrogate broodstock produces salmonids. Nature, 2004, 430: 629–630

46. Okutsu T, Suzuki K, Takeuchi Y, et al. Testicular germ cells can colonize sexually undifferentiated embryonic gonad and produce functional eggs in fish. Proc Natl Acad Sci USA, 2006, 103: 2725–2729

47. Okutsu T, Shikina S, Kanno M, et al. Production of trout offspring from triploid salmon parents. Science, 2007, 317: 1517

48. Liu S. Distant hybridization leads to different ploidy fishes. Sci China Life Sci, 2010, 53: 416–425

49. Gui J F. Genetic Basis and Artificial Control of Sexuality and Reproduction in Fish (in Chinese). Beijing: Science Press, 2007

50. Jalabert B. Particularities of reproduction and oogenesis in teleost fish compared to mammals. Reprod Nutr Dev, 2005, 45: 261–279

51. Xie J, Wen J J, Chen B, et al. Differential gene expression in fully grown oocytes between gynogenetic and gonochoristic crucian carps. Gene, 2001, 271: 109–116

52. Dong C H, Yang S T, Yang Z A, et al. A C-type lectin associated and translocated with cortical granules during oocyte maturation and egg fertilization in fish. Dev Biol, 2004, 265: 341–354

53. Xu H Y, Gui J F, Hong Y H. Differential expression of *vasa* RNA and protein during spermatogenesis and oogenesis in the gibel carp (*Carassius auratus gibelio*), a bisexually and gynogenetically reproducing vertebrate. Dev Dyn, 2005, 233: 872–882

54. Peng J X, Xie J L, Zhou L, et al. Evolutionary conservation of Dazl genomic organization and its continuous and dynamic distribution throughout germline development in gynogenetic gibel carp. J Exp Zool Part B, 2009, 312B: 855–871

55. Xie J, Wen J J, Yang Z A, et al. Cyclin A2 is differentially expressed during oocyte maturation between gynogenetic silver crucian carp and gonochoristic color crucian carp. J Exp Zool, 2003, 295: 1–16

56. Chen B, Gui J F. Identification of a novel C1q family member in color crucian carp (*Carassius auratus*) ovary. Comp Biochem Phys B, 2004, 138: 285–293

57. Mei J, Chen B, Yue H M, et al. Identification of a C1q family member associated with cortical granules and follicular cell apoptosis in *Carassius auratus gibelio*. Mol Cell Endocrinol, 2008, 289: 67–76

58. Mei J, Zhang Q Y, Li Z, et al. *C1q-like* inhibits *p53*-mediated apoptosis and controls normal hematopoiesis during zebrafish embryogenesis. Dev Biol, 2008, 319: 273–284

59. Wu N, Yue H M, Chen B, et al. Histone H2A has a novel variant in fish oocytes. Biol Reprod, 2009, 81: 275–283

60. Wang X L, Sun M, Mei J, et al. Identification of a Spindlin homolog in gibel carp (*Carassius auratus gibelio*). Comp Biochem Physiol B Biochem Mol Biol, 2005, 141: 159–167

61. Sun M, Li Z, Gui J F. Dynamic distribution of spindlin in nucleoli, nucleoplasm and spindle from primary oocytes to mature eggs and its critical function for oocyte-to-embryo transition in gibel carp. J Exp Zool, 2010, 313A: 461–473

62. Mylonas C C, Fostier A, Zanuy S. Broodstock management and hormonal manipulations of fish reproduction. Gen Comp Endo, 2010, 165: 516–534

63. Zohar Y, Muñoz-Cueto J A, Elizur A, et al. Neuroendocrinology of reproduction in teleost fish. Gen Comp Endocrinol, 2010, 165: 438–455

64. Li W S, Lin H R. The endocrine regulation network of growth hormone synthesis and secretion in fish: Emphasis on the signal integration in somatotropes. Sci China Life Sci, 2010, 53: 462–470

65. Li C J, Zhou L, Wang Y, et al. Molecular and expression characterization of three gonadotropin subunits commonα, FSHβ and LHβ in groupers. Mol Cell Endocrinol, 2005, 233: 33–46

66. Cao H, Zhou L, Zhang Y Z, et al. Molecular characterization of Chinese sturgeon gonadotropins and cellular distribution in pituitaries of mature and immature individuals. Mol Cell Endocrinol, 2009, 303: 34–42

67. Li S, Hu W, Wang Y, et al. Cloning and expression analysis in mature individuals of two chicken type-II GnRH (cGnRH-II) genes in common carp (*Cyprinus carpio*). Sci China Ser C-Life Sci, 2004, 47: 349–358

68. Hu W, Li S, Tang B, et al. Antisense for gonadotropinreleasing hormone reduces gonadotropin synthesis and gonadal development in transgenic common carp (*Cyprinus carpio*). Aquaculture, 2007, 271: 498–506

69. Xu J, Huang W, Zhong C, et al. Defining global gene expression changes of the hypothalamic-pituitary-gonadal axis in female sgnrh-antisense transgenic common carp (*Cyprinus carpio*). PLoS ONE, 2011, 6: e21057

70. Brunner B, Hornung U, Shan Z, et al. Genomic organization and expression of the double-sex related gene cluster in vertebrates and detection of putative regulatory regions for DMRT1. Genomics, 2001, 77: 8–17

71. Delvin R H, Nagahama Y. Sex determination and sex

differentiation in fish. Aquaculture, 2002, 208: 191–364

72. Matsuda M, Nagahama Y, Shinomiya A, et al. DMY is a Y-specific DM-domain gene required for male development in the medaka fish. Nature, 2002, 417: 559–663
73. Volff J N, Kondo M, Schartl M. Medaka dmY/dmrt1Y is not the universal primary sex-determining gene in fish. Trends Genet, 2003, 19: 196–199
74. Herpin A, Schartl M. *Dmrt1* genes at the crossroads: a widespread and central class of sexual development factors in fish. FEBS J, 2011, 278: 1010–1019
75. Xia W, Zhou L, Yao B, et al. Differential and spermatogenic cell-specific expression of DMRT1 during sex reversal in protogynous hermaphroditic groupers. Mol Cell Endocrinol, 2007, 263: 156–172
76. Jeong H B, Park J G, Park Y J, et al. Isolation and characterization of DMRT1 and its putative regulatory region in the protogynous wrasse, *Halichoeres tenuispinis*. Gene, 2009, 438: 8–16
77. Huang X, Guo Y, Shui Y, et al. Multiple alternative splicing and differential expression of dmrt1 during gonad transformation of the rice field eel. Biol Reprod, 2005, 73: 1017–1024
78. He C L, Du J L, Wu G C, et al. Differential Dmrt1 transcripts in gonads of theprotandrous black porgy, *Acanthopagrus schlegeli*. Cytogenet Genome Res, 2003, 101: 309–313
79. Wu G C, Chiu P C, Lin C J, et al. Testicular dmrt1 is involved in the sexual fate of the ovotestis in the protandrous black porgy. Biol Reprod, 2012, 86: 41
80. Liarte S, Chaves-Pozo E, Garcia-Alcazar A, et al. Testicular involution prior to sex change in gilthead seabream is characterized by a decrease in DMRT1 gene expression and by massive leukocyte infiltration. Reprod Biol Endocrinol, 2007, 5: 20
81. Graves J A M. Weird animal genomes and the evolution of vertebrate sex and sex chromosomes. Annu Rev Genet, 2008, 42: 565–586
82. Yao B, Zhou L, Wang Y, et al. Differential expression and dynamic changes of SOX3 during gametogenesis and sex inversion in protogynous hermaphroditic fish. J Exp Zool, 2007, 307A: 207–219
83. Huang W, Zhou L, Li Z, et al. Expression pattern, cellular localization and promoter activity analysis of ovarian aromatase (Cyp19a1a) in protogynous hermaphrodite red-spotted grouper. Mol Cell Endocrinol, 2009, 307: 224–236
84. Zhang Y, Zhang W, Yang H, et al. Two cytochrome P450 aromatase genes in the hermaphrodite rice field eel *Monopterus albus*: mRNA expression during ovarian development and sex change. J Endocrinol, 2010, 199: 317–331
85. Guiguen Y, Fostier A, Piferrer F, et al. Ovarian aromatase and estrogens: a pivotal role for gonadal sex differentiation and sex change in fish. Gen Comp Endocrinol, 2010, 165: 352–366
86. Zhou L, Gui J F. Molecular mechanisms underlying sex change in hermaphroditic groupers. Fish Physiol Biochem, 2010, 36: 181–193
87. Arnold A P. The end of gonad-centric sex determination in mammals. Trends Genet, 2012, 28: 55–61
88. Quinn A E, Sarre S D, Ezaz T, et al. Evolutionary transitions between mechanisms of sex determination in vertebrates. Biol Lett, 2011, 7: 443–448
89. Mank J E. Small but mighty: the evolutionary dynamics of W and Y sex chromosomes. Chromosome Res, 2012, 20: 21–33
90. Stöck M, Horn A, Grossen C, et al. Ever-young sex chromosomes in European tree frogs. PLoS Biology, 2011, 9: e1001062
91. Yoshida K, Terai Y, Mizoiri S, et al. B chromosomes have a functional effect on female sex determination in lake victoria cichlid fishes. PLoS Genet, 2011, 7: e1002203
92. Hughes J F, Skaletsky H, Pyntikova T, et al. Chimpanzee and human Y chromosomes are remarkably divergent in structure and gene content. Nature, 2010, 463: 536–539
93. De-santis C, Jerry D R. candidate growth genes in finfish-Where should we be looking? Aquaculture, 2007, 272: 22–38
94. Johnston I A, Bower N I, Macqueen D J. Growth and the regulation of myotomal muscle mass in teleost fish. J Exp Biol, 2011, 214: 1617–1628
95. Li C J, Wei, Q W, et al. Molecular and expression characterization of two somatostatin genes in the Chinese sturgeon, *Acipenser sinensis*. Comp Biochem Physiol A Mol Integr Physiol, 2009, 154: 127–134
96. Ye X, Li W S, Lin H R. Polygenic expression of somatostatin in orange-spotted grouper (*Epinephelus coioides*): molecular cloning and distribution of the mRNAs encoding three somatostatin. Mol Cell Endocrinol, 2005, 241: 62–72
97. Zhang L, Li W S, Hong X, et al. Regulation of preprosomatostatin I (PSSI) gene expression by 17β-estradiol and identification of the PSSI promoter region in orange-spotted grouper (*Epinephelus coioides*). Mol Cell Endocrinol, 2009, 311: 87–93
98. Liu Y, Lu D, Zhang Y, et al. The evolution of somatostatin in vertebrates. Gene, 2010, 463: 21–28
99. Dong H, Li W, Lin H. Comparative analyses of sequence structure, evolution, and expression of four somatostatin receptors in orange-spotted grouper (*Epinephelus coioides*). Mol Cell Endocrinol, 2010, 323: 125–136
100. Yan A F, Zhang L J, Tang Z G, et al. Orange-spotted grouper (*Epinephelus coioides*) orexin: Molecular cloning, tissue expression, ontogeny, daily rhythm and regulation of *NPY*gene expression. Peptides, 2011, 32: 1363–1370
101. Li X, He J, Hu W, et al. The essential role of endogenous ghrelin in growth hormone expression during zebrafish adenohypophysis development. Endocrinol, 2009, 150: 2767–2774
102. Li X, He J, Hu W, et al. Enhanced hyperplasia in muscles of transgenic zebrafish expressing *Follistatin 1*. Sci China Life Sci, 2011, 54: 159–165
103. Zhang Q Y. A review of viral diseases of aquatic in

China. Acta Hydrobiol Sin, 2002, 26: 93–101

104. Qiu T, Lu R H, Zhang J, et al. Complete nucleotide sequence of the S10 genome segment of grass carp reovirus (GCRV). Dis Aqua Org, 2001, 44: 69–74
105. Yang F, He J, Lin X H, et al. Complete genome sequence of the shrimp white spot bacilliform virus. J Virol, 2001, 75: 11811–11820
106. He J G, Deng M, Weng S P, et al. Complete genome analysis of the mandarin fish infectious spleen and kidney necrosis iridovirus. Virology, 2001, 291: 126–139
107. Zhang Q Y, Xiao F, Xie J, et al. Complete genome sequence of lymphocystis disease virus (LCDV-C) isolated from China. J Virol, 2004, 78: 6982–6994
108. Xie X X, Xu L M, Yang F. Proteomic analysis of the major envelope and nucleocapsid proteins of white spot syndrome virus. J Virol, 2006, 80: 10615–10623
109. Zhou Q, Xu L, Li H, et al. Four major envelope proteins of white spot syndrome virus bind to form a complex. J Virol, 2009, 83: 4709–4712
110. Zhao Z, Ke F, Huang Y H, et al. Identification and characterization of a novel envelope protein in *Rana grylio* virus. J Gen Virol, 2008, 89: 1866–1872
111. Dong C F, Xiong X P, Shuang F, et al. Global landscape of structural proteins of infectious spleen and kidney necrosis virus. J Virol, 2011, 85: 2869–2877
112. Zhao Z, Shi Y, Ke F, et al. Constitutive expression of thymidylate synthase from LCDV-C induces a transformed phenotype in fish cells. Virology, 2008, 372: 118–126
113. Wang Z L, Xu X P, He B L, et al. Infectious spleen and kidney necrosis virus ORF48R functions as a new viral vascular endothelial growth factor. J Virol, 2008, 82: 4371–4383
114. Zhang Y B, Zhang Q Y, Xu D Q, et al. Identification of antiviral-relevant genes in the cultured fish cells induced by UV-inactivated virus. Chin Sci Bull, 2003, 48: 581–588
115. Zhang Y B, Gui J F. Molecular characterization and IFN signal pathway analysis of Carassius auratus *CaSTAT*1 identified from the cultured cells in response to virus infection. Dev Comp Immunol, 2004, 28: 211–227
116. Zhang Y B, Gui J F. Identification and expression analysis of two IFN-inducible genes in crucian carp (*Carassius auratus* L.). Gene, 2004, 325: 43–51
117. Hu C Y, Zhang Y B, Huang G P, et al. Molecular cloning and characterization of a fish PKR-like gene from the cultured CAB cells induced by UV-inactivated virus. Fish Shellfish Immunol, 2004, 17: 353–366
118. Zhang Y B, Jiang J, Chen Y D, et al. The Innate Immune Response to Grass Carp Haemorrhagic Virus (GCHV) in Cultured *Carassius auratus* Blastulae (CAB) Cells. Dev Comp Immunol, 2007, 31: 232–243
119. Zhang Y B, Wang Y L, Gui J F. Identification and characterisation of two homologs of interferon stimulated gene ISG15 in crucian carp. Fish Shellfish Immunol, 2007, 23: 52–61
120. Shi Y, Zhang Y B, Zhao Z, et al. Molecular characterization and subcellular localization of*Carassius auratus* interferon regulatory factor-1. Dev Comp Immunol, 2008, 32: 134–146
121. Jiang J, Zhang Y B, Li S, et al. Expression regulation and functional characterization of a novel interferon-inducible gene Gig2 and its promoter. Mol Immunol, 2009, 46: 3131–3140
122. Yu F F, Zhang Y B, Liu T K, et al. Fish virus-induced interferon exerts antiviral function through Stat1 pathway. Mol Immunol, 2010, 47: 2330–2341
123. Sun F, Zhang Y B, Liu T K, et al. Characterization of fish IRF3 as an IFN-inducible protein reveals evolving regulation of IFN response in vertebrates. J Immunol, 2010, 185: 7573–7582
124. Sun F, Zhang Y B, Liu T K, et al. Fish MITA activation serves as a mediator for distinct fish IFN gene activation dependent on IRF3 or IRF7. J Immunol, 2011, 187: 2531–2539
125. Zhu R, Zhang Y B, Zhang Q Y, et al. Functional domains and the antiviral effect of the double-stranded RNA-dependent protein kinase PKR from *Paralichthys olivaceus*. J Virol, 2008, 82: 6889–6901
126. Liu T K, Zhang Y B, Liu Y, et al. Cooperative roles of fish PKZ and PKR in IFN-mediated antiviral response. J Virol, 2011, 85: 12769–12780
127. Jin J Y, Zhou L, Wang Y, et al. Antibacterial and antiviral roles of a fish β-defensin expressed both in pituitary and testis. PLoS ONE, 2010, 5: e12883
128. Wang S, Liu N, Chen A J, et al. TRBP homolog interacts with eukaryotic initiation factor 6 (eIF6) in *Fenneropenaeus chinensis*. J Immunol. 2009, 182: 5250–5258
129. Zhao Z Y, Yin Z X, Xu X P, et al. A novel C-type lectin from the shrimp *Litopenaeus vannamei* possesses anti-white spot syndrome virus activity. J Virol, 2009, 83: 347–356
130. Chen A J, Wang S, Zhao X F, et al. Enzyme E2 from Chinese white shrimp inhibits replication of white spot syndrome virus and ubiquitinates its RING domain proteins. J Virol, 2011, 85: 8069–8079
131. Chang M, Collet B, Nie P, et al. Expression and functional characterization of the RIG-I-like receptors MDA5 and LGP2 in Rainbow trout (*Oncorhynchus mykiss*). J Virol, 2011, 85: 8403–8412
132. Barrett R D, Paccard A, Healy T M, et al. Rapid evolution of cold tolerance in stickleback. Proc Biol Sci, 2011, 278: 233–238
133. Chen Z, Cheng C H, Zhang J, et al. Transcriptomic and genomic evolution under constant cold in Antarctic notothenioid fish. Proc Natl Acad Sci USA, 2008, 105: 12944–12949
134. Xu Q, Cheng C H, Hu P, et al. Adaptive evolution of hepcidin genes in antarctic notothenioid fishes. Mol Biol Evol, 2008, 25: 1099–1112
135. Deng C, Cheng C H, Ye H, et al. Evolution of an antifreeze protein by neofunctionalization under escape from adaptive conflict. Proc Natl Acad Sci USA, 2010, 107: 21593–21598

136. Bickler E, Buck T. Hypoxia tolerance in reptiles, amphibians, and fishes: life with variable oxygen availability. Annu Rev Physiol, 2007, 69: 145–170

137. Lushchak V I. Adaptive response to oxidative stress: Bacteria, fungi, plants and animals. Comp Biochem Physiol C Toxicol Pharmacol, 2011, 153: 175–190

138. Zhong X P, Wang D, Zhang Y B, et al. Identification and characterization of hypoxia-induced genes in *Carassius auratus* blastulae embryonic cells using suppression subtractive hybridization. Comp Biochem Physiol B, 2009, 152: 161–170

139. Wang D, Zhong X P, Qiao Z X, et al. Inductive transcription and protective role of fish heme oxygenase-1 under hypoxic stress. J Exp Biol, 2008, 211: 2700–2706

140. Sun C F, Tao Y, Jiang X Y, et al. IGF binding protein 1 is correlated with hypoxia-induced growth reduce and developmental defects in grass carp (*Ctenopharyngodon idellus*) embryos. Gen Comp Endocrinol, 2011, 172: 409–415

141. Feng X, Liu X, Zhang W, et al. p53 directly suppresses BNIP3 expression to protect against hypoxia-induced cell death. EMBO J, 2011, 30: 3397–3415

142. Guan B, Ma H, Wang Y, et al. Vitreoscilla hemoglobin (VHb) overexpression increases hypoxia tolerance in zebrafish (*Danio rerio*). Mar Biotechnol (NY), 2011, 13: 336–344

143. Liu Z J. Fish genomics and analytical genetic technologies, with examples of their potential applications in management of fish genetic resources. In: Bartley D M, Harvey B J, Pullin R S V, eds. Workshop on Status and Trends in Aquatic Genetic Resources: a basis for international policy. Food and Agriculture Organization of the United Nations, Rome, 2007. 145–178

144. McAndrew B, Napier J. Application of genetics and genomics to aquaculture development: current and future directions. J Agric Sci, 2010, 149: 143–151

145. Abernathy J W, Peatman E, Liu Z J. Basic aquaculture genetics. SRAC Publication No. 5001. 2010, 1–16

146. Wang Z W, Zhu H P, Wang D, et al. A novel nucleo-cytoplasmic hybrid clone formed via androgenesis in polyploid gibel carp. BMC Res Notes, 2011, 4: 82

147. Liu H Q, Cui S Q, Hou C C, et al. YY supermale generated gynogenetically from XY female in *Pelteobagrus fulvidraco* (Richardson). Acta Hydrobiol Sin, 2007, 31: 718–725

148. Wang D, Mao H L, Chen H X, et al. Isolation of Y- and X-linked SCAR markers in yellow catfish and application in the production of all-male populations. Anim Genet, 2009, 40: 978–981

SECTION II: NUTRITION

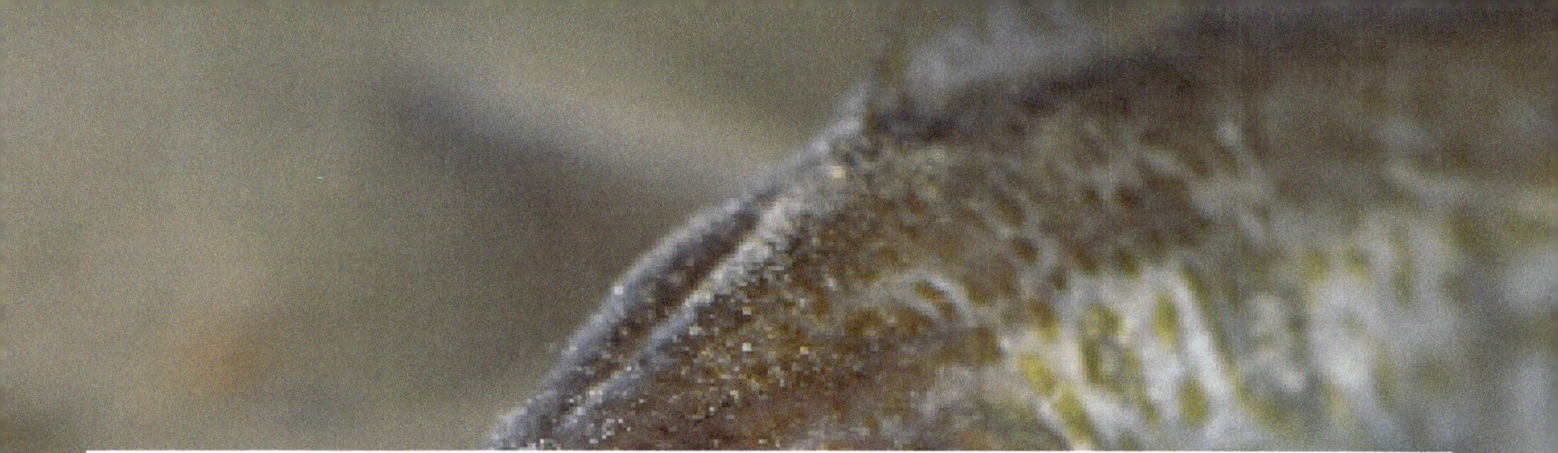

CHAPTER 2

EFFECTS OF DIETARY PROTEIN AND LIPID LEVELS ON GROWTH AND BODY COMPOSITION OF JUVENILE FAR EASTERN CATFISH (SILURUS ASOTUS).

Kyoung-Duck Kim, Sang Gu Lim, Yong Jin Kang, Kang-Woong Kim, and Maeng Hyun Son

Keywords: Dietary Protein and Lipid, Far Eastern Catfish, Growth, Body Composition

INTRODUCTION

Nutrient and energy sources in feed are needed for the growth and maintenance of fish. Fish are known to utilize protein preferentially to lipid or carbohydrate as an energy source (Walton and Cowey, 1982). Dietary protein is the most important factor affecting growth performance of fish and feed cost (Lovell, 1989). Therefore, it is important from a nutritional and economical point of view to improve protein utilization for tissue synthesis rather than for energy use.

Dietary lipids play important roles as a source of energy and essential fatty acids necessary for fish growth and development. Increasing the energy density of diets has been suggested as a strategy to spare protein and limit ammonia production for several fish species including red drum, *Sciaenops ocellatus* (McGoogan and Gatlin, 1999). A supplementation of lipid rather than carbohydrate as a non-protein energy source is generally more effective for increasing dietary energy level because lipid is an energy-dense nutrient that is readily metabolized by fish, especially by carnivorous fish (NRC, 1993).

Far eastern catfish is one of the most important commercial freshwater fish species, because this species has high consumer demand in Korea. This species is distributed in Korea, Japan, China and Taiwan (Chyoung, 1996). Aquaculture production of far eastern catfish has steadily increased during the last decade, and its production in 2010 reached 4,194 metric tons, which is second to eel (7,902 metric tons) among freshwater fish aquaculture in Korea (Statistics Korea, 2010).

Recently, nutritional studies evaluating availability of dietary lipid source and substitution of fish meal with soybean meal for growth in this species have been performed (Kim et al., 2009; 2010). However, information on growth performance of dietary protein and lipid levels is limited to date in this species. Therefore, this study was conducted to investigate the effects of dietary protein and lipid levels on the growth and body composition of juvenile far eastern catfish.

Citation: Kyoung-Duck Kim, Sang Gu Lim, Yong Jin Kang, Kang-Woong Kim, and Maeng Hyun Son, "Effects of Dietary Protein and Lipid Levels on Growth and Body Composition of Juvenile Far Eastern Catfish (Silurus asotus).. doi: 10.5713/ajas.2011.11089.

MATERIALS AND METHODS

Design of the Experimental Diets and Feeding Trial

A 3×2 factorial design with three replicates per dietary treatment was used in this study. Six experimental diets were formulated to contain three levels of protein (20%, 30% and 40%) and two levels of lipid (9% and 17%). Gross energy levels of the diets were calculated based on 4, 9 and 4 kcal/g diets for protein, lipid and nitrogen-free extract, respectively (Garling and Wilson, 1976). Ingredients and nutrient contents of the experimental diets are presented in Table 1. To determine the effects of the three protein levels and two lipid levels, the amounts of fish meal, squid liver oil and soybean oil in the experimental diets were adjusted at the expense of dextrin and cellulose. Diets were prepared with a laboratory pellet machine after 35 to 40 g water was mixed with 100 g mixture of ingredients, and dried over night at room temperature. The experimental diets were stored in freezer at −24°C until use.

Table 1: Ingredients and nutrient contents of the experimental diets

	Diets (% protein/% lipid)					
	20/9	20/17	30/9	30/17	40/9	40/17
Ingredients (%)						
Brown fish meal [1]	27	27	40.5	40.5	54	54
Dextrin	42	34	32	24	22	14
α-Starch	15	15	15	15	15	15
Squid liver oil [2]	3.8	3.8	2.4	2.4	1	1
Soybean oil	3.5	11.5	3.5	11.5	3.5	11.5
α-Cellulose	4.2	4.2	2.1	2.1		
Vitamin premix [3]	2	2	2	2	2	2
Mineral premix [3]	2	2	2	2	2	2
Choline chloride	0.5	0.5	0.5	0.5	0.5	0.5
Nutrient contents (dry matter basis)						
Crude protein (%)	20.0	20.4	29.7	30.3	39.2	39.9
Crude lipid (%)	7.7	14.9	9.0	16.6	9.5	17.6
Ash (%)	4.7	4.9	6.7	6.8	8.6	8.6
Crude fiber (%)	4.6	4.7	3.0	3.2	2.0	2.5
NFE (%) [4]	63.1	55.0	51.5	43.1	40.7	31.3
Gross energy (kcal/g diet)	4.01	4.36	4.06	4.43	4.05	4.43
Protein to energy ratio (mg/kcal)	50	47	73	68	97	90
n-3 HUFA [5]	1.0	1.3	1.2	1.3	1.4	1.5

[1] Imported from Chile. [2] Provided by E-wha Oil and Fat Industry Corporation, Busan, South Korea.
[3] Same as Lee et al. (2003). [4] Nitrogen-free extract calculated by the difference (= 100-crude protein-crude lipid-ash-crude fiber).
[5] Highly unsaturated fatty acids (C≥20), calculated (dietary total lipid %×area %×0.892). (Yoshimatsu et al., 1997).

Table 2. Major fatty acids composition (% of total fatty acids) of the experimental diets

	Diets (% protein/% lipid)					
	20/9	20/17	30/9	30/17	40/9	40/17
C14:0	2.9	1.8	2.8	1.5	2.7	1.4
C16:0	14.3	13.7	14.9	13.7	16.0	13.8
C16:1n-7	5.1	3.5	4.7	2.6	3.8	2.0
C18:0	3.2	2.1	3.8	5.9	5.2	2.8
C18:1n-9	25.4	27.1	24.4	24.5	23.2	28.0
C18:2n-6	24.0	33.6	24.3	35.7	23.2	36.4
C18:3n-3	5.2	3.6	4.0	2.7	2.3	1.8
C20:2n-6	1.5	0.9	1.2	0.8	1.0	0.6
C20:4n-6	0.4	0.3	0.5	0.3	0.6	0.4
C20:5n-3	6.5	4.4	6.6	3.7	6.6	3.4
C22:5n-3	0.8	0.6	1.0	0.6	1.3	0.7
C22:6n-3	6.2	4.1	6.4	3.8	8.0	4.6
n-3HUFA [1]	14.3	10.0	14.8	8.7	16.6	9.7

[1] Highly unsaturated fatty acids (C≥20).

Fish and Rearing Conditions

Juvenile far eastern catfish (*Silurus asotus*) were obtained from a local fish farm (Jinhae, Korea). They were acclimated to laboratory conditions for 2 weeks. Experimental fish (initial body weight 7.6±0.5 g) were randomly distributed into each of eighteen 1,000 L cylindrical plastic tanks with 60 fish to each tank in semi-recirculation system. Partial sedimentation and biofitration was used to maintain stable conditions. Fish were hand-fed to apparent satiation twice a day (0900 and 1700 h, 6 days a week) for 66 days. Freshwater was supplied at a flow rate of 10 L/min in each tank. Mean water temperature and dissolved oxygen, based on weekly measurements, over the experimental periods were 25±1.7°C and 6.9±0.7 mg/L, respectively. Photoperiod was left at natural condition during the feeding trial. All fish in each tank were weighed collectively at the beginning and end of feeding trial being fasted for 48 h. Records were kept of daily feed consumption, mortalities and feeding behavior.

Sample Collections and Chemical Analysis

At the end of the feeding trial, five fish in each tank were collected and stored in a freezer at −70°C for chemical analysis. Crude protein was determined by Kjeldahl method using Auto Kjeldahl System (Gerhardt VAP50OT/TT125, KG, Germany). Crude lipid was determined by ether extraction using Soxhlet extractor (Velp SER 148, Usmate, Italy). Moisture was determined by oven drying at 135°C for 2 h. Ash was determined by muffle furnace at 550°C for 6 h. Crude fiber was determined by an automatic analyzer (Fibertec, Tecator, Hoganas, Sweden). Nitrogen free extract was calculated by difference. Lipid for fatty acid analysis was extracted by mixture of chloroform and methanol according to the method of Folch et al. (1957). The fatty acid composition was determined after methylation with 14% BF_3 methanol (Sigma, St Louis, MO, USA) by a gas chromatography (HP-6890N, Hewlett-Packard, Palo Alto, CA, USA) with a flame ionization detector, equipped with SP™-2560 capillary column (100 m×0.25 mm i.d., film thickness 0.20 μm; Supelco, Bellefonte, PA, USA). Injector and detector temperatures were 260°C. The column temperature was programmed from 140°C to 240°C at a rate of 4°C/min. Helium was used as the carrier gas. Fatty acids were identified by comparison with known standards.

Statistical Analysis

The data were subjected to a two-way ANOVA to test the effects of the dietary protein and lipid levels on fish performance. If significant ($p<0.05$) differences were found in factors, Duncan's multiple range test (Duncan, 1955) was used to rank the means. All statistical analyses were carried out by using the SPSS program Version 11.5 (SPSS, Chicago, Illinois, USA).

RESULTS

Growth Performance

Survival of juvenile far eastern catfish in response to diets containing 20 to 40% protein and 9 to 17% lipid for 66 day was over 88%, and there were no significant difference among all groups (Table 3). Final mean weight, feed efficiency, protein efficiency ratio, daily feed intake and daily protein intake were significantly affected by both dietary protein and lipid levels ($p<0.01$). Final mean weight was improved with increasing dietary protein and lipid levels, and the highest final mean weight was observed in fish fed the 40/17 (% protein/% lipid) diet. No significant difference was observed in final mean weight for fish fed between 30/17 diet and 40/9 diet. Feed efficiency (FE) of fish fed the diets containing over 30% protein levels with 9% and 17% lipid levels were significantly higher than those of fish fed the 20% protein levels ($p<0.05$). FE of fish fed the 30/17 diet was not significantly different from that of fish fed the 40/9 diet or 40/17 diet. Protein efficiency ratio (PER) was significantly decreased with increasing dietary protein level ($p<0.05$). FE and PER of fish fed the 20% protein diet with 17% lipid level were significantly higher than those of fish fed 9% lipid diet ($p<0.05$). Daily feed intake of fish tended to decrease with increasing dietary protein and lipid levels. Daily protein intake of fish was significantly increased with increasing dietary protein level ($p<0.05$), and tended to decrease with increasing dietary lipid level. Significant interactions between protein and lipid levels were founded for final mean weight, PER and daily feed intake.

Table 3: Growth performance of juvenile far eastern catfish fed diets containing three protein and two lipid levels for 66 days

Protein (%)	Lipid (%)	Initial mean weight (g/fish)	Survival (%)	Final mean weight (g/fish)	Feed efficiency (%)[1]	Protein efficiency ratio[2]	Daily feed intake[3]	Daily protein intake[3]
20	9	7.5±0.3 ^{ns}	93±1.1 ^{ns}	43±0.4^a	61±0.9^a	3.20±0.05^c	3.49±0.04^d	0.66±0.01^b
20	17	7.7±0.3	94±1.5	46±1.3^b	71±1.5^b	3.55±0.07^d	3.08±0.02^c	0.61±0.00^a
30	9	7.4±0.4	94±1.5	59±1.1^c	78±0.2^c	2.71±0.01^b	3.02±0.00^{bc}	0.87±0.00^c
30	17	7.6±0.2	94±1.5	64±0.4^d	82±0.9^{cd}	2.78±0.03^b	2.92±0.02^b	0.86±0.01^c
40	9	7.5±0.4	91±4.3	65±0.7^d	84±1.7^d	2.20±0.05^a	2.95±0.05^{bc}	1.13±0.02^e
40	17	7.7±0.2	88±3.5	78±0.7^e	87±3.1^d	2.25±0.08^a	2.76±0.08^a	1.06±0.03^d
Two-way ANOVA								
Protein			p<0.2	p<0.01	p<0.01	p<0.01	p<0.01	p<0.01
Lipid			p<0.8	p<0.01	p<0.01	p<0.01	p<0.01	p<0.01
Interaction			p<0.9	p<0.01	p<0.1	p<0.05	p<0.05	p<0.2

Values (mean±SE of three replications) in the same column not sharing a common superscript letter are significantly different ($p<0.05$).

[1] Fish wet weight gain×100/feed intake (dry matter). [2] Fish wet weight gain/protein intake.

[3] Feed intake (or protein)×100/((initial fish wt.+final fish wt.+dead fish wt.)×days fed/2). ^{ns} Not significant ($p>0.05$).

Body Composition

Moisture content of whole body was significantly affected by dietary protein ($p<0.05$) and lipid ($p<0.01$) levels (Table 4). Crude lipid content of whole body was significantly affected by dietary lipid levels ($p<0.01$). Moisture content of whole body in fish fed the 9% lipid diets were significantly higher than that of fish fed the 17% lipid diets at the same protein level ($p<0.05$), but the opposite trends were found for crude lipid content. No significant differences were observed in crude protein and ash contents. Fatty acids composition of whole body in juvenile far eastern catfish fed the diets containing various protein and lipid levels are shown in Table 5. Significant effects of dietary lipid were observed for most fatty acids, according to their relative values in the diets. The fish fed the 9% the lipid diets showed the significantly higher contents of 14:0, 16:1n-7, 20:2n-6, 20:5n-3, 22:5n-3, 22:6n-3, n-3 poly unsaturated fatty acids (PUFA) and n-3 highly unsaturated fatty acids (HUFA) than those of fish fed the 17% lipid diets ($p<0.05$). However, 18:2n-6 and n-6 PUFA contents of fish fed the 17% lipid diets were significantly higher than those of fish fed the 9% lipid diets at all protein levels ($p<0.05$).

Table 4: Proximate compositions of whole body in juvenile far eastern catfish fed diets containing three protein and two lipid levels for 66 days

Protein (%)	Lipid (%)	Moisture (%)	Crude protein (%)	Crude lipid (%)	Ash (%)
20	9	72.7 ± 0.08^{b}	14.6 ± 0.57^{ns}	8.2 ± 0.30^{ab}	2.5 ± 0.06^{ns}
20	17	70.8 ± 0.38^{a}	14.0 ± 0.11	11.4 ± 1.11^{b}	2.2 ± 0.01
30	9	74.6 ± 0.62^{c}	14.6 ± 0.44	5.5 ± 0.41^{a}	2.3 ± 0.26
30	17	72.0 ± 0.43^{ab}	14.4 ± 0.09	9.4 ± 0.62^{b}	2.2 ± 0.14
40	9	74.7 ± 0.92^{c}	15.0 ± 0.06	6.1 ± 0.92^{a}	2.3 ± 0.05
40	17	71.3 ± 0.53^{ab}	15.2 ± 0.09	10.0 ± 1.70^{b}	2.3 ± 0.11
Two-way ANOVA					
Protein		$p<0.05$	$p<0.1$	$p<0.1$	$p<0.7$
Lipid		$p<0.01$	$p<0.5$	$p<0.01$	$p<0.3$
Interaction		$p<0.5$	$p<0.5$	$p<0.9$	$p<0.6$

Values (mean±SE of three replications) in the same column not sharing a common superscript letter are significantly different ($p<0.05$).
^{ns} Not significant ($p>0.05$).

Table 5: Major fatty acids composition (% of total fatty acids) of whole body in juvenile far eastern catfish fed diets containing three protein and two lipid levels for 66 days

Protein	20		30		40		Pooled	Two-way ANOVA		
Lipid	9	17	9	17	9	17	SEM	Protein	Lipid	Interaction
C14:0	2.5^{b}	1.8^{a}	2.5^{b}	1.6^{a}	2.5^{b}	1.6^{a}	0.10	$p<0.5$	$p<0.01$	$p<0.4$
C16:0	16.0^{a}	15.5^{a}	16.9^{a}	15.8^{a}	18.9^{b}	16.1^{a}	0.33	$p<0.01$	$p<0.01$	$p<0.1$
C16:1n-7	5.9^{b}	4.2^{a}	6.4^{bc}	3.7^{a}	6.8^{c}	4.1^{a}	0.31	$p<0.3$	$p<0.01$	$p<0.1$
C18:0	7.7^{ns}	4.9	5.0	5.9	5.3	6.9	0.42	$p<0.7$	$p<0.9$	$p<0.1$
C18:1n-9	21.5^{ns}	24.9	25.4	24.3	26.0	24.5	0.51	$p<0.2$	$p<0.8$	$p<0.1$
C18:2n-6	20.2^{b}	28.3^{c}	18.8^{b}	29.8^{cd}	16.6^{a}	30.1^{d}	1.37	$p<0.2$	$p<0.01$	$p<0.01$
C18:3n-3	4.1^{d}	3.4^{c}	3.2^{c}	2.7^{b}	2.0^{a}	1.7^{a}	0.20	$p<0.01$	$p<0.01$	$p<0.5$
C20:2n-6	3.9^{d}	2.7^{b}	3.6^{cd}	2.4^{b}	3.3^{c}	1.9^{a}	0.17	$p<0.01$	$p<0.01$	$p<0.8$
C20:4n-6	0.9^{ns}	0.9	0.8	0.9	0.8	0.8	0.02	$p<0.1$	$p<0.7$	$p<0.7$
C20:5n-3	5.3^{c}	3.2^{a}	4.8^{bc}	3.0^{a}	4.5^{b}	2.8^{a}	0.24	$p<0.05$	$p<0.01$	$p<0.6$
C22:2n-6	1.5^{b}	1.2^{b}	1.5^{b}	1.0^{ab}	1.2^{b}	0.6^{a}	0.09	$p<0.05$	$p<0.01$	$p<0.7$
C22:5n-3	1.6^{b}	1.2^{a}	1.7^{bc}	1.1^{a}	1.9^{c}	1.1^{a}	0.07	$p<0.3$	$p<0.01$	$p<0.05$
C22:6n-3	6.6^{b}	4.9^{a}	6.7^{bc}	4.8^{a}	7.5^{c}	4.9^{a}	0.28	$p<0.3$	$p<0.01$	$p<0.3$
n-3 PUFA [1]	18.0^{d}	13.1^{b}	16.9^{cd}	12.0^{ab}	16.3^{c}	10.8^{a}	0.66	$p<0.01$	$p<0.01$	$p<0.8$
n-6 PUFA [1]	27.3^{c}	34.1^{d}	25.8^{b}	35.3^{d}	22.9^{a}	34.6^{d}	1.18	$p<0.01$	$p<0.01$	$p<0.01$
n-3 HUFA [2]	14.3^{b}	10.0^{a}	14.1^{b}	9.6^{a}	14.8^{b}	9.4^{a}	0.60	$p<0.8$	$p<0.01$	$p<0.6$

Values (mean of three replications) in each row not sharing a common superscript are significantly different ($p<0.05$).
[1] Polyunsaturated fatty acids ($C\geq18$). [2] Highly unsaturated fatty acid ($C\geq20$). ^{ns} Not significant ($p>0.05$).

DISCUSSION

Improved growth and feed efficiency with increasing dietary protein levels is well known in fish (NRC, 1993). In the present study, growth performance was increased from 20% to 40% protein levels at the same dietary lipid level. The result indicates that the protein requirement for maximum growth of juvenile far eastern catfish may be higher than 40%. The dietary protein requirement for juvenile far eastern catfish determined in this study is comparable to protein requirements (42–44%) estimated for bagrid catfish *Pseudobagrus fulvidraco* (Kim and Lee, 2005) and Malaysian freshwater catfish *Mystus nemurus* (Khan et al., 1993), but is higher compared with the dietary protein requirements of 32 to 36% reported for channel catfish *Ictalurus punctatus* (Garling and Wilson, 1976).

Growth rate of far eastern catfish fed the 17% lipid diets was significantly higher than that of fish fed the 9% lipid diets at all dietary protein levels, and improvement in weight gain, FE and PER of fish fed the 30/17 diet were compared with those of fish fed the 40/9 diet. These results indicate that increasing the dietary energy level by lipid will provide a more efficient utilization of dietary protein for the growth of fish. Improvement of fish performance at increasing dietary energy levels was observed in some other catfish species (Page and Andrews, 1973; Pei et al., 2004). Protein sparing effect by dietary lipid is well established (Cho and Kaushik, 1990). It is important to provide an adequate level and ratio of dietary protein and non-protein energy in order to reduce catabolism of protein for energy. The optimum protein to energy ratio in diet for maximum growth for several fish species such as rainbow trout, *Salmo gairdneri* and Hybrid bass, *Moroni chrysops*×*M. saxitalis* ranged from 81 to 112 mg/kcal (NRC, 1993). The maximum growth rate was obtained in fish fed the 40/17 diet in this study, corresponding to a protein to energy ratio of 90 mg/kcal, which is similar to that (88 mg/kcal) for channel catfish fingerlings (Garling and Wilson, 1976).

Dietary protein and lipid levels could effect on PER of fish. The PER increased with decrease in dietary protein at the same lipid level in this study. Similar result of improved PER with decreasing dietary protein level was reported in other studies (Lee et al., 2002; Kim and Lee, 2005).

This study showed a lower daily feed intake for fish fed the 17% lipid diets than 9% lipid diets. Similar results of reduced feed consumption of fish fed a high-energy diet compared with a low-energy diet have been reported in other species of fish (Lee and Putnam, 1973; Page and Andrews, 1973; Lee et al., 2002). The high-energy diet caused a reduction in feed intake, probably because fish eat to satisfy their energy requirements. Daily feed intake was negatively correlated with dietary lipid level ($r = -0.64$, $p<0.01$) in this study.

The practice of increasing of dietary lipid levels should be carefully considered as it may affect carcass quality, flavor and storage characteristics, as a result of increased lipid deposition (Cowey, 1993; Hillestad and Johnsen, 1994). In this study, the lipid contents of fish were positively correlated with dietary lipid level. This is an agreement with other studies showing that lipid content of fish fed high energy diets is higher than that of fish fed low energy diets (Kim and Lee, 2005). Despite of increase of body lipid contents in the high lipid diets, considering the improvement of growth performance, high lipid diets seem to have more beneficial effects for fish performance compared to low lipid diets when proper protein is provided (Kim and Lee, 2005; Ahmad, 2008). Moreover, the decrease of feed and protein intakes in high lipid diets in this study appears to have an effect of reduction of ammonia production as described by McGoogan and Gatlin (1999).

The fatty acid composition of fish in this study well reflected the fatty acid composition of dietary lipid. Similar results had been founded in previous catfish study (Kim et al., 2010). Kim et al. (2010) reported that linseed oil or squid liver oil containing n-3 fatty acids are the proper dietary lipid sources for the growth of juvenile far eastern catfish. The importance of essential fatty acids, largely two types, linoleic acid and linolenic acid, for most of freshwater fish has been reported (NRC, 1993). Linolenic acid requirement of fresh water fish species ranged from 0.5% to 2% of the diet (NRC, 1993), and Satoh et al. (1989) reported that dietary essential fatty acids requirement for channel catfish is 1 to 2% linolenic acid or 0.5 to 0.75% n-3 HUFA such as 20:5n-3 and 22:6n-3. In this study, fish meal and squid liver oil contained considerable content of n-3 HUFA were used in experimental diets. Although information on requirement level of n-3 fatty acids for far eastern catfish is unknown, 1.0 to 1.5% n-3 HUFA in diets used in this study was above the requirement for n-3HUFA of channel catfish.

The results of this study suggest that the protein requirement for maximum growth of juvenile far eastern catfish may be higher than 40%, and increase of dietary lipid level from 9% to 17% can improve growth and feed utilization.

ACKNOWLEDGEMENTS

This work was funded by a grant from the National Fisheries Research and Development Institute (RP-2011-AQ-123).

REFERENCES

1. Ahmad MH. Response of African catfish, *Clarias gariepinus*, to different dietary protein and lipid levels in practical diets. J World Aquac Soc. 2008;39:541–548.
2. Cho CY, Kaushik SJ. Nutritional energetics in fish: energy and protein utilization in rainbow trout (*Salmo gairdneri*) World Rev Nutr Diet. 1990;61:132–172.
3. Chyoung MK. The Fishes of Korea. 5th ED. Il Ji Co; Seoul: 1996.
4. Cowey CB. Some effects of nutrition and flesh quality of cultured fish. In: Kaushik SJ, Luquet P, editors. Fish Nutrition in Practice; Proc of the IV Int Symp on Fish Nutrition and Feeding; Paris: Les Colloques INRA; 1993. pp. 227–236.
5. De Silva SS, Gunasekera RM, Shim KF. Interactions of varying dietary protein and lipid levels in young red tilapia: evidence of protein sparing. Aquaculture. 1991;95:305–318.
6. Duncan DB. Multiple-range and multiple F tests. Biometrics. 1955;11:1–42.
7. Folch J, Lees M, Sloane-Stanley GH. A simple method for the isolation and purification of total lipids from animal tissues. J Biol Chem. 1957;226:497–509.
8. Garling DL, Wilson RP. Optimum dietary protein to energy ratios for channel catfish fingerlings, *Ictalurus punctatus*. J Nutr. 1976;106:1368–1375.
9. Hillestad M, Johnsen FT. High-energy/low-protein diets for Atlantic salmon: effects on growth, nutrient retention and slaughter quality. Aquaculture. 1994;124:109–116.
10. Khan MS, Ang KJ, Ambak MA, Saad CR. Optimum dietary protein requirement of a Malaysian freshwater catfish, *Mystus nemurus*. Aquaculture. 1993;112:227–235.
11. Kim LO, Lee SM. Effects of the dietary protein and lipid levels on growth and body composition of bagrid catfish, *Pseudobagrus fulvidraco*. Aquaculture. 2005;243:323–329.
12. Kim KD, Lim SG, Hwang JA, Kim JD, Kang YJ.

Evaluation of soybean meal as a partial substitute for fish meal in diet and experimental practical diet for growth in the far eastern catfish (*Silurus asotus*)Korean J Fish Aquat Sci. 2009;42:349–353.

13. Kim KD, Kim JD, Lim SG, Kang YJ, Son MH. Effects of dietary lipid sources on the growth and body composition of the far eastern catfish, *Silurus asotus*. Kor J Fish Aquat Sci. 2010;43:445–450.
14. Lee SM, Kim DJ, Choi SH. Effects of dietary protein and lipid level on growth and body composition of juvenile ayu (*Plecoglossus altivelis*) reared in seawater. Aquac Nutr. 2002;8:53–58.
15. Lee DJ, Putnam GB. The response of rainbow trout to varying protein/energy ratios in a test diet. J Nutr. 1973;103:916–922.
16. Lee SM, Kim KD, Lall SP. Utilization of glucose, maltose, dextrin and cellulose by juvenile flounder (*Paralichthys olivaceus*) Aquaculture. 2003;221:427–438.
17. Lovell RT. Nutrition and feeding of fish. Van Nostrand Reinhold; New York: 1989. p. 260.
18. McGoogan BB, Gatlin DM. Dietary manipulations affecting growth and nitrogenous waste production of red drum, *Sciaenops ocellatus*: I. Effects of dietary protein and energy levels. Aquaculture.1999;178:333–348.
19. National Research Council (NRC) Nutrient requirements of fish. National Academy Press; Washington, DC, USA: 1993.
20. Page JW, Andrews JW. Interaction of dietary levels of protein and energy on channel catfish (*Ictalurus punctatus*) J Nutr. 1973;103:1339–1346.
21. Pei Z, Xie S, Lei W, Zhu X, Yang Y. Comparative study on the effect of dietary lipid level on growth and feed utilization for gibel carp (*Carassius suratus gibelio*) and Chinese longsnout catfish (*Leiocassis longirostris* Gunther) Aquac Nutr. 2004;10:209–216.
22. Satoh S, Poe WE, Wilson RP. Effect of dietary n-3 fatty acids on weight gain and liver polar lipid fatty acid composition of fingerling channel catfish. J Nutr. 1989;119:23–28.
23. Statistics Korea. Fishery production survey. 2010 Retrieved Mar. 15 2011f rom http://fs.fips.go.kr.
24. Walton MJ, Cowey CB. Aspects of intermediary metabolism in fish. Comp. Biochem. Physiol.1982;73B:59–79.
25. Yoshimatsu T, Imoto H, Hayash M, Toda K, Yoshimura K. Preliminary results in improving essential fatty acids enrichments of rotifer cultured in high diensity. Hydrobiologia. 1997;358:153–157.

CHAPTER 3

GROWTH, FEED UTILIZATION AND BLOOD METABOLIC RESPONSES TO DIFFERENT AMYLOSE-AMYLOPECTIN RATIO FED DIETS IN TILAPIA (OREOCHROMIS NILOTICUS)

Meng-Yao Chen, Ji-Dan Ye, Wei Yang, and Kun Wang

Keywords: Performance, Amylose, Amylopectin, Postprandial Response, Tilapia

INTRODUCTION

Carbohydrate and particular cereal starch are the main sources for energy production of most terrestrial warm-blooded animals including human beings, but its impact is less in fish (Wilson, 1994). Although fish have the same carbohydrate metabolic pathways as mammals, they have a lower carbohydrate tolerance than mammals, as evidenced by a prolonged postprandial hyperglycemia after feeding carbohydrate-rich diets (Wilson, 1994; Moon, 2001). As a result, the use of carbohydrate for energy is limited and a high dietary inclusion level of carbohydrates could lead to growth retardation in fish (Wilson, 1994). Fish depend more than most other animals on amino acids as precursors to glucose for tissue protein and glucose synthesis as well as energy (Mommsen and Plisetskaya, 1991; Moon, 2001). Nevertheless, from an economic point of view, it is desirable to reduce the protein and lipid use for energy consumption by increasing the carbohydrate use in the aquafeeds where carbohydrates are more readily available and less expensive than other macro-nutrients. The nutritional value of starches varies widely depending on its sources (Granfeldt et al., 1994;Gutierrez Del Alamo et al., 2009; Glencross et al., 2012), processing (Mahadevamma and Tharanathan, 2007), hardness (Denardin et al., 2012), non-starch polysaccharides (Glencross et al., 2012) and fish species (Wilson, 1994) as well as its complexity (Jun et al., 2010). Furthermore, previous studies have shown that the metabolic responses induced by dietary starches differ (Hemre et al., 2002; Stone, 2003; Enes et al., 2009). Starch is composed of two types of glucose polymers, amylose, a linear chain of glucose units joined by α-1,4-glucosidic linkages, and amylopectin, a highly branched amylose in which the branches are tied by α-1,6 bonds (Zobel, 1988). Amylopectin-rich starches are easily digested, which may induce a faster and higher glycemic and insulinemic responses compared to those high in amylose (Behall et al., 1989; Weststrate and van Amelsvoort, 1993; Granfeldt et al., 1995; Behall et al., 2002; Rawles and Lochmann, 2003). Although these metabolic responses were previously observed both in mammals and fish, the mechanisms behind these responses are not well understood.

In terms of growth effects in relation to starch type, most of the research has focused on terrestrial animals, but contradictory results have occurred in different studies. The

Citation: Meng-Yao Chen, Ji-Dan Ye, Wei Yang, and Kun Wang, "Growth, Feed Utilization and Blood Metabolic Responses to Different Amylose-amylopectin Ratio Fed Diets in Tilapia (Oreochromis niloticus). doi: 10.5713/ajas.2013.13022.

positive effects on growth in response to dietary amylose were observed in pigs (Cohen and Tanksley, 1973), poultry (Pirgozliev at al., 2000;Gutierrez Del Alamo et al., 2009) and goats (Kyuma and Ishida, 1989), but the negative effects were observed in rats (Aziz et al., 2009; Denardin et al., 2012), pigs (Perez and Aumaitre, 1979; Camp et al., 2003), and poultry (Ertl and Dale, 1997; Bergh et al., 1999; Pirgozliev et al., 2002; Collins et al., 2003;Pirgozliev at al., 2010). No effect of starch type on growth was also observed in rats (Mongeau et al., 1995;Zhou and Kaplan, 1997; Kabir et al., 1998) and pigs (Shelton et al., 2004). In fish, Rawles and Lochmann (2003) first investigated the starch utilization related to starch type and observed an enhanced growth of sunshine bass (Morone chrysops ♀×M. saxatilis ♂) when fed with a diet at the higher ratio of amylose-amylopectin starch. To the best of our knowledge, the impact of dietary amylose-amylopectin ratio in the starch fraction of the diet on other fish species has not yet been determined. The omnivorous fish such as tilapia (Oreochromis niloticus) have greater ability to digest starch than carnivorous fish (Krogdahl et al., 2005). We hypothesized that the tilapia fed diets with increasing amylose-amylopectin ratio would exhibit better growth. Therefore, we investigated the effects of dietary amylose-amylopectin ratio on the growth, feed utilization, digestibility, digestive enzyme activity, and blood intermediary metabolites in the tilapia.

MATERIALS AND METHODS

Diets

High-amylose corn starch (Hi-Maize 260, National Starch and Chemical Co., Ltd., China, 66.9% amylose) and waxy rice (the local glutinous rice, 81.0% amylopectin and 0.1% amylose) were used as the main amylose and amylopectin sources, respectively. Steam dried fish meal was obtained from Shine Biotech Co., Ltd., China. Fish oil, soybean oil, and other feed ingredients were obtained from Baisuihang Feed Co. Ltd., China. Based on the proximate composition of feed ingredients, five isonitrogenous (32% crude protein) and isolipidic (6.5% crude fat) diets containing an equal starch level (35% starch) were formulated. Diets differing in ratios of amylose-amylopectin (diet 1, 0.11; diet 2, 0.24; diet 3, 0.47; diet 4, 0.76; and diet 5, 0.98) were achieved by adjusting the proportion of corn starch to waxy rice, i.e. by incremental levels of amylose from 2.58% to 14.12% at the expense of amylopectin. The ingredients and proximate composition of the experimental diets are shown in Table 1. The coarse dry ingredients were ground using a hammer mill and then passed through a 60 mesh sieve, weighed and blended together into a homogeneous mixture. The liquid ingredients (mixed oil and water) were then added, and the mixture was made into a mash. The dough was pelleted through a 2.5-mm die using multifunctional spiral extrusion machinery (CD4XITS, South China University of Technology, China). The pellets were dried in a forced-air oven at 50°C for 24 h and then sealed in plastic bags and kept at −20°C until further use for feeding.

Table 1: Ingredients and composition of the experimental diets (as-fed basis)

Ingredients (%)	Diets at different amylose-amylopectin ratios				
	Diet 1 (0.11)	Diet 2 (0.24)	Diet 3 (0.47)	Diet 4 (0.76)	Diet 5 (0.98)
Fish meal	45.77	46.28	46.9	47.34	47.67
Fish and soybean oil (1:1)	2.43	2.39	2.34	2.31	2.28
High-amylose corn starch[1]	3.85	8.47	14.12	18.15	21.17
Waxy rice	31.15	26.53	20.88	16.85	13.83
Microcrystalline cellulose[2]	14.70	14.23	13.66	13.25	12.95
Choline chloride	0.1	0.1	0.1	0.1	0.1
Vitamin premix[3]	0.4	0.4	0.4	0.4	0.4
Mineral premix[4]	0.5	0.5	0.5	0.5	0.5
Sodium alginate	0.1	0.1	0.1	0.1	0.1
$Ca(H_2PO_4)_2$	1.0	1.0	1.0	1.0	1.0
Analyzed composition					
Dry matter (%)	91.59	91.21	91.93	91.18	91.30
Crude protein (%)	32.77	32.54	32.80	32.58	32.71
Crude fat (%)	6.85	6.41	6.83	6.56	6.72
Ash (%)	7.30	7.49	7.59	7.72	7.66
Amylose (%)	2.58	5.56	9.17	12.17	14.12
Amylopectin (%)	25.81	23.18	19.51	16.02	14.41
Amylose-amylopectin	0.11	0.24	0.47	0.76	0.98

[1] High-amylose corn starch (Hi-Maize 260) obtained from National Starch and Chemical Co., Ltd., Shanghai, China.

[2] Microcrystalline cellulose obtained from Baisuihang Feed Co. Ltd., Xiamen, China.

[3] Vitamin premix (per kg diet): vitamin C (as Stay-C 35%) 400 mg; vitamin A (as vitamin A palmitate) 15,000 IU; vitamin D_3 7,500 IU; vitamin E (as D, L-a-tocopheryl acetate) 150 mg; vitamin K_3 (as menadione sodium bisulphite) 7.5 mg; vitamin B_1 (as thiamin hydrochloride) 75 mg; riboflavin 60 mg; vitamin B_6 (as pyridoxine hydrochloride) 100 mg; D-calcium pantothenate 120 mg; nicotinic acid 150 mg; biotin 1.0 mg; inositol 450 mg.

[4] Mineral premix (per kg diet): Fe (from Ferrous sulfate) 80 mg; Zn (from Zinc sulfate) 40 mg; Cu (from Copper sulfate) 5 mg; Mn (from Manganese sulfate) 20 mg; Se (from Sodium selenite) 0.2 mg; I (from Potassium iodine) 30 μg; Co (from Cobalt chloride) 50 μg; Mg (from Magnesium sulphate) 300 mg; P (from Sodium dihydrogen phosphate) 600 mg; K (from Potassium chloride) 740 mg.

Feeding Procedures

The tilapia juveniles obtained from a local commercial farm in Xiamen were transported to the aquaculture laboratory of Jimei University. The fish were stocked into a closed recirculating system consisting of two circular fiber glass tanks (0.85 m height×1.22 m upper diameter, 1.04 m lower diameter) with a polygeyser bead filter (Aquaculture Systems Technologies, LLC., USA) and were initially maintained on a commercial tilapia feed for 15 d. A total of four hundred and fifty animals, initially weighing an average of 9.81±0.42 g (means±SD) were randomly divided into five experimental groups, each with six replicate 100–l tanks at a density of 15 fish per tank. During a feeding period of 42 d, the fish in each group were hand-fed on one of the diets to satiation at 08:00, 13:00 and 18:00 h per day under a natural photoperiod. After 7 d of feeding, feces were collected from wk 2, until the end of the feeding trial. Excess feed was collected one hour after each feeding and then feces were collected by siphoning. To minimize the nutrient leaching in feces, only fresh and intact feces were collected. Feces collected from a single tank were pooled each day and dried at 70°C. Dissolved oxygen was measured daily and nitrite-N was monitored twice a week with a multi-parameter photome (HI83200, Hanna Instruments, USA). During the feeding trial, the water temperature, dissolved oxygen and nitrite-N ranged from 27.5 to 32°C, 4.9 to 5.6 mg/L and 0 to 0.2 mg/L, respectively.

Sampling Procedures

At the termination of the feeding trial, fish in each tank were bulk weighed and counted after 24 h starvation. Three fish were randomly taken from each of four replicate tanks per treatment and killed with an overdose of MS 222 (tricaine methanesulfonate) solution, pooled, weighed, minced and then dried at 70°C for whole-body composition analysis. Additionally, sixteen fish from four replicate tanks in each treatment were randomly sampled, anesthetized (MS 222, 100 mg/L) and weighed individually. Blood was collected via venipuncture using a heparinized syringe and immediately centrifuged at 3,000×g for 10 min at 4°C. Plasma samples from each tank were pooled and stored at −80°C for subsequent analysis of blood biochemical parameters. After blood collection, the liver and intestine were immediately removed and weighed. Liver and intestine samples were frozen in liquid nitrogen and stored at −80°C before use. The remaining fish in each treatment were used for the subsequent postprandial blood response test.

Postprandial Blood Metabolic Response Test

After sample collection, the remaining fish in each treatment were readjusted and distributed into the same tanks as used in the growth trial with three tanks of 22 fish per tank. Fish were continued to be maintained on their respective diets for 7 d. The next morning, fish received a single meal to apparent satiety within 30 min. Blood samples were collected from three fish in each tank with a total of nine fish for each treatment at each sampling time, according to the above described method and blood were collected at 0, 1, 3, 6, 9, 12 and 24 h after a meal. Plasma aliquots were stored at −80°C for subsequent analysis of blood biochemical parameters.

Plasma Components and Liver Glycogen Analysis

Plasma glucose, triglyceride (TG), total cholesterol (T-CHO), high-density lipoprotein cholesterol (HDL-C) and low-density lipoprotein cholesterol (LDL-C) were measured by using kits (BioSino Bio-technology and Science, Inc., China). All the blood parameters were determined by using an automatic biochemical analyzer (Hitachi 7020, Japan) as described by Ye et al. (2011). Total plasma amino acid and liver glycogen were analyzed using the L-amino acid assay kit and glycogen assay kit, respectively (Nanjing Jiancheng Bioengineering Institute, China) according to the manufacturer's protocols by using UV-2800S spectrophotometer (UNICO Instruments Co., Ltd., China).

Digestive Enzyme Analysis

The pooled intestines of each sample were homogenized in chilled saline at a ratio of 1:9 (w/v) for 10 min. The supernatant was collected after centrifugation at 6,261×g for 15 min at 4°C. Aliquots of the supernatant, designated as the crude extract, were used to estimate the activities of amylase and protease, according to the procedure described by Ye et al. (2011), and the lipase activity was assayed with the lipase activity assay kit (NJBI, China) using a UV-2800S spectrophotometer.

Proximate Composition Analysis

Crude protein, crude fat, moisture and ash contents in feed ingredients, diets, feces and whole-body fish samples were determined according to the methods of the Association of Official Chemists (AOAC, 1995). Crude protein content was determined by the Kjeldahl method (N×6.25) using Kjeltec™ 8400 Auto Sample Systems (Foss Tecator AB, Sweden). Crude fat content was determined by the Soxhlet method by using Soxtec Avanti 2050 (Foss Tecator AB). Moisture content was determined by the drying method, using an oven at 105°C. Ash content was determined by the combustion method, using a muffle furnace at 550°C. Acid-insoluble ash content was determined by using the same muffle furnace at 550°C, after acid digestion. Acid-insoluble ash was used as an endogenous indicator, instead of chromium oxide as an exogenous marker, to determine the nutrient digestibility because the dietary chromium oxide as a glucose tolerance factor is believed to affect starch utilization (Shiau and Chen, 1993). The total starch content in corn starch, waxy rice, diets and feces was calorimetrically determined by using the total starch assay kit (Megazyme International Ireland Ltd., Ireland). Using the amylose assay kit (Megazyme International Ireland Ltd., Ireland), the amylose content in high-amylose corn starch, sticky rice and diets was analyzed.

Calculations

Percent weight gain, feed conversion ratio, protein efficiency ratio, feeding rate, survival, hepatosomatic index and condition factor were calculated as follows:

Percent weight gain (WG, %) = $100\times(W_f-W_i)/W_i$

Feed efficiency (FE, %) = $100\times(W_f-W_i)/W_c$

Protein efficiency ratio (PER) = $(W_f-W_i)/(W_c\times P)$

Feeding rate (FR, %) = $100\times W_c/[(W_i+W_f)/2]/t$

Hepatosomatic index (HSI, %) = $100\times W_l/W_b$

Condition factor (CF) = $100\times W_b/L^3$

where W_i (g) and W_f (g) are the initial body weight (IBW) and final body weight (FBW) respectively; W_c(g) is the average feed intake per fish (on an as-fed basis); t is the time duration of the experiment (days); P is the protein content of the diet; W_l (g) is the liver weight of the sampled fish; W_b (g) is the final body weight of the sampled fish and L (cm) is the final body length of the sampled fish.

The apparent digestibility coefficients (ADCs) of the experimental diets for dry matter, crude protein, crude fat, ash and starch were calculated according to the following equation (Spyridakis et al., 1989):

$$ADC(\%)=100-100\times(N_f)/(N_d)\times(M_d)/(M_f)$$

where N_f is the percentage of a specific nutrient in feces, N_d is the percentage of the same nutrient in the diet, M_d is the percentage of acid-insoluble ash in the diet and M_f is the percentage of acid-insoluble ash in feces.

Statistical Analysis

Data are presented as mean±SD. The normality and variance's homogeneity of all data were systematically checked before applying One-way ANOVA, and the mean values were compared by using the Student-Neuman-Keuls multiple comparison tests, at the 5% level of significance. The results of the measured variables were examined by regression. A repeated measures ANOVA was carried out to examine mean differences in postprandial concentrations of blood glucose, TG and total amino acids among treatments at time intervals. All statistical tests were calculated using SPSS13.0 software (SPSS Inc., Chicago, USA).

RESULTS

Growth Performance

The fish in each treatment exhibited proper growth and survived across the feeding period. Fish fed diet 2 exhibited significantly ($p<0.05$)higher WG (%), FE (%) and PER compared with fish fed diet 5 (Table 2). A strong negative correlation was observed between WG%, FE% and PER and dietary amylose-amylopectin ratio (R = −0.8907, −0.9347 and −0.9031, respectively) and the values for WG (%), FE (%) and PER in fish fed diet 2 were significantly higher than those in fish fed diet 5 ($p<0.05$). Although HSI (%) did not differ across treatments, the parameter was linearly (R = −0.9728) correlated with dietary amylose-amylopectin ratio. No influences of dietary amylose-amylopectin ratio on FR% and CF could be detected ($p>0.05$).

Table 2: Effects of dietary amylose-amylopectin ratios on the growth performance of tilapia over a feeding period of 6 wks

Diets[1]	Growth parameters						
	FAW (g)[2]	WG (%)[2]	FE (%)[2]	PER[2]	FR (%)	HSI (%)[3]	CF[3]
Diet 1 (0.11)	69.48[ab]	606.7[ab]	79.4[ab]	2.42[ab]	5.28	2.64	3.98
Diet 2 (0.24)	75.38[a]	659.7[a]	82.6[a]	2.55[a]	5.13	2.49	4.03
Diet 3 (0.47)	62.56[b]	530.0[b]	76.3[ab]	2.33[ab]	5.26	2.39	3.96
Diet 4 (0.76)	58.69[b]	490.9[b]	73.5[ab]	2.25[ab]	5.59	2.20	3.83
Diet 5 (0.98)	57.63[b]	479.7[b]	69.4[b]	2.15[b]	5.57	2.18	3.97
Pooled SE	2.045	20.541	1.399	0.043	0.083	0.078	0.038
p-value	0.009	0.006	0.015	0.015	0.334	0.028	0.592

WG (Y)-amylose-amylopectin ratio (X) correlation: Y= 651.62-191.84X; R = -0.8907; p = 0.0427.
FE (Y)-amylose-amylopectin ratio (X) correlation: Y= 83.05-13.29X; R = -0.9347; p = 0.0199.
PER (Y)-amylose-amylopectin ratio (X) correlation: Y= 2.54-0.386X; R = -0.9031; p =0.0357.
HSI (Y)-amylose-amylopectin ratio (X) correlation: Y= 2.65-0.527X; R = -0.9728; p = 0.0054.

Values with different superscripts in each column indicate significant difference ($p<0.05$).
FAW = Final average weight; WG % = Percent weight gain; FE = Feed efficiency; PER = Protein efficiency ratio; FR = Feeding rate; HIS = Hepatosomatic index; CF = Condition factor.
[1] Values in parentheses are amylose-amylopectin ratios in diets. [2] Data are means of four replicate tanks per dietary treatment.
[3] Data are means of 24 fish per dietary treatment.

Nutrient Digestibility

As shown in Table 3, the digestibility values for dry matter, crude protein and lipid did not differ among dietary treatments ($p>0.05$). However, there was a negatively linear correlation between dry matter digestibility and dietary amylose-amylopectin ratio (R = −0.8948; p = 0.0403). The starch digestibility showed a decreasing trend with the increase in dietary amylose-amylopectin ratio (R = −0.9893, p = 0.0013), and the values of fish fed diets 1–2 were significantly ($p<0.05$) higher than those of fish fed with other diets. When considering WG%, FE% and PER in relation to nutrient digestibility, the FE% and PER were linearly correlated with the values for dry matter (R = 0.973, p = 0.005; R = 0.928, p = 0.023, respectively) and for starch (R = 0.956, p = 0.011; R = 0.903, p = 0.036, respectively). The WG% was also correlated linearly with apparent digestibility of starch (R = 0.911, p = 0.031), but protein and lipid digestibility did not show this relationship (Table 3).

Table 3: Effects of dietary amylose-amylopectin ratios on nutrient digestibility for tilapia over a feeding period of 6 wks

	Apparent digestibility coefficients (%) [2]			
	Dry matter	Crude protein	Crude fat	Starch
Diet 1 (0.11)	82.29	90.85	94.44	72.53[a]
Diet 2 (0.24)	82. 95	91.53	93.06	70.76[a]
Diet 3 (0.47)	82.19	91.14	93.96	66.47[b]
Diet 4 (0.76)	81.59	90.98	94.89	64.25[b]
Diet 5 (0.98)	80.49	90.45	93.97	61.59[b]
Pooled SE	0.755	0.607	0.618	1.191
p-value	0.912	0.992	0.940	0.001

Dry matter (Y)-amylose-amylopectin ratio (X) correlation: Y = 83.08-2.30X; *R* = 0.8948; p = 0.0403.
Starch (Y)-amylose-amylopectin ratio (X) correlation: Y = 73.48-12.43X; *R* = -0.9893; p = 0.0013.
Dry matter (Y)-feed efficiency (X) correlation: Y= 068.49+0.18X; *R* = 0.973; p = 0.005.
Dry matter (Y)-protein efficiency ratio (X) correlation: Y = 4.59+0.82X; *R* = 0.928; p = 0.023.
Starch (Y)-feed efficiency (X) correlation: Y = 68.46+5.75X; *R* = 0.956; p = 0.011.
Starch (Y)-protein efficiency ratio (X) correlation: Y = 5.10+26.51X; *R* = 0.903; p = 0.036.
Starch (Y)-percent weight gain (X) correlation: Y = 37.0+0.05X; *R* = 0.911; p = 0.031.

Values with different superscripts in each column indicate significant difference (p<0.05).
[1] Values in parentheses are amylose-amylopectin ratios in diets. [2] Data are means of four replicate tanks per dietary treatment.

Digestive Enzyme Activity

Highest protease activity was observed in fish fed diet 2, and was significantly (p<0.05) higher than those in fish fed the other diets (Table 4). A decreasing trend of lipase (R = −0.9080, p = 0.0330) and amylase (R = −0.8617, p = 0.0604) activities was observed as dietary amylose-amylopectin ratio increased and the values of amylase and lipase in fish fed with diet 1 were significantly (p<0.05) higher than those of fish fed other diets. The amylase activity was linearly correlated (R = 0.910; p = 0.032) with starch digestibility in response to dietary amylose-amylopectin ratio (Table 4).

Table 4: Effects of dietary amylose-amylopectin ratios on digestive enzyme activity in the gut of tilapia over a feeding period of 6 wks

Diets[1]	Digestive enzyme activity in gut (U/mg protein)[2]		
	Protease	Lipase	Amylase
Diet 1 (0.11)	1,803[b]	207. 9[a]	6.51[a]
Diet 2 (0.24)	2,020[a]	152.2[b]	4.93[b]
Diet 3 (0.47)	1,221[c]	137.9[b]	3.43[c]
Diet 4 (0.76)	1,253[c]	123.2[bc]	3.09[c]
Diet 5 (0.98)	1,361[c]	104.2[c]	3.24[c]
Pooled SE	76.21	8.50	0.35
p-value	<0.001	<0.001	<0.001

Lipase (Y)-amylose-amylopectin ratio (X) correlation: Y = 195.94-99.34X; *R* = -0.9080; p = 0.0330.
Amylase (Y)-amylose-amylopectin ratio (X) correlation: Y = 6.04-3.51X; *R* = -0.8617; p = 0.0604.
Amylase (Y)-starch digestibility (X) correlation: Y = -15.89+0.3X; *R* = 0.910; p = 0.032.

Values with different superscripts in each column indicate significant difference (p<0.05).
[1] Values in parentheses are amylose-amylopectin ratios in diets. [2] Data are means of four replicate tanks per dietary treatment.

Whole-Body Composition

The values of moisture, crude protein, crude fat and ash of the fish did not vary across dietary treatments (Table 5). In case of the relationship between body lipid and dietary amylose-amylopectin ratio, lipid retention was inversely related to the ratio (R = −0.9642, p = 0.0081).

Table 5: Effects of different dietary amylose-amylopectin ratios on whole body composition of tilapia over a feeding period of 6 wks

Diets	Whole-body composition (%)[1]			
	Moisture[2]	Crude protein[2]	Crude fat[2]	Ash[2]
Diet 1 (0.11)	74.03	12.68	8.02	3.66
Diet 2 (0.24)	73.83	13.11	7.78	3.66
Diet 3 (0.47)	72.96	13.66	7.68	3.93
Diet 4 (0.76)	74.77	13.07	6.89	4.11
Diet 5 (0.98)	74.47	13.39	6.86	3.86
Pooled SE	0.398	0.212	0.217	0.089
p-value	0.702	0.705	0.326	0.495

Crude fat (Y)-amylose-amylopectin ratio (X) correlation: $Y = 8.18-1.44X$; $R = -0.9642$; $p = 0.0081$.

Values with different superscripts in each column indicate significant difference ($p<0.05$).
[1] Crude protein, crude fat and ash are expressed on a wet weight basis. [2] Data are means of four replicate tanks per dietary treatment.

Plasma Components and Liver Glycogen

There were no significant differences in plasma glucose, TG, T-CHO, LDL-C, and HDL-C concentrations between dietary treatments (Table 6). However, liver glycogen content showed significant difference between different dietary treatments ($p<0.05$), and the highest value was seen in fish fed diet 4 and lowest in fish fed diet 3.

Table 6: Effects of dietary amylose-amylopectin ratios on the concentrations of plasma components and liver glycogen in tilapia over a feeding period of 6 wks

Diets[1]	Plasma components (mmol/L)[2]					Liver glycogen (mg/g)
	Glucose	TG	T-CHO	LDL-C	HDL-C	
Diet 1 (0.11)	7.22	4.75	5.19	2.21	2.42	70.3^{b}
Diet 2 (0.24)	7.36	4.70	5.79	2.56	2.72	56.7^{b}
Diet 3 (0.47)	6.45	4.58	5.56	2.52	2.61	54.6^{b}
Diet 4 (0.76)	6.17	4.42	4.50	2.14	2.57	98.3^{a}
Diet 5 (0.98)	6.87	4.53	4.43	2.38	2.63	97.7^{a}
Pooled SE	0.132	0.037	0.151	0.047	0.034	4.856
p-value	0.121	0.339	0.158	0.627	0.983	0.003

TG (Y)-amylose-amylopectin ratio (X) correlation: $Y = 4.470-0.135\log(X)$; $R = 0.9054$; $p = 0.0344$.

Values with different superscripts in each column indicate significant difference ($p<0.05$).
TG = Triglyceride; T-CHO = Total cholesterol; HDL-C = High-density lipoprotein cholesterol; LDL-C = Low-density lipoprotein cholesterol, LDL-C.
[1] Values in parentheses are amylose-amylopectin ratios in diets. [2] Data are means of four replicate tanks per dietary treatment.

Postprandial Blood Metabolic Response

Plasma glucose concentrations of fish fed diets 1, 2 and 5 reached a peak 3 h after a meal (Figure 1), but the peak was reached in fish fed diets 3–4 only at 6 h, and thereafter the values gradually decreased to about the preprandial levels till 24 h. However, the fish fed diet 1 exhibited a significant difference ($p<0.05$), when compared with the lower peaks achieved in diets 3–5.

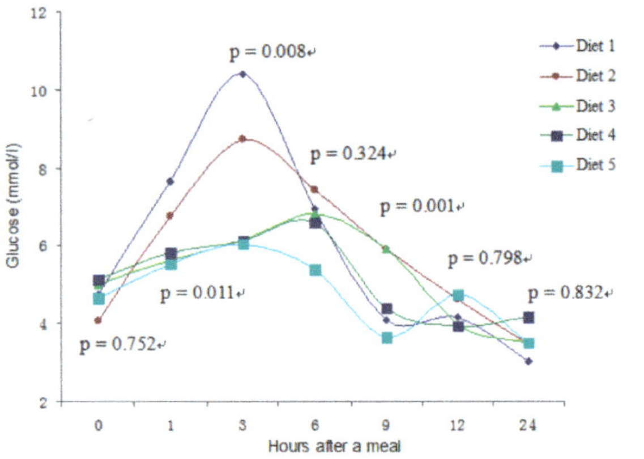

Figure 1: Effect of amylose-amylopectin ratios on the plasma glucose concentrations in the fish at different diets. Plasma glucose concentrations were measured at various time intervals after a meal in fish fed diets varying in amylose-amylopectin ratios (0.11, 0.24, 0.47, 0.76 and 0.98, respectively) over the 6-wk feeding period. Data are means±SD ($n = 3$), represents three replicate tanks used at each sampling time. Plasma samples of three fish were pooled for tank. Plasma glucose was significantly affected by dietary treatments and time after a meal.

A decreasing trend was observed with the increase in dietary amylose-amylopectin ratio, and the fish fed diet 1, showed the highest value, which was significantly different ($p<0.05$) from the fish fed diets 3–5 at 1, 3 or 9 h, while lowest value was observed in the fish fed diet 5. Highest peak was observed in fish fed diets 1–3 and diet 5 3 h after a meal, but it reached its peak in fish fed diet 4 only after 6 h (Figure 2). However, the peak in fish fed diet 1 was significantly ($p<0.05$) higher than fish fed diets 3 and 5 at 3 h. The values in fish fed diet 1 maintained at a higher level than those of fish fed other diets for the 9 h after a meal. The values across all treatments were reduced to below the preprandial value till 24 h.

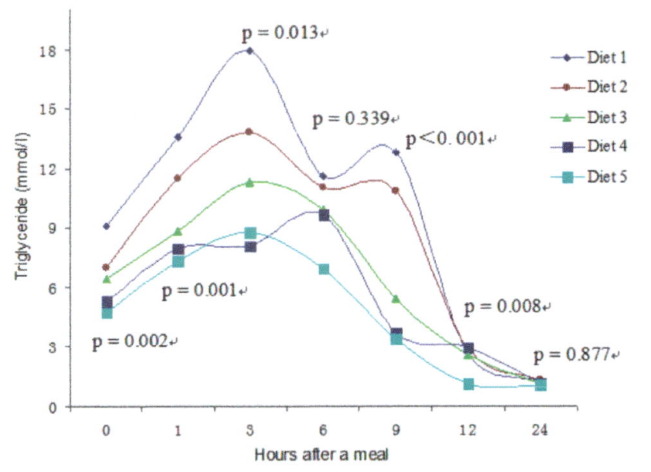

Figure 2: Effect of amylose-amylopectin ratios on the plasma triglyceride concentrations in the fish at different diets. Plasma triglyceride concentrations were measured at various time intervals after a meal in fish fed diets varying in amylose-amylopectin ratios (0.11, 0.24, 0.47, 0.76 and 0.98, respectively) over the 6-wk feeding period. Data are means±SD (n = 3), represents three replicate tanks used at each sampling time. Plasma samples of three fish were pooled for tank. Plasma triglyceride was significantly affected by dietary treatments and time after a meal.

Contrary to the plasma glucose changes, total amino acid concentrations of fish fed diets 1 and 2 reached a peak 6 h after a meal, whereas the peak of fish fed diets 3–5 was reached at 3 h (Figure 3). The peak of fish fed diet 5 at 3 h was at the lowest and was not different from the initial value and those at other sampling times.

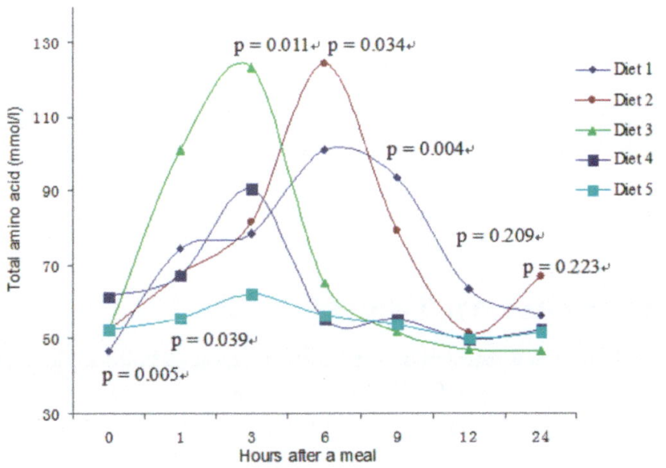

Figure 3: Effect of amylose-amylopectin ratios on the plasma total amino acid concentrations in the fish at different diets. Plasma total amino acid concentrations were measured at various time intervals after a meal in fish fed diets varying in amylose-amylopectin ratios (0.11, 0.24, 0.47, 0.76 and 0.98, respectively) over the 6-wk feeding period. Data are means±SD (n = 3), represents three replicate tanks used at each sampling time. Plasma samples of three fish were pooled for tank. Plasma total amino acid was significantly affected by dietary treatments and time after a meal.

DISCUSSION

The results of our study demonstrated that amylose-amylopectin ratio in the starch fraction of the diet affects the growth performance of the tilapia. The fish fed with the diet containing amylose-amylopectin at ratio of 0.24 had superior performance (WG% and FE%) relative to those fed the diets at higher or lower ratio, which was not accordance with the results of Rawles and Lochmann (2003) who observed enhanced growth of sunshine bass that were fed a 70% amylose diet. Our finding is consistent with what has been reported for broilers (Dai et al., 2008) and piglets (Deng et al., 2010) wherein the dietary amylose-amylopectin ratio at 0.23 is suitable for this production. The observation in our study and the previous reports are closer to the average value of amylose-amylopectin in cereal starch fraction (Shibanuma et al., 1994; Hruby, 2005). The digestion of starch depends on its amylose-amylopectin ratio and their molecular weight (Granfeldt et al., 1994; Englyst and Hudson, 1996). Amylose has more hydrogen bonds and undergoes more rapid retrogradation than does amylopectin (Zhou and Kaplan, 1997), while the amylopectin, having a branched chain, allows a greater access to digestive enzymes. For this reason, amylose is more resistant to hydrolytic enzymes compared with amylopectin (Kabir et al., 1998; Weurding et al., 2001), which may be partly attributed to the presence or development of resistance starch (Behall and Hallfrisch, 2002). Therefore, the ratio of amylose to amylopectin affects the digestion of starch and other macronutrients and its metabolic responses (Le Leu et al., 2009; Deng et al., 2010; Fugui et al., 2010; Jun et al., 2010). In the present study, a linear correlation was found between amylase activity and starch digestibility in response to dietary amylose-amylopectin ratio, which reflects the direct impact of dietary amylose-amylopectin ratio on starch digestion at a higher level. Furthermore, FE%, PER or/and WG% correlated linearly with the apparent digestibility of dry matter and starch. Our results agreed with what was reported by Denardin et al. (2012) for rats. However, the mismatch between protease activity and protein digestibility and between lipase activity and lipid digestibility in response to dietary amylose-amylopectin ratio indicated that protein and lipid digestibility was less affected by starch type compared with starch digestibility. This indicates that nutrient digestibility is not the only reason for the differences in starch utilization in terms of starch type. Previous studies showed that rats and poultry were able to adapt to dietary changes by increasing their feed intake and dietary energy consumption to compensate the lower digestibility values of diets caused by starch type (Zhou and Kaplan, 1997;Pirgozliev et al., 2010). Similar responses also exist in some carnivorous fish (Bergot, 1979; Yamamoto et al., 2001; Boujard et al., 2004), that is, such fish have the ability to compensate for imbalances in nutrient density through increasing feed intake. The tilapia, however, exhibited a partial ability in this regard in our current study, because although the feed intake increased with an increasing dietary ratio of amylose-amylopectin, the fish fed higher amylose still exhibited lower nutrient digestibility and growth. Berger and Halver (1987) showed that liver enlargement is associated with increased lipid or glycogen deposition. An increased HSI was found in high amylose-fed rats (Zhou and Kaplan, 1997). However, the opposite result was observed in high amylose-fed sunshine bass (Rawles and Lochmann, 2003), consistent with our current results. This indicates that liver was in a good state, though liver glycogen concentration was higher in fish fed a diet at a high amylose-amylopectin ratio compared with those at the low ratio. In the present study, liver glycogen deposition indicates that starch not used for energy may accumulate in the liver of the tilapia fed higher-amylose-amylopectin ratio diets, supporting the observations for rainbow trout Oncorhynchus mykiss (Brauge et al., 1994). These observations may partly account for the inferior growth and feed utilization of fish that were fed higher-amylose diets.

The whole-body composition was not affected by dietary treatments, which was similar to the observations by Solomon et al. (2007). Although not significant, there was a progressive reduction in whole-body lipid retention to dietary amylose-amylopectin ratio. The inverse relationship between dietary amylose-amylopectin ratio and whole-body lipid retention appears interesting, since an increased dietary amylose level did not lead to undesirable lipid accumulation in the body of the fish, though the whole-body protein retention was not related with the dietary amylose-amylopectin ratio.

Rawles and Lochmann (2003) reported a reduced liver glycogen deposit in higher amylose-fed sunshine bass. The opposite response was observed in our current study and in previous work with gibel carp (Carassius auratus gibelio; Tan et al., 2006). Regardless of starch type, high dietary starch levels could elevate liver glycogen deposits in rainbow trout (Bergot, 1979; Suárez et al., 2002), eel (Angilla Anguilla; Suárez et al., 2002) and Atlantic cod (Gadus morhua; Hemre et al., 1993), indicating the relative low ability of these carnivorous fish to use glucose by the glycolytic pathway. The variations may be caused by the difference in fish species.

The plasma intermediary metabolites were not affected by dietary treatments, which is consistent with those of Camp et al. (2003), who reported no effect of starch type on plasma metabolites (TG, LDL-C and HDL-C) in growing pigs. However, decreased serum T-CHO and TG levels in mammals was observed by Behall et al. (1988) and Denardin et al. (2012), whereas the opposite effect was observed by Yin et al. (2011) in weaned pigs after the consumption of a diet rich in amylose compared to a diet rich in amylopectin. The reason for this difference is unclear in that the dietary starch level was identical for each experiment besides protein and lipid level. Some previous studies have also shown that starch type is not the sole determinant factor affecting starch digestion and postprandial glycemic responses (Zhou and Kaplan, 1997) as well as growth performance (Pirgozliev et al., 2010), but there are other physiochemical characteristics of cereals like genotype (Granfeldt et al., 1994; Gutierrez Del Alamo et al., 2009), processing (Mahadevamma and Tharanathan, 2007; Glencross et al., 2012), hardness (Denardin et al., 2012) and non-starch polysaccharides (Glencross et al., 2012), which could also exert some effect on the growth.

Because of the high enzyme resistance of amylose, mammals ingesting high dietary amylose have low glycemic and insulinemic postprandial responses compared with amylopectin (Granfeldt et al., 1994; Noakes et al., 1996; Zhou and Kaplan, 1997; Behall and Hallfrisch, 2002; Aziz et al., 2009). This effect is beneficial for the control of diabetes and also to reduce the risk of diabetes development (Behall and Hallfrisch, 2002;Aziz et al., 2009). Contrary to terrestrial animals, tolerance to carbohydrate is somewhat lower in fish (Wilson, 1994; Enes et al., 2010), and its underlying mechanism remains unclear (Moon, 2001; Polakof et al., 2010). Few studies have observed fish glycemic changes in response to amylose-amylopectin ratio. In the present study, we measured the postprandial changes in several blood intermediary metabolites of fish fed diets varying in amylose-amylopectin ratios. The weak peak level of blood glucose in the tilapia in our study and in sunshine bass in the previous study conducted by Rawles and Lochmann (2003) was observed after the high-amylose load when compared with those with the low-amylose load, a fact suggesting that feeding higher-amylose diet might lead to higher tolerance to starch. Meanwhile, peak blood TG paralleled with the glucose change of the tilapia in response to dietary amylose-amylopectin ratio. In contrast with peak blood glucose, peak total amino acid time was delayed when fish were fed the low-amylose diet when compared with those fed the high-amylose diet. We also observed that the highest dietary amylose-amylopectin ratio produced the weakest metabolic responses, regardless of blood glucose, TG or total amino acids. The lowered metabolic responses in the presence of high-amylose starch could be associated with a lowered rate of amylolysis and this may be due to the increased time of gastrointestinal transit after ingestion of high-amylose (Goddardet al., 1984; Granfeldt et al., 1994; Weurding et al., 2001).

CONCLUSIONS

There is considerable variation in growth and feed utilization with the dietary use of starch type in terrestrial animals and fish. Our results in the present study showed that there were negative linear relationships between percent weight gain, feed efficiency, protein efficiency ratio, starch digestibility, the activities of protease, lipase and amylase and the dietary amylose-amylopectin ratio. After a meal, peak blood glucose and TG were lower in fish fed higher-amylose diets than those in fish fed lower-amylose diets, and peak total amino acid time was delayed in the fish fed the low-amylose diet when compared with those fed the high-amylose diet. The decreased growth performance, along with the lower-peak blood glucose and TG as well as the delayed peak total amino acid time in fish fed higher-amylose-amylopectin diets let us to conclude that higher dietary amylose compromises the growth but helps in reducing the blood glucose stress on the fish caused by postprandial starch load.

ACKNOWLEDGMENTS

This study was funded by the Fujian Provincial Natural Science Foundation (2012J01141), the Foundation for Innovative Research Team of Jimei University (2011A001), and the Research Foundation of Jimei University (F04008).

REFERENCES

1. AOAC . Association of Official Analytical Chemists. In: Cunniff PA, editor. Official Methods of Analysis. 16th edn. AOAC International; Arlington, VA, USA: 1995.
2. Aziz AA, Kenney LS, Goulet B, Abdel-Aal E. Dietary starch type affects body weight and glycemic control in freely fed but not energy-restricted obese rats. J Nutr. 2009;139:1881–1889.
3. Behall KM, Hallfrisch J. Plasma glucose and insulin reduction after consumption of breads varying in amylose content. Eur J Clin Nutr. 2002;56:913–920.
4. Behall KM, Scholfield DJ, Canary J. Effect of starch structure on glucose and insulin responses in adults. Am

J Clin Nutr. 1988;47:428–432.

5. Behall KM, Scholfield DJ, Yuhaniak I, Canary J. Diets containing high amylose vs amylopectin starch: effects on metabolic variables in human subjects. Am J Clin Nutr. 1989;49:337–344.
6. Berger A, Halver JE. Effect of dietary protein, lipid and carbohydrate content on the growth, feed efficiency and carcass composition of striped bass, Morone saxatillis (Walbaum), fingerlings. Aquac Res. 1987;18:345–356.
7. Bergh M, Razdan A, Aman P. Nutritional influence of broiler chicken diets based on covered normal, waxy and high amylose barleys with or without enzyme supplementation. Anim Feed Sci Technol.1999;78:215–226.
8. Bergot F. Carbohydrate in rainbow trout diets: effects of the level and source of carbohydrate and the number of meals on growth and body composition. Aquaculture. 1979;18:157–167.
9. Boujard T, Gelineau A, Coves D, Corraze G, Dutto G, Gasset E, Kaushik S. Regulation of feed intake, growth, nutrient and energy utilisation in European sea bass (Dicentrarchus labrax) fed high fat diets. Aquaculture. 2004;231:529–545.
10. Brauge C, Medale F, Corraze G. Effect of dietary carbohydrate levels on growth, body composition and glycaemia in rainbow trout, Oncorhynchus mykiss, reared in seawater. Aquaculture.1994;123:109–120.
11. Camp LK, Southern LL, Bidner TD. Effect of carbohydrate source on growth performance, carcass traits, and meat quality of growing-finishing pigs. J Anim Sci. 2003;81:2488–2495.
12. Cohen RS, Tanksley TD., Jr Energy and protein digestibility of sorghum grains with different endosperm textures and starch types by growing swine. J Anim Sci. 1973;37:931–935.
13. Collins NE, Moran ET, Jr, Stilborn HL. Performance of broilers fed normal and waxy corn diets formulated with chick and rooster derived apparent metabolizable energy values for the grains. J Appl Poult Res. 2003;12:196–206.
14. Dai Q, Li X, Zhang S, Jiang G, Hu Y, Chen F. Effects of different dietary amylose-amylopectin ratios on production performance and nutrient availability of yellow broilers. Chinese J Anim Nutr.2008;20:249–255.
15. Denardin CC, Boufleur N, Reckziege P, da Silva LP, Walter M. Amylose content in rice (Oryza sativa) affects performance, glycemic and lipidic metabolism in rats. Ciênc Rural. 2012;42:381–387.
16. Deng J, Wu X, Bin S, Li TJ, Huang R, Liu Z, Liu Y, Ruan Z, Deng Z, Hou Y, Yin YL. Dietary amylose and amylopectin ratio and resistant starch content affects plasma glucose, lactic acid, hormone levels and protein synthesis in splanchnic tissues. J Anim Physiol Anim Nutr. 2010;94:220–226.
17. Enes P, Panserat S, Kaushik S, Oliva-Teles A. Nutritional regulation of hepatic glucose metabolism in fish. Fish Physiol Biochem. 2009;35:519–539.
18. Enes P, Peres H, Couto A, Oliva-Teles A. Growth performance and metabolic utilization of diets including starch, dextrin, maltose or glucose as carbohydrate source by gilthead sea bream (Sparus aurata) juveniles. Fish Physiol Biochem. 2010;36:903–910.
19. Englyst HN, Hudson GJ. The classification and measurement of dietary carbohydrates. Food Chem.1996;57:15–21.
20. Ertl D, Dale N. The metabolizable energy of waxy vs. normal corn for poultry. J Appl Poult Res.1997;6:432–435.
21. Fugui Y, Zhenzhen Z, Ju H, Yulong Y. Digestion rate of dietary starch affects systemic circulation of amino acids in weaned pigs. Br J Nutr. 2010;103:1404–1412.
22. Glencross B, Blyth D, Tabrett S, Bourne N, Irvin S, Anderson M, Fox-Smith T, Smullen R. An assessment of cereal grains and other starch sources in diets for barramundi (Lates calcarifer) – implications for nutritional and functional qualities of extruded feeds. Aquacult Nutr. 2012;18:388–399.
23. Goddard MS, Young G, Marcus R. The effect of amylose content on insulin and glucose responses to ingested rice. Am J Clin Nutr. 1984;39:388–392.
24. Granfeldt Y, Drews A, Björck I. Arepas made from high amylose corn flour produce favorably low glucose and insulin response in healthy humans. J Nutr. 1995;125:459–465.
25. Granfeldt Y, Liljeberg H, Drews A, Newman R, Björck I. Glucose and insulin responses to barley products: influence of food structure and amylose-amylopectin ratio. Am J Clin Nutr. 1994;59:1075–1082.
26. Gutierrez Del, Alamo A, Verstegen MWA, den Hartog LA, Perez de Ayala P, Villamide MJ. Wheat starch digestion rate affects broiler performance. Poult Sci. 2009;88:1666–1675.
27. Hemre GI, Lie Ø, Sundby A. Dietary carbohydrate utilization in cod (Gadus morhua): metabolic responses to feeding and fasting. Fish Physiol Biochem. 1993;10:455–463.
28. Hemre G-I, Mommsen TP, Krogdahl Å. Carbohydrates in fish nutrition: effects on growth, glucose metabolism and hepatic enzymes. Aquacult Nutr. 2002;8:175–194.
29. Hruby M. Challenge of corn variability. Feed Int. 2005;26:5–11.
30. Jun H, Daiwen C, Bing Y. Metabolic and transcriptomic responses of weaned pigs induced by different dietary amylose and amylopectin ratio. PLoS One. 2010;5:e15110. [PMC free article]
31. Kabir M, Rizkalla SW, Champ M, Luo J, Boillot J, Bruzzo F, Slama G. Dietary amylose-amylopectin starch content affects glucose and lipid metabolism in adipocytes of normal and diabetic rats. J Nutr.1998;128:35–43.
32. Krogdahl Å, Hemre G-I, Mommsen TP. Carbohydrates in fish nutrition: digestion and absorption in postlarval stages. Aquac Nutr. 2005;11:103–122.
33. Kyuma T, Ishida M. Growth and digestion of goats fed high-amylose corn and waxy corn. Asian-Aust J Anim Sci. 1989;2:359–360.

34. Le Leu RK, Hu Y, Brown IL, Young GP. Effect of high amylose maize starches on colonic fermentation and apoptotic response to DNA-damage in the colon of rats. Nutr Metab. 2009;6:11.[PMC free article]

35. Mahadevamma S, Tharanathan RN. Processed rice starch characteristics and morphology. Eur Food Res Technol. 2007;225:603–612.

36. Mommsen TP, Plisetskaya EM. Insulin in fishes and agnathans: history, structure and metabolic regulation. Rev Aquat Sci. 1991;4:225–259.

37. Mongeau R, Sarwar G, Brassard R, Botting HG. Effects of amylose and wheat bran on the levels of blood serum urea nitrogen (BUN), other blood parameters, growth and fecal characteristics in rats.Plant Foods Hum Nutr. 1995;48:95–105.

38. Moon TW. Glucose intolerance in teleost fish: fact or fiction? Comp. Biochem. Physiol. B, Biochem Mol Biol. 2001;129:243–249.

39. Noakes M, Clifton PM, Nestel PJ, Leu RL, McIntosh G. Effect of high-amylose starch and oat bran on metabolic variables and bowel function in subjects with hypertriglyceridemia. Am J Clin Nutr.1996;64:944–951.

40. Perez JM, Aumaitre A. Waxy versus regular maize: Energy value for growing pigs and utilization in piglet diets. Anim Feed Sci Technol. 1979;4:109–115.

41. Pirgozliev VR, Rose SP, Bedford MR. The effect of amylose-amylopectin ratio in dietary starch on growth performance and gut morphology in broiler chickens. Arch Geflügelkd. 2010;74:S21–S29.

42. Pirgozliev VR, Rose SP, Graybosch RA. Energy and amino acid availability to chickens of waxy wheat. Arch Geflügelkd. 2002;66:108–113.

43. Pirgozliev VR, Rose SP, Kettlewell PS, Bedford MR. Relationship between chemical composition of wheat and broiler chicken growth performance. Brit Poult Sci. 2000;41:S697–S698.

44. Polakof S, Álvarez R, Soengas JL. Gut glucose metabolism in rainbow trout: implications in glucose homeostasis and glucosensing capacity. Am. J. Physiol.-Regul Integr Comp Physiol. 2010;299:R19–R32.

45. Rawles S, Lochmann R. Effects of amylopectid-amylose starch ratio on growth, body composition and glycemic response of sunshine bass Morone chrysops ♀×M. saxatilis ♂ J World Aquacult Soc.2003;34:278–288.

46. Shelton JL, Matthews JO, Southern LL, Higbie AD, Bidner TD, Fernandez JM, Pontif JE. Effect of nonwaxy and waxy sorghum on growth, carcass traits, and glucose and insulin kinetics of growing-finishing barrows and gilts. J Anim Sci. 2004;82:1699–1706.

47. Shiau S-Y, Chen M-J. Carbohydrate utilization by tilapia (Oreochromis niloticus×O. aureus) as influenced by different chromium sources. J Nutr. 1993;123:1747–1753.

48. Shibanuma K, Takeda Y, Hizukuri S, Shibata S. Molecular structure of some wheat starches.Carbohydr Polym. 1994;25:111–116.

49. Solomon SG, Tiamiyu LO, Agaba UJ. Effect of feeding different grain sources on the growth performance and body composition of tilapia, (Oreochromis niloticus) fingerlings fed in outdoor hapas.Pakistan J Nutr. 2007;6:271–275.

50. Spyridakis P, Metailler R, Gabaudan J, Riaza A. Studies on nutrient digestibility in European sea bass (Dicentrarchus labrax): 1. Methodological aspects concerning faeces collection. Aquaculture.1989;77:61–70.

51. Stone DA. Dietary carbohydrate utilization by fish. Rev Fish Sci. 2003;11:337–369.

52. Suárez MD, Sanz A, Bazoco J, García-Gallego M. Metabolic effects of changes in the dietary protein: carbohydrate ratio in eel (Angilla anguill a) and trout (Oncorhynchus mykiss) Aquacult Int.2002;10:143–156.

53. Tan Q, Xie S, Zhu X, Lei W, Yang Y. Effect of dietary carbohydrate sources on growth performance and utilization for gibel carp (Carassius auratus gibelio) and Chinese longsnout catfish (Leiocassis longirostris Günther) Aquacult Nutr. 2006;12:61–70.

54. Weststrate JA, van Amelsvoort JMM. Effects of the amylose content of breakfast and lunch on postprandial variables in male volunteers. Am J Clin Nutr. 1993;58:180–186.

55. Weurding RE, Veldman A, Veen WAG, van der Aar PJ, Verstegen MWA. In vitro starch digestion correlates well with rate and extent of starch digestion in broiler chickens. J Nutr. 2001;131:2336–2342.

56. Wilson RP. Utilization of dietary carbohydrate by fish. Aquaculture. 1994;124:67–80.

57. Yamamoto T, Shima T, Furuita H, Suzuki N, Sánchez-Vázquez FJ, Tabata M. Self-selection and feed consumption of diets with a complete amino acid composition and a composition deficient in either methionine or lysine by rainbow trout Oncorhynchus mykiss (Walbaum) Aquac. Res. 2001;32(Suppl. 1):83–91.

58. Ye J, Wang K, Li F, Sun Y. Single or combined effects of fructo- and mannan oligosaccharide supplements and Bacillus clausii on the growth, feed utilization, body composition, digestive enzyme activity, innate immune response and lipid metabolism of the Japanese flounder Paralichthys olivaceus.Aquac. Nutr. 2011;17:e902–e911.

59. Yin F, Yin Y, Zhang Z, Xie M, Huang J, Huang R, Li T. Digestion rate of dietary starch affects the systemic circulation of lipid profiles and lipid metabolism-related gene expression in weaned pigs. Br J Nutr. 2011;106:369–377.

60. Zhou X, Kaplan ML. Soluble amylose cornstarch is more digestible than soluble amylopectin potato starch in rats. J Nutr. 1997;127:1349–1356.

61. Zobel HF. Molecules to granules: a comprehensive starch review. Starch-Stärke. 1988;40:44–50.

CHAPTER 4 ISOENERGETIC REPLACEMENT OF FAT BY STARCH IN DIETS FOR AFRICAN CATFISH (CLARIAS GARIEPINUS)

Till S. Harter, Johan A. J. Verreth, Leon T. N. Heinsbroek, Johan W. Schrama

Keywords: Performance, Amylose, Amylopectin, Postprandial Response, Tilapia

INTRODUCTION

Many physiological functions in fish, like respiration and absorption of nutrients demand large surfaces of permeable epithelia to enable active and diffusive transport processes. But this also increases the exposure of the animal to environmental conditions, such as salinity and pH, which may differ from the physiologically desired values. Electrochemical and osmotic gradients across the semipermeable membranes lead to unidirectional passive diffusion processes, which need to be counteracted actively to maintain homeostasis. Osmoregulation is the sum of regulatory processes that maintain the osmotic concentration of body fluids within the range of the physiological tolerance of the species [1]. Unlike their marine counterparts, freshwater (FW) fish are hyperosmotic compared to their environment. This means that they constantly lose ions to the environment and take up water by diffusion and intercellular leakage [1]. Processes of water balance take place in the skin, the gills, the kidneys and contrary to former beliefs [2], also in the gastro intestinal tract (GIT).

In FW species the role of the GIT in osmoregulation only becomes apparent during feeding, in relation to the uptake of dietary ions [3] and postprandial drinking [4], [5], [6], [7]. Due to its attractive simplicity, most studies on osmoregulation have been performed on starved animals and in FW fish this has led to the underestimation of the role of the GIT in osmoregulation. Under these circumstances it was acknowledged, that FW fish would cover their ion demand solely by branchial and renal uptake. Today it is recognized that this is only one part of the picture and that the uptake of dietary ions is an important pillar of the hyperosmotic strategy of FW fish [3].

In modern aquaculture, fish are fed extremely concentrated dry pelleted diets. It is assumed that from an evolutionary point of view, fish are not adapted to process such high dietary dry matter (DM) contents and that this places a physiological strain on the GIT [8]. Previous studies have reported a longer retention time of chyme in the stomach upon consumption of dry diets and that a high DM content of chyme would be counteracted by an increased addition of water to the GIT [7], [9]. These fluxes have been reported to be of both, endogeneous and exogeneous origin [6]. Chyme characteristics have been suggested to influence the water balance in the GIT.

Citation: Harter TS, Verreth JAJ, Heinsbroek LTN, Schrama JW (2013) Isoenergetic Replacement of Fat by Starch in Diets for African Catfish (Clarias gariepinus) doi:10.1371/journal.pone.0055245

Bucking and Wood [10] investigated the water fluxes in the digestive tract of FW rainbow trout (*Oncorhynchus mykiss*) during the processing of a single meal, and detected large additions of water to the stomach and proximal intestine. The osmotic pressure generated in the stomach, upon consumption of the dry diet, has been proposed to drive the addition of endogeneous water from the interstitial fluid to the lumen. This assumption requires further substantiation. Furthermore, the effect of dietary characteristics on the kinetics of DM and ion concentration of the chyme, along the GIT has been described in previous research [11]. Therefore, in the present study, the water balance in the GIT was assessed in response to different dietary characteristics, which were expected to yield a difference in chyme osmolality and DM content during digestion. It was expected, that in line with previous findings, these differences in chyme characteristics would be reflected in altered water balance processes.

In the quest for sustainability, increasing amounts of plant ingredients are being included in aquaculture diets, as energy and protein carriers. For many farmed fish species, starch is being employed as a cheap source of non-protein energy. Its digestion in the GIT yields large amounts of osmotically active mono- and disaccharides. If the osmotic gradient across the epithelium drives the addition of water to the GIT, as proposed by Bucking and Wood [10], changes in the osmolality of chyme should be reflected in altered water fluxes. Therefore, the replacement of fat by starch is expected to increase the water additions to the GIT. Based on the methodology employed by Bucking and Wood [10], the present study assessed the postprandial water fluxes in the GIT by dissection of fish and chyme collection, using yttrium oxide (Y_2O_3) as an inert marker to account for the absorption of DM. Contrasts in dietary characteristics were created by the isoenergetic replacement of dietary fat by starch.

Large additions of endogeneous water to the GIT may lead to the rectal excretion of water with the faeces [10]. This is considered advantageous for hyperosmoregulators, but the implication of an increased osmotic challenge to the animals is yet unknown. The present study assessed, whether the replacement of dietary fat by starch led to an altered rectal excretion of water.

In light of the present gaps in knowledge, two diets with different sources of non-protein energy, i.e. by replacing fat versus starch in an isoenergetic way, were used to induce contrasts in dietary characteristics. The aim of the present research was to detect a causal relationship, between dietary characteristics, the resulting development of chyme characteristics in the GIT and processes of water balance.

MATERIALS AND METHODS

Ethics

This study was carried out in strict compliance with the Dutch legislation on animal experiments and the protocol was approved by the ethics committee for animal experiments of Wageningen University (DEC: 2009121.b).

Animals and Housing

Adult African catfish (*Clarias gariepinus*) (n=180), of mixed sex and average individual body weight of 941±195 g (Mean±SD.) were obtained from the experimental hatchery of Wageningen University, "De Haar Vissen" (Wageningen, The Netherlands). Fish were randomly stocked into twelve 120-litre aquaria, all connected to a single recirculation system (comprising a common water reservoir, a plate separator for solids removal and a trickling filter for gas exchange and NH_4^+ removal). Initial stocking density was approximately 118 kg m^{-3}. The flow rate in the tanks was set to 7–8 L min^{-1}; temperature was 27–28°C and a photoperiod of 12:12 h was maintained, with daybreak at 7.00 am. Water quality was monitored daily and measured values did not exceed the predefined limits: NH_4-N, 7 mg L^{-1}; NO_2-N, 4 mg L^{-1}; NO_3-N 250 mg L^{-1}; conductivity 3000–3500 μS cm^{-1} and pH 7–7.5. Whenever necessary, low pH was corrected by adding $NaHCO_3$.

Diet Preparation

Two extruded pelleted dry feeds were produced by Research Diet Services B.V. (Wijk bij Duurstede, The Netherlands). The experimental diets differed only in the source of non-protein energy. In the starch diet (SD) 300 g of gelatinized maize starch was added to 700 g of basal ingredient mixture, while in the fat diet (FD) this starch was replaced by an isoenergetic amount of vegetable oil (125 g) (Table 1).

Table 1. Formulation and measured nutrient composition of experimental starch (SD) and fat diet (FD)

	Diet	
Diet formulation	**SD[1]**	**FD[2]**
	[g kg^{-1}]	
Gelatinized maize starch	300.0	–
Palm oil	–	75.8
Soy oil	–	75.8
Fish meal (CP>68%)	150.0	181.8
Pea protein concentrate	150.0	181.8
Soy protein concentrate	150.0	181.8
Rapeseed meal	150.0	181.8
Wheat	66.0	80.0
Fish oil	10.0	12.1
DL-methionine	3.9	4.7
Yttrium oxide	0.15	0.18
Diamol	10.0	12.1
Mineral premix	10.0	12.1
Measured nutrient content [g kg^{-1} DM]		
Dry matter [g kg^{-1}]	961	975
Crude protein (CP)	421	514
Fat	56	213
Carbohydrates	467	204
Ash	57	69
AIA	9.20	9.51
Gross energy (GE) [MJ kg^{-1} DM]	20.27	24.17
CP GE^{-1} [mg kJ^{-1}]	20.76	21.25

[1]Starch diet.
[2]Fat diet.
doi:10.1371/journal.pone.0055245.t001

Therefore, the ratio of nutrients (including protein) to gross energy (GE) was the same for both diets. Due to the higher energy density of fat, FD was more concentrated compared to SD on a DM basis. It was chosen, not to overcome this by adding a filler such as cellulose, since an effect on water fluxes is difficult to exclude. After extrusion, the starch in SD expanded considerably, while no increase in volume was observed in FD. Therefore, the presented formulation resulted in floating pellets for SD and sinking pellets for FD. Yttrium oxide (Y_2O_3) was included in both diets to account for the absorption of DM along the GIT in subsequent calculations. Feeds were kept in cold storage (4°C) and were sieved (>1.5 mm), to remove dust and smaller particles prior to use. A representative sample of both diets was taken and stored at 4°C for analysis.

Experimental Design

The aim of the present study was to determine whether dietary composition (fat vs. starch), had an effect on chyme characteristics and the postprandial water fluxes in the gastro intestinal tract (GIT). The calculation of water fluxes was based on chyme analysis by dissection of experimental animals at three different points in time after feeding (2, 5 and 8 h). The twelve aquaria were randomly assigned to one of the two dietary treatments and within diets randomly to one of the three sampling times; resulting in a 2×3 factorial design, with two replicates per treatment (n=2). Since the collected chyme was pooled per tank prior to analysis, the experimental unit in this study was tanks with15 fish.

Experimental Protocol

Animals were fed restrictively during the entire experiment at approximately 80% of *ad libitum*, based on the data presented by van de Nieuwegiessen et al. [12]. Due to the difference in nutrient density of the two diets, the feeding level was based on GE instead of DM. Therefore, animals were fed isoenergetically, isonitrogeneously and received the same amount of basal ingredients, but SD fed animals had a 19.18% higher feed ration based on DM, compared to FD. The initial daily feeding ration per tank was calculated from the average animal body weight over tanks. The daily increase in feeding ratio was calculated from the expected daily animal growth, assuming a FCR of one. To ensure complete feed intake (FI) and to account for differences in feeding speed, animals were hand fed and the feeding time was recorded. To estimate FI per tank, feed refusal was weighed. Furthermore, feed spillage was collected by settling according to the procedure described by Amirkolaie et al. [11] or by netting out of the tank. The pellets were counted and their dry weight was estimated from the average pellet weight.

Chyme sampling took place on two consecutive days (34–35) and on each day, six tanks were sampled; i.e. one tank per treatment. Animals were fed a single meal and groups of 15 fish were euthanized (2- phenoxy-ethanol at 4 ml L^{-1}) at 2, 5 or 8 h after feeding. Animals were weighed individually and dissected for chyme collection. Chyme was collected in four compartments of the GIT: Stomach, proximal-, mid- and distal intestine. The distinction of the gastro intestinal compartments was standardized prior to sampling, according to a subjective assessment of physiological traits (including diameter and coloration). Samples were taken quantitatively, meaning that all chyme was stripped out of each compartment. The collected samples were pooled per tank and stored in aluminium trays for the compartments stomach and mid intestine and in 50 ml plastic containers for the compartments proximal- and distal intestine. From these volumes a 0.5 ml subsample was taken for the analysis of osmolality. The wet weight of the samples was recorded before and after subsampling, to account for this loss of chyme. Prior to analysis samples were freeze-dried and ground (1.2 mm screen grinder).

Pellet Water Uptake

Water uptake of pellets during feeding has been described in several previous studies [6], [10]. A separate trial was performed in order to estimate the absorption kinetics for the two experimental diets. Therefore, 8 g of diet was placed in a spherical metal tea sieve and was submerged into 250 ml of tap water (Wageningen, The Netherlands) at room temperature (20°C) for 15, 30, 60, 120, 180 or 300s. The sieves were hung for 30 s to allow dripping and were then weighed. Samples were dried at 103°C for 3 h, cooled in a desiccator for 1 h and DW was determined. Measurements were performed in triplicate.

Calculations

FI [g] per tank was calculated from the daily feeding ration, subtracting the feed refusal and the feed spillage. GE intake [kJ] was calculated by multiplying the FI with the energy content of the respective diet [kJ g^{-1}]. The water content of chyme [g] was calculated from the chyme wet weight and DM content and was then converted into a volumetric measure [ml], assuming that one g of water had a volume of 1 ml under the tested conditions. To calculate water fluxes it was necessary to account for the absorption of DM along the GIT. Therefore, the water content (W) [ml] of chyme samples was expressed relative to its marker content (M) [mg] as:

$$RW = \frac{W}{M} \tag{1}$$

Since water fluxes in the GIT were expected to be dependent on the meal size, the relative amount of ingested DM (RDM_i) [g DM mg^{-1} yttrium] was calculated from the ingested DM (DM_i) on the sampling day and the marker content [mg] of the ingested feed (M) as:

$$RDM_i = \frac{DM_i}{M} \tag{2}$$

The relative water fluxes (RWF) [ml water g^{-1} ingested DM] were calculated from the relative water content (Eqn. 1) of chyme in one compartment (RW_2), the relative water content in the previous compartment (RW_1) and the relative ingested DM (Eqn. 2) as:

$$RWF = \frac{(RW_2 - RW_1)}{RDM_i} \tag{3}$$

The water absorption (WA) of pellets [% of pellet weight] was calculated from the wet weight of the sample (WW_{fin}) after water exposure and the initial sample wet weight (WW_{ini}) as:

$$WA = \frac{(WW_{fin} - WW_{ini})}{WW_{ini}} \times 100 \tag{4}$$

The leaching (L) of pellet DM [% of pellet weight] was calculated from the initial dry weight of the sample (DW_{ini}) and the final dry weight of the sample (DW_{fin}) as:

$$L = \frac{(DW_i - DW_{fin})}{DW_{ini}} \times 100 \quad (5)$$

Analytical Techniques

Due to insufficient sample material, chyme DM was assessed as freeze DW. For the measurements of osmolality chyme samples (0.5 ml) were centrifuged at 12,000 G for 10 min. The osmolality of the supernatant was determined by freezing point osmometry, in an Advanced Model 3320 Micro-Osmometer (Advanced Instruments, Norwood, MA, USA). For the analysis of yttrium, dry samples (0.3 g) were prepared by wet acid digestion. Analysis was by inductively coupled plasma atomic emission spectroscopy (ICP-OES) [13].

Statistics

Data was analyzed using the IBM SPSS statistics software v19.0.0.1 (SPSS Inc., an IBM company, Copyright 1989–2010, Chicago, USA). Feeding time and FI data was analyzed by means of One-Way ANOVA (n=6, Tukey, p<0.05). Pellet water uptake, leaching and relative water fluxes were analyzed using the linear regression function in SPSS (p<0.05). Slopes and intercepts of the regression analysis were compared using an ANCOVA (p<0.05) with the independent variable as covariate. Chyme characteristics and relative water fluxes were analyzed by means of F-tests using a split plot model (repeated measurements ANOVA in SPSS) for the effects of diet (D), time (T), compartment of the GIT (C) and their interactions (p<0.05). The applied model was specified as:

$$Y_{ijkl} = \mu + D_i + T_j + (D \times T)_{ij} + e_{1ijk} + C_l + (C \times D)_{il} + (C \times T)_{jl} + (C \times D \times T)_{ijl} + e_{2ijkl} \quad \textbf{(6)}$$

Where Y_{ijkl} = value of the chyme characteristic in diet i, at time j, replicate tank k and GIT compartment l; μ= overall mean; D_i = effect of diet i (i=1,2); T_j = effect of time j (j=2,5,8); e_{1ijk} = error term 1, which represents the random effect of tank k within diets and time (k=1,2); C_l = effect of compartment l (l=1,...,4), e_{2ijkl} = error term 2, which represents the random effect within tanks between compartments. Furthermore, main effects and interactions, were tested separately, within each compartment, by 2-way ANOVA (n=2, Tukey, p<0.05). The assumption of normal distribution was tested with the Shapiro-Wilkinson test (p<0.05). All results are given as Means±SE.

RESULTS

Overall, both diets were readily consumed and all fish were feeding frenetically in the corner, in which the feed was supplied. Feed refusal was reported mainly in SD and at the beginning of the trial. Since animals were hand fed, feed spillage was generally low. According to subjective observations, SD groups showed a higher level of activity and aggressive behavior. On the sampling day, FI was significantly higher (p<0.001) in SD (144±2 g), compared to FD (117±0 g). This is a consequence of the different energy density of the diets and the isoenergetic feeding. However, GE intake was similar for both diets (p>0.05) and on average 303±4 kJ. SD fed animals consumed their feeding ration significantly slower (p<0.001) compared to FD fed animals. Feeding times were 57.5±6.8 s and 17.7±1.3 s for SD and FD, respectively.

Pellet Water Absorption

Results for pellet water absorption of the two experimental diets are given by the linear relationships between soaking time (t) [s] and the water absorption (WA) of pellets [% of pellet weight] as:

$$\text{SD} : \text{WA} = 65.55(\text{SE}3.93) + 0.289(\text{SE}0.03) \times \text{t}(R^2 = 0.918)$$

$$\text{FD} : \text{WA} = 17.30(\text{SE}0.99) + 0.068(\text{SE}0.01) \times \text{t}(R^2 = 0.918)$$

Water absorption was higher in SD compared to the FD for all measured soaking times and the intercepts were tested significantly different (p<0.001). The absorption of water increased with soaking time, for both diets. This increase was higher in SD compared to FD, as indicated by the significant difference in the slopes of the regression lines (p<0.001). Since the animals were hand fed, adapted to the feeding speed, it was assumed, that single pellets were consumed within 5 s. By extrapolation of the two linear models, previously described, the amount of water absorption during this soaking time was estimated. In SD 66.99% of the pellet weight was taken up as water, compared to 17.64% in FD. Results of the pellet water absorption trial revealed that pellet leaching was not significantly different from zero or between diets (p>0.05). Therefore, leaching within the considered 300 s period can be regarded as negligible and no adjustment of ingested DM or nutrient content was required.

Chyme Osmolality

The results for chyme osmolality are depicted in Figure 1. Chyme osmolality was significantly different between compartments of the GIT (p<0.001), but not dependent on the diet (p>0.05). A time effect on chyme osmolality was detected (p=0.002) and dependent on the compartment (p<0.001). The average chyme osmolality per compartment was 369±22 mOsm kg^{-1} in the stomach and increased slightly towards the proximal intestine with 383±10 mOsm kg^{-1}. Thereafter, chyme osmolality decreased to 343±4 mOsm kg^{-1} in the mid intestine and 321±8 mOsm kg^{-1} in the distal intestine.

In the stomach, within compartment analysis detected significant main effects of diet (p=0.008), time (p<0.001) and their interaction (p=0.001). At 2 h after feeding chyme osmolality in the stomach was significantly higher in FD (498±2 mOsm kg^{-1}), compared to SD (424±10 mOsm kg^{-1}). Over time, chyme osmolality in the stomach decreased for both diets. A non-significant trend suggests that at 8 h after feeding, stomach chyme osmolality in FD was still higher compared to SD (p<0.10). Average chyme osmolality over diets in the stomach at 8 h after feeding was 295±12 mOsm kg^{-1}. In the proximal intestine, a significant time effect was detected (p=0.027), which was not influenced by the diet. Averaged over diets chyme osmolality in the proximal intestine decreased from 420±18 mOsm kg^{-1} at 2 h after feeding to 362±5 mOsm kg^{-1} at 8 h after feeding. No significant main effects or interactions were detected in the compartments mid- and distal intestine.

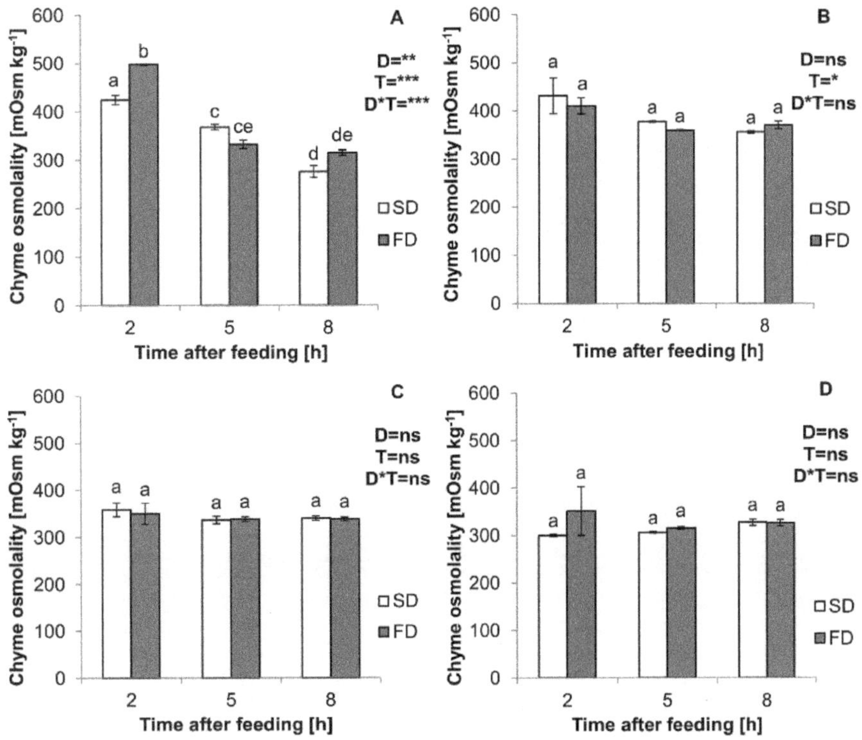

Figure 1. Osmolality of chyme [mOsm kg^{-1}] (Means±SE). Measured in the stomach (A), proximal intestine (B), mid intestine (C) and distal intestine (D) of African catfish (*Clarias gariepinus*) at 2, 5 and 8 h after feeding a starch (SD) or fat (FD) diet. A 2-Way ANOVA ($p<0.05$, Tukey) was used to detect the effects of diet (D), time (T) and the diet*time interaction (D*T) within compartments. Effects are marked with ns for non-significant; # $p<0.10$; * $p<0.05$; ** $p<0.01$; *** $p<0.001$. Means ($n=2$) within compartments marked with different letters are significantly different from each other ($p<0.05$).

Chyme DM

Results for the DM content of chyme are depicted in Figure 2. Chyme DM was influenced by the compartment ($p<0.001$), the diet ($p=0.034$) and the time after feeding ($p=0.006$). Furthermore, the compartment had an influence on the effects of diet ($p<0.001$) and time ($p<0.001$). Average DM content was highest in the stomach (243±19 g kg^{-1}); decreased towards the proximal- (105±3 g kg^{-1}) and mid-intestine (110±4 g kg^{-1}) and increased slightly in the distal intestine (153±4 g kg^{-1}).

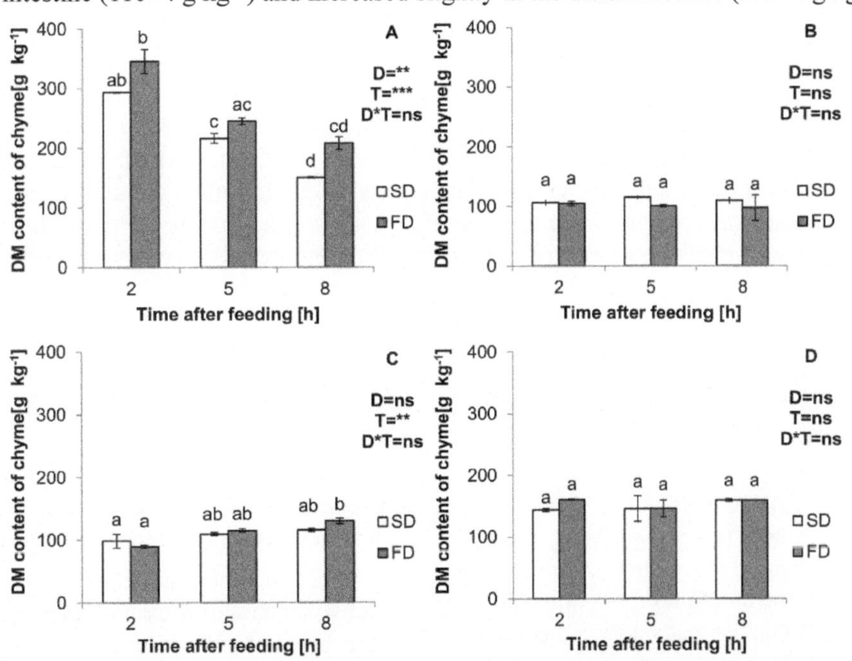

Figure 2. Dry matter (DM) content of chyme [g kg^{-1}] (Means±SE). Measured in the stomach (A), proximal intestine (B), mid intestine (C) and distal intestine (D) of African catfish (*Clarias gariepinus*) at 2, 5 and 8 h after feeding a starch (SD) or fat (FD) diet. A 2-Way ANOVA ($p<0.05$, Tukey) was used to detect the effects of diet (D), time (T) and the diet*time interaction (D*T) within compartments. Effects are marked with ns for non-significant; #$p<0.10$; * $p<0.05$; ** $p<0.01$; *** $p<0.001$. Means ($n=2$) within compartments marked with different letters are significantly different from each other ($p<0.05$).

Chyme DM in the stomach was influenced by the diet ($p=0.002$) and averaged over time, the DM content in the stomach was 220±26 g kg^{-1} and 266±27 g kg^{-1} for SD and FD respectively. Averaged over diets chyme DM in the stomach at 2 h after feeding was 319±17 g kg^{-1} and decreased ($p=0.001$) to 179±17 g kg^{-1} at 8 h after feeding. In all other compartments, chyme DM was similar between diets. In the mid intestine chyme DM, averaged over diets, increased ($p=0.004$) from 93±5 g kg^{-1} at 2 h after feeding, to 123±5 g kg^{-1} at 8 h after feeding.

Water Fluxes

Results for the relative water fluxes (RWF) (Eqn. 3) in the GIT are depicted in Figure 3. Water fluxes were influenced by the compartment ($p<0.001$) and this effect was dependent on diet ($p=0.032$) and time after feeding ($p=0.047$). The diet did not have an influence on the calculated water fluxes ($p>0.05$). Averaged over diets and time, 3.01±0.32 ml water g^{-1} DM were added to the stomach and 3.80±0.53 ml g^{-1} DM to the proximal intestine. In the mid- and distal intestine instead, a net absorption of water took place; and relative water fluxes were −3.71±0.33 ml g^{-1} DM and −1.83±0.33 ml g^{-1} DM, respectively. The sum of relative water influx to stomach and proximal intestine was similar for both diets (One-way ANOVA, $n=12, p<0.05$) and on average 6.52±0.85 ml water g^{-1} DM.

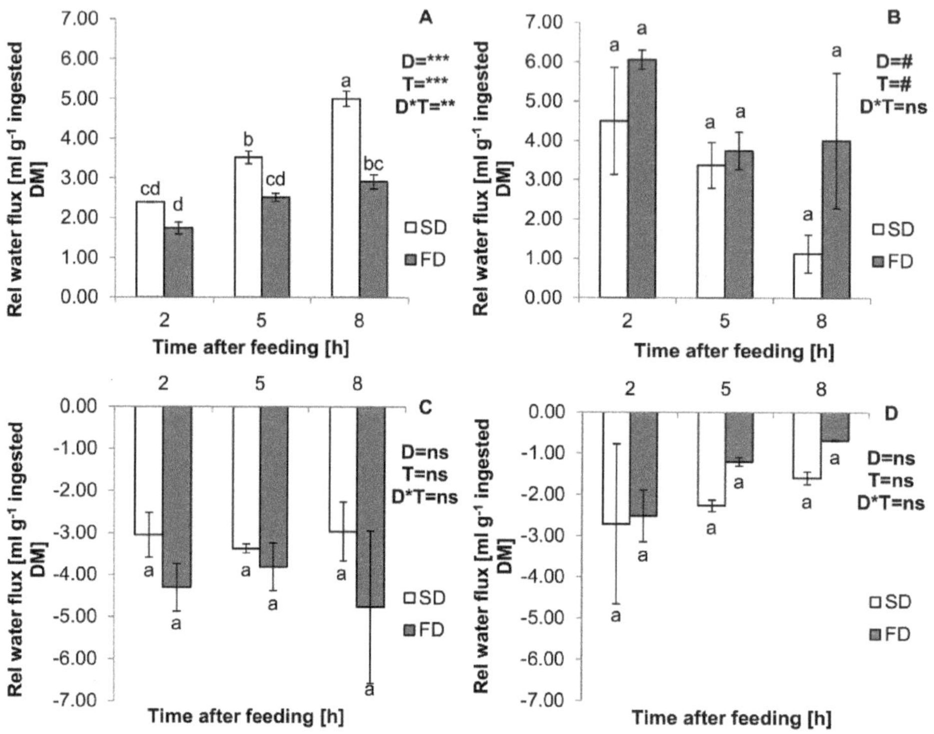

Figure 3. Relative water flux [ml g^{-1} **ingested DM**] (Means±SE). Measured in the stomach (A), proximal intestine (B), mid intestine (C) and distal intestine (D) of African catfish (*Clarias gariepinus*) at 2, 5 and 8 h after feeding a starch (SD) or fat (FD) diet. A 2-Way ANOVA ($p<0.05$, Tukey) was used to detect the effects of diet (D), time (T) and the diet*time interaction (D*T) within compartments. Effects are marked with ns for non-significant; # $p<0.10$; * $p<0.05$; ** $p<0.01$; *** $p<0.001$. Means ($n=2$) within compartments marked with different letters are significantly different from each other ($p<0.05$).

Cumulative water additions to the stomach were significantly higher ($p<0.001$) in SD fed animals (3.64±0.48 ml g^{-1} DM), compared to FD (2.39±0.23 ml g^{-1} DM). Averaged over diets, the water flux in the stomach increased ($p<0.001$) from 2.07±0.20 g g^{-1} DM at 2 h after feeding to 3.96±0.61 ml g^{-1} DM at 8 h after feeding. This time effect was dependent on the diet ($p=0.006$), and a stronger increase in water additions to the stomach was observed in SD. When correcting the water fluxes in the stomach for the cumulative effect of this compartment, results suggest that the water addition decreased over time ($p=0.053$). Averaged over diets the corrected water influx to the stomach decreased from 2.07±0.20 ml g^{-1} DM at 2 h after feeding, to 0.94±0.32 ml g^{-1} DM at 8 h after feeding. In the proximal intestine a non-significant trend indicates, that more water was added to FD fed animals (4.60±0.66 ml g^{-1} DM), compared to SD (3.00±0.74 ml g^{-1} DM) ($p=0.091$). A non-significant trend suggests that water fluxes in the proximal intestine decreased over time ($p=0.081$). Averaged over diets the water influx decreased from 5.27±0.72 ml g^{-1} DM at 2 h after feeding to 2.56±1.11 ml g^{-1} DM at 8 h after feeding. In the mid- and distal intestine, water fluxes were constant between diets and over time, since no significant main effects or interactions were detected ($p>0.05$).

To better depict the relationship between the water fluxes in the stomach and proximal intestine, the relative water flux of the proximal intestine (RWF_{prox}) was plotted against the relative water flux of the stomach (RWF_{stom}) for both diets and a regression line was fitted (Figure 4). The linear relationship between RWF_{prox} and RWF_{stom} [ml g^{-1} DM] per diet was:

$$SD: RWF_{prox} = 7.635(SE1.651) - 1.274(SE0.435) \times RWF_{stom}(R^2 = 0.682)$$

$$FD: RWF_{prox} = 7.801(SE3.128) - 1.339(SE1.281) \times RWF_{stom}(R^2 = 0.215)$$

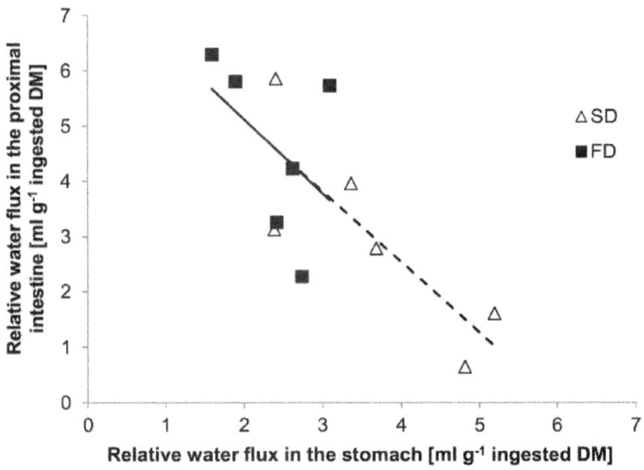

Figure 4. Relative water flux in proximal intestine and stomach [ml g^{-1} **DM]**. Relative water flux in the proximal intestine vs. Relative water flux in the stomach of African catfish (*Clarias gariepinus*) measured at 2, 5 and 8 h after feeding a starch (SD) or fat (FD) diet. The estimated linear models are given in the text. No significant differences were detected between intercepts (p=0.962) or slopes (p=0.959) of the linear models (One-Way ANCOVA, n=2, p<0.05).

No significant differences were detected between slopes or intercepts of the two linear models (p>0.05). The linear relationship of the pooled data between RWF_{prox} and RWF_{stom} [ml g^{-1}DM] was:

$$\text{Pooled data}: RWF_{prox} = 7.672(SE1.092) - 1.285(SE0.342) \times RWF_{stom}(R^2 = 0.585)$$

DISCUSSION

The aim of the present study was to detect a relationship, between dietary characteristics, i.e. the replacement of dietary fat by starch, the resulting chyme characteristics in the GIT and processes of water balance. The present results show that chyme osmolality, DM content and water fluxes were affected by the planned contrasts in dietary composition. Overall, compartments of the GIT differed significantly in chyme characteristics, which validates the division of segments made during dissection. Largest differences in chyme characteristics, between diets and over time, were observed in the stomachs of the fish. In subsequent compartments, these differences were much smaller. In general, it was observed, that the smaller compartments, proximal- and distal intestine, displayed a larger variability in the measured parameters, compared to the stomach and mid intestine. This was mainly caused by the lower amount of chyme present in these compartments, which made it increasingly difficult to sample in a quantitative way. In many fish, the distal intestine was empty. Therefore, fewer individuals contributed to the collected chyme and the effect of inter-individual differences was larger; thereby increasing variability.

The observed high initial osmotic pressure in the stomach (498±2 mOsm kg^{-1} in FD and 424±10 mOsm kg^{-1} in SD) is in line with the results obtained by Bucking and Wood [10] in their study on rainbow trout. However, they described extreme osmotic values in the stomach of up to 2.8 times plasma osmolality at 2 h after feeding. In the present study, osmolality of stomach chyme at 2 h after feeding was only 1.5 fold higher, assuming a blood plasma osmolality of approximately 300 mOsm in African catfish. A subsequent decline in chyme osmolality due to dilution by large additions of water was described in both studies. Yet, after 24 h the stomach osmolality in trout was still 1.3 fold higher than plasma values, while in the present study, stomach osmolality was close to plasma values, already at 8 h after feeding. Initial differences in stomach chyme osmolality between these two studies are likely the results of differences in dietary composition. However, the DM content of the chyme suggests that the present diets were more diluted at 2 h after feeding compared to the results presented by Bucking and Wood[10]. As described by Ruohonen et al. [7] trout experience a delay in gastric evacuation after the consumption of dry pelleted diets, which first need to be moistened in the stomach. The same is expected for African catfish, but the present results suggest that moistening of the chyme was faster compared to trout and therefore gastric evacuation was delayed to a lesser extent. However, interspecific differences in gastric evacuation rate cannot be excluded and the lower rearing temperature of trout (10–13°C) [10] likely contributed to a slower gastric evacuation.

In the proximal intestine chyme osmolality increased for both diets (to 383±10 mOsm kg^{-1}), and resembles the observations of Bucking and Wood [14] on rainbow trout. An increase in chyme osmolality in the proximal intestine can be explained by the secretion of bile but also pancreatic digestive fluids. Bile is known to have a higher osmolality compared to the blood plasma in trout [15] and the same is expected for African catfish. Furthermore, the enzymatic breakdown of polysaccharides in the proximal intestine yields mono- and disaccharides such as glucose and maltose. These substances are highly osmotically active and contribute to the observed increase in chyme osmolality. In fish, lipolysis in the duodenum releases mainly free fatty acids [16]. These molecules are larger than mono- or disaccharides and therefore the increase in osmolality upon hydrolysis of fat is smaller compared to starch. However, the original hypothesis that a higher dietary inclusion of starch would lead to a higher osmotic pressure in the proximal intestine was not confirmed. An explanation might be found in a higher secretion of bile in FD fed animals, due to the higher lipid content of their diet. In SD instead, it seems that free glucose (and other mono- and disaccharides) from enzymatic starch hydrolysis was readily absorbed. More herbivorous species have a higher digestive and absorptive capacity for glucose [17] and it appears that, for African catfish, this point was not exceeded by the applied starch inclusion level. The observed decrease (p<0.05) in chyme osmolality within the proximal intestine can be explained by a decreasing addition of bile over time [10] and a decreasing release of osmotically active molecules from the chyme.

Dissection of the experimental animals at different time points after feeding allowed calculating the relative water fluxes (Eqn. 3) along the GIT in response to feeding. Present results show that large amounts of water were added to the stomach and proximal intestine, while in the mid- and distal intestine reabsorption of water took place. Similar results were reported by Bucking and Wood [10], who calculated the postprandial water fluxes in the GIT of rainbow trout; but in addition, the present setup allowed assessing the response of these water fluxes to dietary composition. Results indicate, that during the first 8 h after feeding, significantly more water was added to the stomach of animals fed a higher dietary starch content (Figure 3 A). Interestingly, a trend ($p<0.10$) suggests that in the proximal intestine, the opposite was the case, and the reported water influx was numerically higher in FD (Figure 3 B). By plotting the relative water flux of the proximal intestine versus the relative water flux of the stomach a significant negative relationship was detected (Figure 4). The present data suggests that 59% of the water influx to the proximal intestine can be explained by the previous addition to the stomach. In other words, a surplus water addition to the stomach was compensated by a decreased addition to the proximal intestine. And in fact, the sum of water additions to both compartments, was not influenced by the consumed diet ($p>0.05$) and on average 6.52±0.85 ml water were added to every g ingested DM.

In the present study, it was not possible to differentiate between endogeneous and exogeneous water additions to the stomach, since no marker was added to the system water (as performed by Kristiansen and Rankin [6]). Nevertheless, the addition of water to the stomach until 2 h after feeding was not significantly different between diets. Furthermore, the cumulative water influx to the stomach increased faster for SD, compared to FD ($p<0.01$). This illustrates that the observed differences in water fluxes in the stomach, between diets, can be attributed to postprandial water additions. According to Bucking and Wood [10], the osmotic pressure in the stomach, is the main driver for endogeneous water additions to this compartment. Also in the present study, indications were found that substantiate this assumption and in fact, most water was added to the stomachs at the time of highest chyme osmolality. Interestingly, the surplus addition of water to the stomachs in SD was not accompanied by a higher osmolality of chyme. However, it seems possible that an increase in the concentration of osmotically active substances in the chyme, due to hydrolysis of macromolecules, drew an equivalent amount of water into the stomach and thereby prevented an increase in chyme osmolality. This is substantiated by the lower ($p<0.01$) DM content of the stomach chyme in SD.

Considering that the outcome of the digestive processes in the stomach was a chyme of similar characteristics, it can be speculated that fish were able to actively control the addition of fluid to the stomach. Due to the large osmotic gradient available to drive an endogeneous water influx, it is questionable whether active transport mechanisms are a sound pathway. Therefore, it is suggested that in the same way as marine species drive water uptake in the intestine, both marine and FW species could regulate the influx of interstitial water to the stomach. By actively controlling the osmolality of the lateral interspace, the osmotic gradient across the epithelium could be adapted to the required osmotic water influx to the stomach; lateral Na^+/K^+-ATPase, Na^+/H^+-exchangers and Cl^--channels are suggested as regulating units. The involvement of some type of mitochondria rich cell seems likely and could be substantiated in immunocytochemical studies. According to recent findings, the water flux is likely via the cellular pathway, involving aquaporins [18].

Postprandial drinking is another way, in which the animals can regulate the water addition to the chyme [4], [5], [6], [7]. However, it remains unclear, what triggers these drinking events. In euryhaline fish, drinking represents the most immediate reaction to changes in osmotic conditions and is strongly regulated by anti-dipsogenic (thirst supressing) hormones [19]. In fish, reflex swallowing is triggered by the hindbrain, often in response to hypovolemia (state of decreased blood volume) [20]. Bucking and Wood [10] described an increase in the plasma osmolality of trout, after feeding. This was attributed to a combination of osmotic water additions to the chyme and the absorption of nutrients in the intestine. A higher plasma osmolality could be an indicator for an on-setting hypovolemia, which concentrates the blood plasma and could therefore trigger drinking. This mechanism would function as a feedback loop between endogeneous and exogeneous water additions to the stomach, likely via the action of angiostenin II, a strong dipsogen, which acts on the area postrema, as part of the renin-angiostenin system [21], [22]. The result is a combination of endogeneous and exogeneous, postprandial water additions to the stomach, which Bucking and Wood [10] considered a likely explanation for their results. In a study on juvenile trout, Kristiansen and Rankin [6]incorporated a non-absorbable marker (^{51}Cr-EDTA) into the ambient water during feeding and were able to discriminate between different water flows. They identified 25–35% of stomach water influx, to be exogeneous, while 34–44% was endogeneous. Despite the large inter individual differences described, convincing evidence remains that fish are able to combine different water sources, in order to establish the required water content of chyme, for stomach evacuation and subsequent digestion.

Unlike in the stomach, the measurement of water fluxes in the proximal intestine is not immediately subjected to a bias from exogeneous water fluxes. Water additions to the proximal intestine are expected, mainly from those digestive fluids secreted by the pancreas and the gall bladder [10]. The higher water flux in FD is likely due to an increased bile secretion in response to the higher dietary lipid level [16]. The decrease ($p<0.10$) in water additions over time can be explained by a decreasing secretion of bile and an increased absorption of water in the proximal intestine [23],[10]. Another source of water additions to the proximal intestine is the evacuation of liquid from the stomach [10]. Since the used yttrium oxide is an insoluble marker, a basic assumption for the estimation of water fluxes was that water and chyme display similar passage rates. However, it is possible that the chyme of the two dietary treatments differed in water uptake behavior, as a results of the developed chyme characteristics. The higher lipid content in FD might have restricted the mixing of water and chyme in the stomach. The formation of a distinct liquid phase and its evacuation would lead to an underestimation of the water addition to the stomach and an overestimation of the water addition to the proximal intestine. In FD fed animals, the chyme absorbed most water in the proximal intestine, which is striking for such a short compartment, in which the chyme will only reside for a limited time. This was likely due to the emulsifying action of the added bile, which mediated a water uptake into the hydrophobic chyme. In contrast, in SD, a higher viscosity and water retention capacity of extruded maize [24] made a separation of the stomach chyme into a solid and a liquid phase less likely. Consequently, a more accurate estimation of the real water fluxes is expected in SD. Bucking and Wood [10] combined ballotini beads with a water-soluble marker (^{3}[H]-PEG 4000) to detect different passage rates of the solid and liquid phase of the chyme. Their results indicate that at determined points in time, the soluble marker was evacuated from the stomach at a higher rate than the beads and similar results have been described in studies on mammals[25], [26], [27]. For their results on trout, Bucking and Wood [10] concluded that the slightly faster evacuation of the soluble marker from the stomach had only a minor impact on the outcome of the calculated water fluxes. Due to the different dietary composition and species, it is difficult to make comparisons to the present study. Further research on the discrimination between endogeneous and exogeneous water additions to the chyme and the dissociation of chyme into fractions with different passage rates, in relation to dietary composition and resulting chyme characteristics, is required.

In the mid- and distal intestine, water absorption took place at a similar rate for both diets (Figure 3 C and D). When balancing the water fluxes over the entire GIT, it was observed that for both diets, more water was added to the chyme compared to reabsorbed amounts. This led to the rectal excretion of 1.32±0.34 ml water g^{-1} DM in SD and 1.23±0.24 ml g^{-1} DM in FD. Considering the implications of a hyperosmoregulatory strategy, rectal water excretion can generally be considered advantageous. But a net excretion will only take place, if water additions to the GIT were largely from endogeneous sources. Bucking and Wood [10] found indications for a net excretion of water with the faeces in rainbow trout. In light of the results obtained by Kristiansen and Rankin [6] on trout, it seems likely that water additions are at least partly of exogeneous origin. When assuming that only 45% of the stomach influx was endogeneous, the net water balance for the present study would be close to zero. Yet, this does not mean that no osmotic challenge is imposed on the fish, when consuming dry pelleted feeds, since the times of highest water secretion and absorption can differ significantly [10].

CONCLUSION

In the present study, the planned contrasts in dietary composition, i.e. the replacement of dietary fat by starch had an influence on chyme characteristics and processes of water balance in the GIT. Largest differences in chyme characteristics between diets were observed in the stomach and decreased towards the distal GIT. A high osmotic pressure in the stomach was detected for both diets and is suggested to drive the addition of endogenous water to the chyme. Large additions of water were observed in the stomach and proximal intestine for both diets, while absorption of water took place in the mid and distal intestine. Water balance in the GIT was altered by the consumption of the two experimental diets and present results suggest that the isoenergetic replacement of fat by starch induced a shift in water secretion from the proximal intestine to the stomach. Nevertheless, the overall addition of water to the GIT was not affected by the dietary treatment. For future studies on water fluxes in the GIT of fish, it is recommended to perform feeding trials with a combination of inert markers, as applied by other authors. This will allow accounting for the origin of the water added to the stomach, as well as for differences in passage rate of the liquid and solid phase of the chyme.

ACKNOWLEDGMENTS

Thanks are due to Menno ter Veld, Ronald, Booms, Jade Joubert and Saravanan Subramanian for their help during dissection. Furthermore, Ronald Booms and Tino Leffering are acknowledged for their support during the chemical analysis.

AUTHOR CONTRIBUTIONS

Conceived and designed the experiments: TH JS LH JV. Performed the experiments: TH JS LH. Analyzed the data: TH JS. Contributed reagents/materials/analysis tools: TH JS LH. Wrote the paper: TH JS LH JV.

REFERENCES

1. Marshall WS, Grosell M (2006) Ion transport, osmoregulation and acid–base balance. In: Claiborne DEaJB, editor. The Physiology of Fishes. 3 ed. BocaRaton: CRC Press. pp. 177–230.
2. Heisler N (1984) Acid-base regulation in fish. In: Hoar WS, Randall DJ, editors. Fish Physiology. New York: Academic Press. pp. 315–401.
3. Wood CM, Bucking C (2010) The role of feeding in salt and water balance. In: Grosell M, Farrell AP, Brauner CJ, editors. Fish Physiology The multifunctional gut of fish. London, UK: Academic Press. pp. 165–212.
4. Smith HS (1930) The absorption and excretion of water and salts by marine teleosts. Am J Physiol 93: 480–505.
5. Fuentes J, Eddy FB (1997) Drinking in marine, euryhaline and freshwater teleost fish. In: Hazon N, Eddy FB, Flik G, editors. Ionic Regulation in Animals. New York: Spring-Verlag. pp. 135–149.
6. Kristiansen HR, Rankin JC (2001) Discrimination between endogeneous and exogeneus water sources in juvenile rainbow trout fed extruded dry feed. Aquatic Living Ressources 14: 359–366.
7. Ruohonen K, Grove DJ, McIlroy JT (1997) The amount of food ingested in a single meal by rainbow trout offered chopped herring, dry and wet diets. Journal of Fish Biology 51: 93–105. doi: 10.1111/j.1095-8649.1997.tb02516.x
8. Buddington RK, Krogdahl A, Bakke-McKellep AM (1997) The intestines of carnivorous fish: Structure and functions and the relations with diet. Acta Physiologica Scandinavica 161.
9. Windell JT, Norris DO, Kitchell JF, Norris JS (1969) Digestive response of rainbow trout, Salmo gairdneri, to pellet diets. Journal of the Fisheries Research Board of Canada 26: 1801–1812. doi: 10.1139/f69-164
10. Bucking C, Wood CM (2006) Water dynamics in the digestive tract of the freshwater rainbow trout during the processing of a single meal. Journal of Experimental Biology 209: 1883–1893. doi: 10.1242/jeb.02205
11. Amirkolaie AK, Verreth JAJ, Schrama JW (2006) Effect of gelatinization degree and inclusion level of dietary starch on the characteristics of digesta and faeces in Nile tilapia (Oreochromis niloticus (L.)). Aquaculture 260: 194–205. doi: 10.1016/j.aquaculture.2006.06.039
12. van de Nieuwegiessen PG, Olwo J, Khong S, Verreth

JAJ, Schrama JW (2009) Effects of age and stocking density on the welfare of African catfish, Clarias gariepinus Burchell. Aquaculture 288: 69–75. doi: 10.1016/j.aquaculture.2008.11.009

13. van Bussel W, Kerkhof F, van Kessel T, Lamers H, Nous D, et al. (2010) Accurate Determination of Titanium as Titanium Dioxide for Limited Sample Size Digestibility Studies of Feed and Food Matrices by Inductively Coupled Plasma Optical Emission Spectrometry With Real-Time Simultaneous Internal Standardization. Atomic Spectroscopy 31: 81–88.

14. Bucking C, Wood CM (2006) Gastrointestinal processing of monovalent ions (Na+,Cl−, K+) during digestion: implications for homeostatic balance in freshwater rainbow trout. Am J Physiol 291: R1764–R1772. doi: 10.1152/ajpregu.00224.2006

15. Grosell M, O'Donnell MJ, Wood CM (2000) Hepatic versus gallbladder bile composition – in vivo transport physiology of the gallbladder in the rainbow trout. Am J Physiol 278: R1674–R1684.

16. Bakke AM, Glover C, Krogdahl A (2010) Feeding Digestion and Absorption of Nutrients. In: Grosell M, Farrell AP, Brauner CJ, editors. Fish Physiology The multifunctional gut of fish. London, UK: Academic Press. pp. 56–108.

17. Stone DAJ (2003) Dietary carbohydrate utilization by fish. Rev Fish Sci 11: 337–369. doi: 10.1080/10641260390260884

18. Wood CM, Grosell M (2012) Independence of net water flux from paracellular permeability in the intestine of Fundulus heteroclitus, a euryhaline teleost. Journal of Experimental Biology 215: 508–517. doi: 10.1242/jeb.060004

19. Takei Y (2002) Hormonal control of drinking in eels: an evolutionary approach. In: Hazon N, Flik G, editors. Osmoregulation and Drinking in Vertebrates. Oxford: BIOS Scientific Publishers Ltd. pp. 61–82.

20. Takei Y, Balment RJ (2008) The neuroendocrine regulation of fluid intake and fluid balance. In: J. Bernier N, Van Der Kraak G, Farrell AP, Brauner CJ, editors. Fish Neuroendocrinology. San Diego: Academic Press. 365–419.

21. Takei Y, Hirano T, Kobayashi H (1979) Angiotensin and water intake in the Japanese eel, Anguilla japonica. Gen Comp Endocrinol 38: 446–475. doi: 10.1016/0016-6480(79)90155-2

22. Nobata S, Takei Y (2011) The area postrema in hindbrain is a central player for regulation of drinking behavior in Japanese eels. American Journal of Physiology – Regulatory Integrative and Comparative Physiology 300: 1569–1577. doi: 10.1152/ajpregu.00056.2011

23. Bogé G, Lopez L, Peres G (1988) An in vivo study of the role of the pyloric caeca in water absorption in rainbow trout (Salmo gairdneri). Comp Biochem Physiol 91: 9–13. doi: 10.1016/0300-9629(88)91584-8

24. Anguita M, Gasa J, Martín-Orúe SM, Pérez JF (2006) Study of the effect of technological processes on starch hydrolysis, non-starch polysaccharides solubilization and physicochemical properties of different ingredients using a two-step in vitro system. Animal Feed Science and Technology 106: 99–115. doi: 10.1016/j.anifeedsci.2005.12.004

25. Faichney GJ, Beever DE, Black JL (1980) Prediction of the fractional rate of outflow of water from the rumen of the sheep. Agric Syst 6: 261–264. doi: 10.1016/0308-521x(81)90062-7

26. Faichney GJ, White GA (1988) Rates of passage of solutes, microbes and particulate matter through the gastrointestinal tract of ewes fed at a constant rate throughout gestation. Aust J Agric Res 39: 481–488. doi: 10.1071/ar9880481

27. Owens FN, Goetsch AL (1988) Ruminal fermentation. In The Ruminant Animal: Digestive Physiology and Nutrition; Church DC, editor. Englewood Cliffs NJ: Prentice-Hall. 145–167 p.

CHAPTER 5

A COMPARISON OF THE BENEFICIAL EFFECTS OF LIVE AND HEAT-INACTIVATED BAKER'S YEAST ON NILE TILAPIA: SUGGESTIONS ON THE ROLE AND FUNCTION OF THE SECRETORY METABOLITES RELEASED FROM THE YEAST

Chao Ran, Lu Huang, Zhi Liu, Li Xu, Yalin Yang, Philippe Tacon, Eric Auclair, Zhigang Zhou

INTRODUCTION

Aquaculture remains the fastest growing agri-business sector with total global production (inland and marine, without aquatic plants) increasing from 47.3 million tons in 2006 to 70.2 million tons in 2013 [1]. Antibiotics have been widely used for growth promotion and disease control in aquaculture practices. However, antibiotics usage is associated with problems such as bioaccumulation and development of antibiotic-resistant pathogens [2]. As a result, the European Union has banned the use of antibiotics in animal feed as growth promoters in 2006, and the FDA has introduced measures to ensure the judicious use of antibiotics in food-producing animals in 2012. Thus there has been an increasing interest in developing novel agents or functional dietary supplements to improve growth and health in fish.

Probiotics have received a great deal of attention as novel dietary supplements to replace antibiotics in aquaculture [3, 4]. In particular, yeast, Saccharomyces cerevisiae, has been used as a probiotic in various aquatic species and the beneficial effects reported include growth promotion, enhancement of innate immune response, as well as protection against pathogen infection [5–10]. The beneficial effects of yeast have often been attributed to the structural components in the yeast cell wall, such as β-glucans and mannan oligosaccharides (MOS), due to their immunostimulating or gut-health promoting effects [6, 11, 12]. Yeast also produces various metabolites such as enzymes, oligosaccharides, amino acids, peptides, organic acids, vitamins, and other soluble factors. Various metabolites of yeast have been proposed to contribute to improved growth performance of the broiler chicks [13]. Effects of amino acids, peptides, and vitamins on immune function have been reported in mammals [14–16]. In addition, the soluble fraction of yeast cultures exhibited anti-inflammatory effects in swine intestinal epithelial cell lines and immune-modulating effects in NK cells and B-lymphocytes in vitro [17, 18]. The secretion of metabolites by yeast was suggested to rely on yeast viability [19], and the influence of yeast viability was used to deduce the contribution of secreted factors linked to the anti-inflammatory effect of yeast [17]. Dietary inactive yeast has been investigated on different fish species, with beneficial effects reported in growth performance, disease resistance, and gut microbiota [20–22]. However, no direct comparison of the

Citation: Ran C, Huang L, Liu Z, Xu L, Yang Y, Tacon P, et al. (2015) A Comparison of the Beneficial Effects of Live and Heat-Inactivated Baker's Yeast on Nile Tilapia: Suggestions on the Role and Function of the Secretory Metabolites Released from the Yeast. doi:10.1371/journal.pone.0145448.

probiotic effects of live and inactivated yeast in fish has been conducted.

In this study, heat-inactivated baker's yeast was prepared. The effect of live and heat-inactivated yeast on growth performance, gut morphology, immune response, intestinal microbiota, as well as disease resistance of the economically important Nile tilapia was compared to evaluate the contribution of secretory metabolites to the mode of action of the yeast. Two basal diets with different levels of fishmeal and soybean meal similar to those used in the industry were used to investigate the potential interactions of the yeast (live vsinactivated) with diet.

EXPERIMENTAL METHODS

Experimental Diets

Two basal diets were formulated, representative of current practices aimed at reducing fishmeal incorporation rates: one diet with relatively higher fishmeal and lower soybean content (diet A) and the other with relatively lower fishmeal and higher soybean content (diet B). Each basal diet was then kept as control (CK) or supplemented with dietary live yeast (LY) or its heat-inactivated counterpart (HIY). The live yeast, Saccharomyces cerevisiae, was obtained as commercial preparation, Actisaf® (Lesaffre, France). The experimental diet was incorporated with 1 g/kg of the Actisaf® yeast for a final concentration of 10^7 CFU/g of feeds. Heat-inactivated yeast was prepared by incubating Actisaf® yeast at 80°C for 30 min. After the heat treatment, the yeast preparation was washed with PBS for three times to get rid of potential secretory metabolites present in the heat-inactivated yeast preparation. The effectiveness of the heat treatment was confirmed by attempting to culture on YPD agar, with no yeast colonies growing on the plates. Heat-inactivated yeast was also supplemented as 1 g/kg of the feed. Consequently, a total of six experimental diets were formulated, i.e., CKA, LYA, HIYA, CKB, LYB, and HIYB, representing control, live yeast supplemented diet, and heat-inactivated yeast supplemented diet for diet A and B, respectively. The formulation and chemical composition of the experimental diets (%) are presented in Table 1. The dietary ingredients were blended with 100 ml of water per 1 kg of diet to form a paste which was then passed through a meat grinder equipped with a 3-mm die to obtain uniform pellets. The pelleted diets were air-dried till the moisture content was reduced to less than 10%. The levels of S.cerevisiae in diets LYA and LYB were verified by pellet homogenization, serial dilution followed by culturing of the dilutions on YPD plates. All the diets were stored in plastic bags at 4°C until use.

Table 1. Formulations and chemical compositions of the experimental diets

Ingredients (%)	Basal diet of soybean meal type					
	CKA	LYA	HIYA	CKB	LYB	HIYB
Soybean meal	14.0	14.0	14.0	28.0	28.0	28.0
Cottonseed meal	10.0	10.0	10.0	10.0	10.0	10.0
Zeolite powder	2.95	2.85	2.85	1.48	1.38	1.38
Adhesives	0.20	0.20	0.20	0.20	0.20	0.20
VC phosphate ester	0.05	0.05	0.05	0.05	0.05	0.05
Mineral Premix	0.20	0.20	0.20	0.20	0.20	0.20
Vitamin Premix	0.20	0.20	0.20	0.20	0.20	0.20
Lysine Sulphate	0.22	0.22	0.22	0.19	0.19	0.19
Calcium methionine hydroxy	0.16	0.16	0.16	0.28	0.28	0.28
Calcium dihydrogen phosphate	2.00	2.00	2.00	2.00	2.00	2.00
Bean oil	4.02	4.02	4.02	4.40	4.40	4.40
Rapeseed meal	28.0	28.0	28.0	20.0	20.0	20.0
Flour	24.0	24.0	24.0	24.0	24.0	24.0
Domestic fish meal (mixed)	10.0	10.0	10.0	5.00	5.00	5.00
Corn gluten meal	4.00	4.00	4.00	4.00	4.00	4.00
Live yeast	0	0.10	0	0	0.10	0
Heat-inactivated yeast	0	0	0.10	0	0	0.10
Estimated chemical compositions (%)						
Crude protein	31.9	31.9	31.9	32.4	32.4	32.4
Crude lipid	6.73	6.73	6.73	6.83	6.83	6.83
Crude ash	6.38	6.38	6.38	5.50	5.50	5.50
Crude fiber	5.95	5.95	5.95	6.84	6.84	6.84

doi:10.1371/journal.pone.0145448.t001

Fish and Rearing Conditions

All experimental and animal care procedures were approved by the Feed Research Institute of Chinese Academy of Agricultural Sciences Animal Care Committee, under the auspices of the China Council for Animal Care (Assurance # 2012 ZZGCC01). MS-222 was used as the anaesthetic. Nile tilapia, Oreochromis niloticus (L.) fingerlings were obtained from the local aquaculture farm in Beijing, China. After disinfection with 2‰ salt solution, the fish were acclimatized in aquaria (50 ×30 × 38 cm) in a recirculation system (at 0.5 L/min) for at least 2 weeks before the feeding trial. Fish of a similar size (0.66 g) were randomly chosen and distributed into 24 tanks at a density of 12 fish per tank. Each diet was randomly assigned to four tanks and hand-fed to apparent satiation three times daily (8:00, 11:30 and 17:30). During the feeding trial, water temperature and pH were maintained in the range of 25.4 ± 1.68°C and 7.7 ± 0.1, respectively; dissolved oxygen was higher than 5.0 mg O/L; NH_4^+-N and NO_2^--N were lower than 0.50 mg N/L and 0.05 mg N/L, respectively. The photoperiod was 12 h light and 12 h dark, with the light period from 08:00 to 20:00.

Growth Performance and Sampling

After 8 weeks feeding, all the fish were batch weighed and counted. Two fish were randomly chosen from each tank and anaesthetized with MS-222 (50.0 mg/l). The intestine was sampled. A piece of the midgut (~1 cm) was cut and gently agitated three times in PBS (pH 7.2) for 1 min to remove the digesta for intestinal mucosal morphology investigation; the rest of midgut was washed the same way and used for RT-PCR analysis. The hindgut was used for analysis of autochthonous and allochthonous microbiota. The head-kidney was also sampled for RT-PCR analysis. Weight gain (WG) and Feed conversion ratio (FCR) were calculated as follows:

$$\text{Weight gain (WG\%)} = \frac{\text{final weight(g)} - \text{initial weight(g)}}{\text{initial weight(g)}} \times 100$$

$$\text{Feed conversion ratio (FCR)} = \frac{\text{total feed consumption(total feed casting} - \text{total food residue)(g)}}{\text{total final weight(g)} - \text{total initial weight(g)} + \text{total mora;ity weight(g)}}$$

Intestinal Microvilli Morphology

Intestine samples were processed for transmission electron microscope (TEM) scanning electron microscopy (SEM) analysis according to the methods described by Liu et al. [23]. In brief, the midgut intestine was fixed with 2.5% glutaraldehyde. The segments were washed twice in phosphate buffer (pH 7.2) and then post-fixed in OsO4 (1% in pH 7.2 phosphate buffer, 2 h). After serial dehydration in alcohol from 30% to 100%, the segments were critical-point dried (BAL-TEC, CPD030, Balzers, Liechtenstein), mounted on aluminum stubs, sputter-coated with gold by using a high-resolution fine coater (BAL-TEC, SCD005, Balzers, Liechtenstein), and then examined under a Jeol JSM-6301 (JEOL, Tokyo, Japan) scanning microscope. For each group, samples of eight replicate fish were processed and analyzed, with each fish giving one TEM/SEM image. TEM images (magnification × 30,000) were analyzed to measure the microvilli length. SEM images (magnification × 20,000) were used to measure the microvilli density. For the density calculation, ten 1μm ×1μm zoomed squares were randomly taken from each SEM image by PhotoShop CS6 (Adobe, USA), and the microvilli number in each zoomed square was counted by ImageJ (National Institutes of Health, USA). The density data for each fish was the average of the counted microvilli numbers for the ten zoomed squares.

Expression of *hsp70* and Inflammation-Related Cytokine Genes

Total RNA from the intestine and kidney was extracted using a TRIzol Reagent RNA kit (Promega, Germany), and the integrity of total RNA was verified by visualization on a 1.2% agarose gel. RNA was dissolved in 50 μl RNase-free water and stored at − 70°C until use. cDNA was synthesized for quantitative reverse transcription PCR (RT-qPCR) using the ReverTra Ace-a-RT-PCR kit (TRAN, Beijing, China) according to the manufacturer's instructions. The qPCR primers were designed using Primer 5.0 software based on the available cytokine sequences in GenBank (S1 Table). Additional dissociation curve analysis was performed and showed a single melting curve in all cases. qPCR was performed with the SYBR Green SuperReal PreMix Plus (TIANGEN, Beijing, China) in ABI Step OnePlus fluorescence ration PCR instrument (ABI, USA). The total volume of the PCR reactions was 20 μl and consisted of: 10 μl SYBR Green SuperReal PreMix Plus (2 ×), 1μl primer of each, 2μl cDNA, and 6μl RNase-free H_2O. The cycling conditions were as follows: 95°C for 10 min and then 40 cycles of 95°C for 20 s, 58°C for 20 s, and 72°C for 20 s. All qPCRs were performed at least two times. Data analysis was conducted using the $2^{-\Delta\Delta CT}$method [24], and the β-actin gene (S1 Table) was chosen as the internal standard.

Gut Alkaline Phosphatase Activity after Challenge

After the feeding trial, four replicate tanks of each dietary treatment were infused with a pathogenic strain A. hydrophila NJ-1 (a gift from Dr. Yongjie Liu from Nanjing Agricultural University) at a concentration of 10^8 CFU/ml. After 24h, 2 fish each tank were randomly chosen and anaesthetized with MS-222 (50.0 mg/l). The whole intestine was sampled and gently agitated three times in PBS (pH 7.2) for 1 min to remove the digesta. The alkaline phosphatase activity was determined according to Qin et al. [25]. In brief, the whole intestine was homogenized in PBS and centrifuged at 8000 rpm, 4°C to obtain the supernatant, which was incubated in a p-nitrophenylphosphate liquid substrate system (Sigma) for 30 min following the manufacturer's specifications. The absorbance was measured at 405 nm. The concentration of total protein was measured using a BCA Protein Assay Kit (Novagen, Darmstadt, Germany) following the manufacturer's specifications.

16S rRNA Gene Pyrosequencing and Analysis

The gut wall and content of the hindgut were separated for analysis of autochthonous and allochthonous microbiota, respectively. The DNA was extracted from samples according to Liu et al. [26]. Bacterial 16S rRNA genes were amplified by PCR from DNA samples using barcoded primers targeting the V3 -V4 region of 16S rRNA gene of bacteria (S1 Table). All forward core primers ('core' refers to the original unmodified 16S rRNA amplification primer) were modified by the addition of a PGM sequencing adaptor, a 'GT' spacer and unique error correcting Golay barcode [27], to allow multiplex analyses when necessary. Amplification conditions employed were according to previously published protocols [28]. We amplified a standard 450 bp V3-V4 region using the modified primers 349F and 806R. All PCR products were checked for size and specificity by electrophoresis on 1.2% w/v agarose, gel purified and adjusted to 10ng/μl using molecular grade water and pooled equally for subsequent sequencing. V3-V4 amplicons were sequenced using pair-end method by Illumina Hiseq2500. The sequence data are available from the European Nucleotide Archive (ERA) (accession number PRJEB9953).

The raw data was processed and trimmed by FastQC and NGS toolkits. For V3-V4 pair-end reads, only sequences that overlap longer than 10bp and without any mismatch were assembled according to their overlap sequence. Reads with an average base score higher than 30 and base number lower than 80% of this sequence reads were removed by QIIME1.8. Chimera produced in PCR process was removed by USEARCH v6.1. All the reads from fish fed with the same diet (n = 8) were pooled, making 6 final groups, i.e., CKA, LYA, HIYA, CKB, LYB, and HIYB. Sequences were analyzed in UCLUST v1.2.22 to determine Operational Taxonomic Units (OTU), which was defined as comprising sequences with less than a 3% difference by the furthest-neighbor method. Taxon annotation was conducted using one representative sequence from an OTU by RDP classifier referring to GreenGenes database. The relative abundance of OTUs was calculated, and microbial composition in different taxonomic levels was analyzed for each group. Heatmap was constructed for each sample. Species alpha diversity index including Shannon, PD (phylogenetic diversity), good's coverage was calculated based on normalized data.

Statistical Analysis

Data were subjected to two way ANOVA to test the effects of yeast supplementation, basal diets as well as their interaction. The pyrosequencing data was analyzed with two way ANOVA without replication, and only the main effects of yeast supplementation and basal diets were tested. When the main effect of yeast supplementation was significant or a trend was observed, post hoc testing was performed using Tukey's test. The effects of yeast supplementation under each basal diet were separately analyzed with one way ANOVA when the interaction was significant. Variance homogeneity of the data was examined with Levene's test. Differences were considered significant when $P < 0.05$. Instances in which $0.05 < P < 0.1$ were discussed as trends. All the statistical analysis was conducted in SPSS 17.0.

RESULTS

Growth and Feed Utilization

The initial body weights in all treatments were very similar. Two way ANOVA showed that both yeast supplementation and basal diet had significant effect on the final body weight, weight gain and feeding conversion ratio ($P < 0.05$) (Table 2). Live yeast significantly improved the FBW and WG of fish compared to the control ($P < 0.05$). Also, a trend for lower FCR was observed in live yeast group relative to the control ($P = 0.06$). The inactive yeast had no significant influence on FBW, WG and FCR of fish, although with numerical improvement compared to the control. No significant difference was observed between the live and inactivated yeast groups. Fish fed diet B showed significantly improved FBW, WG and FCR, compared with the diet A fed fish ($P < 0.001$).There were no significant interactions between yeast and basal diet ($P > 0.50$).

Table 2. Effects of yeast supplementation and basal diets on the growth performance and feed utilization of Nile tilapia

	Yeast			Basal diet			*P* value		
	CK	LY	HIY	A	B	SEM	Yeast	Basal diet	Interaction
FBW/g	12.1^{a}	12.6^{b}	12.5^{ab}	12.0	12.8	0.3	0.04	<0.001	0.97
WG/%	1876^{a}	1958^{b}	1942^{ab}	1866	1984	47	0.04	<0.001	0.98
FCR	1.52	1.46	1.47	1.53	1.44	0.04	0.049	<0.001	0.97

Values represent means of data obtained from 8 and 12 tanks for 'Yeast' and 'Basal diet' unit cells, respectively. Means sharing a common superscript letter (a, b, ab) were not significantly different ($P > 0.05$). CK, control check; LY, live baker's yeast; HIY, heat-inactivated yeast; FBW, final body weight; WG, weight gain; FCR, feed conversion ratio.

doi:10.1371/journal.pone.0145448.t002

Intestinal Microvilli Morphology

Both yeast and basal diet significantly influenced gut microvilli morphology of tilapia ($P < 0.05$) (Table 3, Fig 1, S1 Fig). Live yeast significantly improved microvilli length ($P < 0.001$) and density ($P < 0.05$) compared with control, while inactive yeast failed to exert any beneficial effect on the microvilli morphology. Microvilli length was significantly increased in the live yeast group compared with the inactivated yeast group ($P < 0.001$). Similar with the growth performance results, interactions between yeast supplementation and basal diets were not significant for the microvilli morphology ($P > 0.10$).

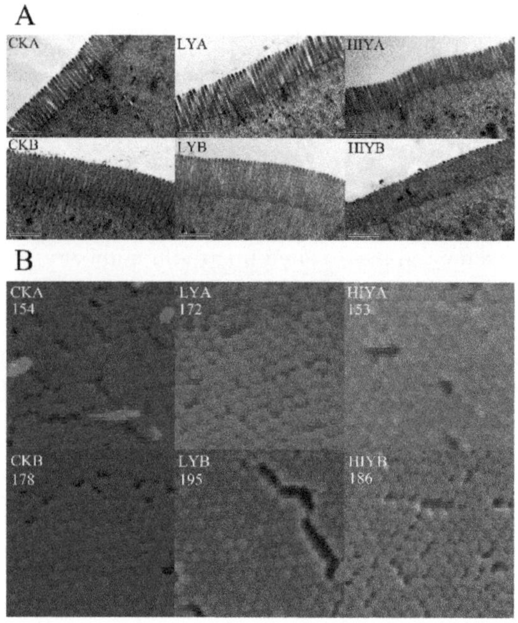

Figure 1. Electron microscope images of the gut microvilli. (A) The

TEM images for microvilli length; (B) The 1μm × 1μm zoomed squares from the original SEM images reflecting the calculation process of the microvilli density. The counted microvilli number in the zoomed square was put on top left. Each zoomed image was selected such that the average density of the corresponding group can be represented, which involved 8 replicate fish and counting of 10 randomly selected 1μm × 1μm squares for the image of each fish. The original SEM images are inS1 Fig.

Table 3. Effects of yeast supplementation and basal diets on intestinal microvilli length and density of Nile tilapia

	Yeast			Basal diet			*P* value		
	CK	LY	HIY	A	B	SEM	Yeast	Basal diet	Interaction
Length (μm)	0.99[a]	1.10[b]	0.96[a]	0.97	1.07	0.03	<0.001	<0.001	0.17
Density (count/μm²)	164.3[a]	181.6[b]	167.8[ab]	156.4	186.0	10.2	0.04	<0.001	0.71

Values represent means of 16 or 24 fish for 'Yeast' and 'Basal diet' unit cells, respectively. Means sharing a common superscript letter (a, b, ab) were not significantly different (*P* > 0.05). CK, control check; LY, live baker's yeast; HIY, heat-inactivated yeast.

doi:10.1371/journal.pone.0145448.t003

Expression of *hsp70* in Intestine and Head Kidney

Live yeast significantly reduced intestinal hsp70 expression compared to the control (P < 0.05). However, the inactive yeast had no significant effect on hsp70 expression (P > 0.05), although the value was numerically lower (Table 4). In head kidney, hsp70 expression was significantly decreased in the live yeast group compared to the control (P < 0.05), while no significant difference was observed in the inactive yeast group (Table 4). In both the gut and head kidney there was no significant interaction between yeast supplementation and basal diet (P > 0.20).

Table 4. Effects of yeast supplementation and basal dies on the relative expression of hsp70 **and inflammation-related cytokines in the intestine** and head kidney of Nile tilapia

		Yeast			Basal diet			*P* value		
		CK	LY	HIY	A	B	SEM	Yeast	Basal diet	Interaction
	hsp70	1.02[a]	0.66[b]	0.80[ab]	0.94	0.72	0.17	0.02	0.04	0.20
Gut	*tnfα*	1.10[a]	0.73[b]	1.12[a]	0.90	1.08	0.20	0.02	0.14	0.42
	il1β	1.03	0.72	1.00	1.02	0.81	0.21	0.05	0.05	0.39
	tgfβ	1.04	0.82	0.89	0.93	0.90	0.17	0.15	0.86	0.02
	hsp70	1.02[a]	0.46[b]	0.77[ab]	0.79	0.70	0.20	0.001	0.43	0.44
HK	*tnfα*	1.03	0.79	0.85	0.89	0.88	0.10	0.10	0.84	0.40
	il1β	1.06	0.72	1.04	0.77	1.09	0.26	0.12	0.04	0.17
	tgfβ	1.01[a]	0.69[b]	0.73[b]	0.77	0.86	0.13	0.002	0.18	0.47

Values represent means of 16 or 24 fish for 'Yeast' and 'Basal diet' unit cells, respectively. The units of the values are arbitrary units (a.u.). Means sharing a common superscript letter (a, b, ab) were not significantly different (*P* > 0.05). CK, control check; LY, live baker's yeast; HIY, heat-inactivated yeast; HK, head kidney.

doi:10.1371/journal.pone.0145448.t004

Expression of Inflammation-Related Cytokine Genes in Intestine and Head Kidney

Live yeast significantly reduced intestinal tnfα expression compared to the control and inactivated yeast group (P < 0.05). A trend towards decreased intestinal expression of il1β (P = 0.08) was observed in the live yeast group compared to the control (Table 4). A significant interaction between yeast supplementation and basal diet was observed for the intestinal expression of tgfβ (P < 0.05). The expression was significantly lower in the live yeast group compared with control and inactivated yeast group when diet A was fed (P < 0.05), while no significant effect was observed when diet B was fed (S2 Fig). Inactivated yeast had no significant effects on the cytokine genes expression.

In head kidney, yeast supplementation had no effect on the expression of tnfα nor il1β (Table 4). The expression of tgfβ was significantly down-regulated by both live and inactivated yeast compared to the control (P < 0.05). No significant interaction between yeast and basal diet was observed (P > 0.10).

Intestinal Autochthonous and Allochthonous Microbiota

For the autochthonous microbiota, the observed OTUs ranged from 1428 to 1678, and good's coverage for all the samples was higher than 0.99. Neither yeast supplementation nor basal diet exerted a significant influence on the diversity of the autochthonous microbiota, as reflected by observed OTUs, Shannon index, and PD (S2 Table). Cluster analysis showed that the basal diet A and B formed two well-separated clusters, while live yeast and inactivated yeast exerted comparable effects on the microbiota with each basal diet (Fig 2). Fusobacteria, Proteobacteriaand Bacteroidetes were the dominant groups in the microbiota. Yeast supplementation and basal diets had no significant effect on the abundance of each detected phylum. At the genus level, the abundance of Deefgea spp. was reduced in the live and inactive yeast groups, as compared to the control (P < 0.05) (Table 5).

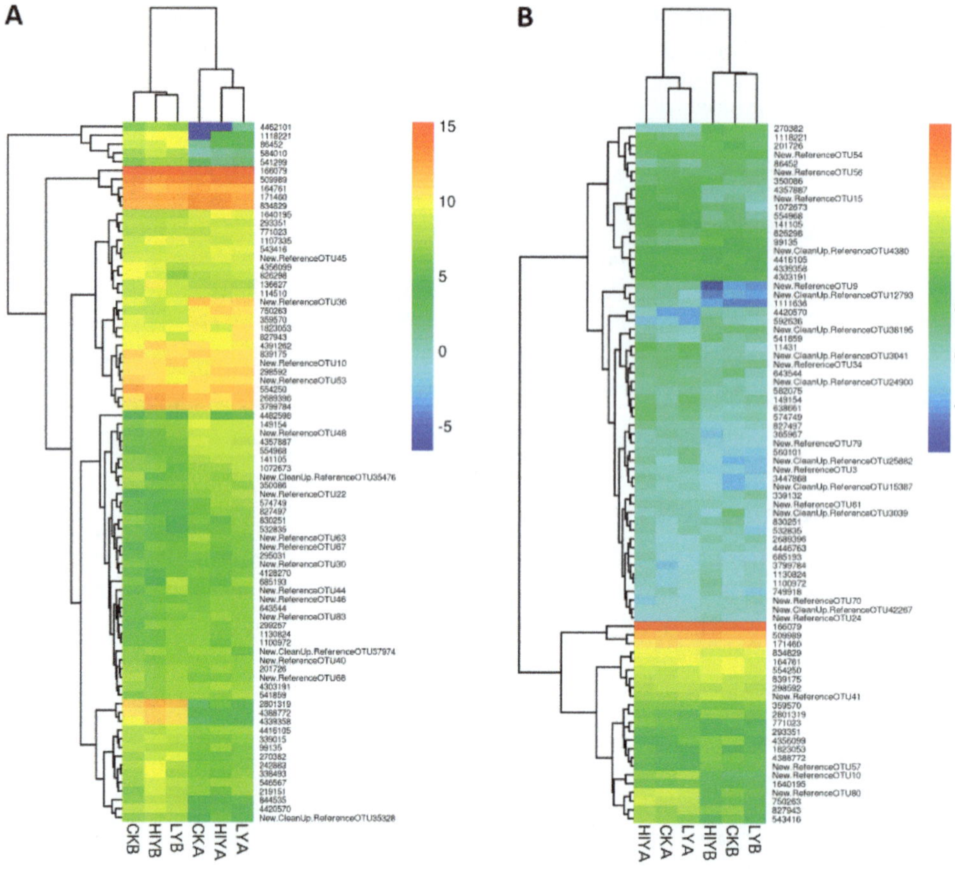

Figure 2. Heatmap showing the relative abundance of the top 80 OTUs of the microbiota. The figure describes autochthonous microbiota (A) and autochthonous microbiota (B) of Nile tilapia after 8 weeks of feeding with different diets. The microbial profiles of the 6 groups were clustered by complete linkage method.

Table 5. Genuses significantly changed in abundance by yeast supplementation

	CK	LY	HIY	*P* value
Deefgea†	0.97[a]	0.78[b]	0.78[b]	0.02
Dechloromonas‡	0.80[a]	0.61[b]	0.90[a]	0.02
Lactococcus‡	0.52[a]	0.64[b]	1.03[c]	0.003

Values are the mean relative abundance of a genus under two basal diets. Means sharing a common superscript letter (a, b, c) were not significantly different (P > 0.05).

†Genus from autochthonous microbiota.

‡Genus from allochthonous microbiota. CK, control check; LY, live baker's yeast; HIY, heat-inactivated yeast.

doi:10.1371/journal.pone.0145448.t005

Similar results were obtained for the allochthonous microbiota, with the observed OTUs ranging from 1295 to 1582 and good's coverage higher than 0.99 for all the samples. The observed species richness was significantly lower in fish fed diet B compared to fish fed diet A ($P < 0.05$). Accordingly, a significantly lower PD was observed in the diet B group ($P < 0.05$), with the Shannon index marginally reduced ($P = 0.097$) (S3 Table). However, yeast supplementation had no significant influence on the diversity of allochthonous microbiota. Basal diet A and B again formed two separated clusters for the allochthonous microbiota. Under each basal diet, the alteration of microbiota induced by inactivated yeast was more pronounced than the one by live yeast (Fig 2). Similar to the autochthonous microbiota, Fusobactetia, Proteobacteria and Bacteroidetes were the dominant groups. Yeast supplementation did not have a significant effect on the abundance of the detected phyla. At the genus level, the abundance of Dechloromonas spp. and Lactococcus spp. was significantly influenced by yeast supplementation. Lactococcus spp. was enriched in both of the yeast supplementation groups compared to the control ($P < 0.05$), and the relative abundance in the inactivated yeast group was further up-regulated compared to the live yeast group ($P < 0.05$) (Table 5).

Protection against *A. hydrophila* NJ-1 Infection

Both live and inactive yeast significantly reduced alkaline phosphatase activity in the gut of tilapia after challenge by A. hydrophila NJ-1 ($P < 0.001$) (S3 Fig). Alkaline phosphatase activity was numerically lower in the live yeast group compared to the inactivated yeast group. Basal diet had no influence on the gut alkaline phosphatase activity, and no significant interaction between yeast supplementation and basal diet was observed ($P > 0.10$).

DISCUSSION

In this study, the contribution of secretory metabolites to the beneficial effects of baker's yeast was investigated through comparison of fish fed diets supplemented with equal doses of live or heat-inactivated yeast, the latter being less susceptible to produce/secrete metabolites, as previously described [17]. It has been reported that some Saccharomyces cerevisiae strains may colonize the intestine of rainbow trout [29]. We conducted a preliminary experiment to investigate the fate of ingested live yeast in the intestine of tilapia by spreading plate method. Results showed that the yeast strain we used was only transient and the population of live yeast did not increase in the intestine of Nile tilapia (data not shown). Thus, we may rule out the proliferation of yeast in the intestine of fish, and the resulting higher concentration as a potential factor of differentiation between live and inactivated yeast. The difference between live and inactivated yeast therefore could be attributed to secretory metabolites.

Dietary supplementation of live baker's yeast improved growth and feed utilization of Nile tilapia. Similar results have been reported previously in various fish species fed live yeast [5, 7–10], with the benefits attributed to improved diet and protein digestibility [5]. The growth promotion effect of yeast structural components like MOS, β-glucan and nucleic acids has been demonstrated in various fish species [12, 30, 31]. In contrast, the contribution of the secretory metabolites of yeast to the growth promotion effect has rarely been studied. Gao et al. suggested that phytase and oligosaccharides, as secretory metabolites in the yeast products, may contribute to improved Ca and P digestibility in broiler chicks [13]. In the present study, inactivated yeast did not significantly improve the growth and feed utilization of fish, indicating advantages of live yeast and the contribution of secretory metabolites to the growth of fish. However, the increased growth observed in the live yeast group relative to the inactivated one was marginal, suggesting that the secretory metabolites did not play a major role in the growth promotion effect. Although there was no direct comparison between live and inactive yeast, significantly improved growth performance with diets supplemented with inactivated yeast has previously been reported in fish [20, 22], supporting the major role of yeast structural components in the growth promotion effect. The lack of a significant growth improvement in fish fed inactivated yeast in our study might be due to the lower dose (0.1%) used here compared to previous studies (1%-2%) [20, 22].

Electron microscopy demonstrated that live yeast significantly improved microvilli length and density in the intestine of tilapia. As a component in the yeast cell wall, Mannan Oligosaccharides (MOS) were reported to improve villi and microvilli morphology in broiler chickens and pigs [32–35], as well as in various fish species [11, 36, 37]. MOS may prevent harmful bacteria from attaching to mannose residues on intestinal epithelial cells, therefore alleviating atrophy of (micro) villi due to inflammation from pathogens and their toxins. However, the present study showed that the inactive yeast, which contains equal amount of MOS to the live yeast, failed to exert beneficial effect on the microvilli morphology, suggesting that the yeast secretory metabolites acted as a main contributory factor to the improved microvilli morphology.

Live yeast supplementation significantly reduced the expression of hsp70 in both intestine and head kidney of tilapia. HSP70 brings appropriate protection of protein structures, strengthens the immune system and stops apoptotic mechanisms [38], and its expression has been found to be related to stressful conditions in aquaculture, such as temperature oscillations, crowding, insufficient water quality, and improper diet [39]. Generally, hsp70 expression decreases in fish given probiotics and coincides with improved welfare and growth [40–43]. In contrast to the live yeast feeding group, only a marginal reduction in hsp70 expression was observed in the inactivated yeast group, indicating the importance of viability of yeast in the relief of stress status in fish.

Live yeast supplementation significantly reduced the expression of tnfα and tended to lower il1βexpression in the intestine of tilapia, pointing to a down-regulation of inflammation in the intestine. Intestinal expression of tgfβ was significantly down-regulated by live yeast in fish fed diet A. TGFβ is known as an anti-inflammatory cytokines and down-regulated tgfβ expression is normally accompanied with up-regulation in proinflammatory cytokines [44]. However, simultaneous down-regulation in the expression of pro- and anti-inflammatory cytokines has also been reported, reflecting a generally down-regulated inflammation response [45]. Therefore, live yeast reduced the inflammation status in the intestine of tilapia. In contrast, the effect of inactivated yeast on expression of cytokines was minimal, suggesting that the secretory metabolites produced by the live yeasts contributed to the anti-inflammatory function. In accordance with our result, live yeast inhibited the expression of pro-inflammatory cytokines by enterotoxigenic E. coli treated porcine intestinal epithelial cells in vitro, with the effect attributed to secretory factors in the yeast [17]. The differential effect of live yeast on the expression of tgfβ in fish fed the two basal diets might be due to the difference in the metabolites produced by live yeasts, and needs further investigation. In head-kidney, the expression of tnfα and il-1β showed a down-regulated tendency in the yeast supplementation groups compared to the control, and the expression of tgfβ was significantly decreased in both live and inactivated yeast groups. Generally down-regulated expression of inflammation-related cytokines in head-kidney was also observed in common carp fed with Saccharomyces cerevisiae derived β-glucans [45]. Nevertheless, a stronger inflammation response was registered in the head-kidney of β-glucans fed common carp following the challenge by A. salmonicida, indicating an enhanced immunity [45]. Increased expressions of pro-inflammatory cytokines in the head-kidney of fish have been correlated with enhanced innate immunity upon probiotics feeding [46, 47, 48]. However, results in the present study suggest that other mechanisms beyond these cytokines may be involved in the immunity enhancement capacity of probiotics, as proposed by Panigrahi et al. [47].

Recent studies suggest that when the physiology of fish is altered through functional feeds the host becomes energy efficient rather than immunologically responsive, and a stronger immune response is achieved upon subsequent pathogen encounter [45, 49–50]. This rationale agrees with the observed expression of inflammation-related cytokines in response to live yeast in our study, where the host was not under the threat of pathogen attack. Also, reduced inflammation may direct more energy towards growth [51] and contribute to an enhanced transport of nutrients and metabolic processes in the intestine [45], leading to improved growth performance.

Beneficial effects of yeast products in ruminants are due to increased concentration of total and cellulolytic ruminal bacteria [52, 53], which may increase the availability of ME from diets, thereby increasing production. However, effects of yeast products on the intestinal microbiota of monogastric animals have been variable [13, 54]. Based on the results of 16S rRNA gene pysosequencing, supplementation of baker's yeast, either alive or inactivated, exerted a subtle influence on autochthonous and allochthonous intestinal microbiota of tilapia fed with both diet A and B. The basal diets formed well-separated clusters for both autochthonous and allochthonous microbiota. The relative abundance of most genera detected was not significantly different between the yeast groups and control for both the autochthonous and allochthonous microbiota. Only Deefgea spp., Dechloromonas spp. and Lactococcus spp. showed significant difference in relative abundance due to yeast treatment. Deefgea spp. is a new genus including two species, i.e., Deefgea rivuli and Deefgea chitinilytica, both of which were isolated from aquatic environments, with Deefgea chtinilytica a chitinolytic bacteria that has been reported to be regularly isolated from freshwater ornamental fish [55]. Dechloromonas spp. is a genus from the phylum Proteobacteria, with some members having the ability to reduce perchlorate or degrade benzene. The potential function of these two genera in the intestinal microbiota of fish, as well as the implication of their abundance variation for the probiotic effect of baker's yeast requires further investigation. The relative abundance of allochthonous Lactococcus spp. was significantly elevated in the yeast groups compared to the control, which is interesting because lactic acid bacteria including Lactococcus spp. are considered as beneficial bacteria [56, 57]. Increased lactic acid bacteria after yeast products/subcomponents supplementation has also been reported in other fish species [11, 20]. The extent to which this elevated abundance ofLactococcus spp. contributes to the observed beneficial effect of yeast deserves further assessment.

Gut alkaline phosphatase activity has been shown to be positively correlated with the number ofA. hydrophila cells in the intestine after challenge and may reflect the resistance level of host against the pathogen (Liu et al. unpublished data). Protection of the host against pathogen was reflected by the significantly lower gut alkaline phosphatase activity in the baker's yeast supplement groups. Similarly, increased diseases resistance after yeast supplementation has been reported in various fish species [7, 8, 10, 21]. The mechanism has been mainly attributed to the immunostimulating structural components, i.e., β-glucans, mannan oligosaccharides, chitin, as well as nucleotides [58]. In our study, the protection effect was marginally higher in the live yeast group compared with the inactivated yeast group, implying a minor involvement of the yeast secretory metabolites in the protection.

The formulations of the two basal diets were both extensively used in the aquaculture practice. In both diets, the inclusion of fishmeal was relatively low, which was replaced by soybean meal and rapeseed meal. Although the level of soybean meal was higher in diet B, no intestinal inflammation was observed, which may be attributed to the low inclusion of soybean meal in this diet. Moreover, diet B generally gave improved growth, feed utilization, microvilli morphology, as well as intestinal stress status (lower hsp70 expression), indicating a more balanced nutrition of diet B compared with diet A. Nevertheless, there were no difference in disease resistance between fish groups fed diet A and diet B, implying that the advantage of diet B mainly involves nutrition, with no effect on the host immunity. Apart from the tgfβexpression in intestine, there were no interactions between yeast treatment and basal diets for all the parameters evaluated, indicating that the beneficial effects of yeast, as well as the advantage of live yeast, don't depend on the protein profiles of the basal diets and the different intestinal microbiota they induced.

In conclusion, the present study showed the advantages of live yeast as a dietary supplement for Nile tilapia, as indicated by the improved gut microvilli morphology, reduced hsp70expression level, as well as the generally down-regulated intestinal inflammation status, which suggests a role for yeast secretory metabolites. However, the observed improvement at the microscopic/molecular level did not translate into evident improvement in growth performance and disease resistance, and only a marginal increase was observed. Therefore, the secretory metabolites might not play a major role in growth promotion and disease protection effects of yeast, as compared to the structural components, which might be due to the limited dose provided by the live yeast in the host or the duration of feeding. Nevertheless, the beneficial effects and modes of action of yeast secretory metabolites deserve further study, which may lead to the development of novel functional dietary supplements for the replacement of antibiotics.

SUPPORTING INFORMATION

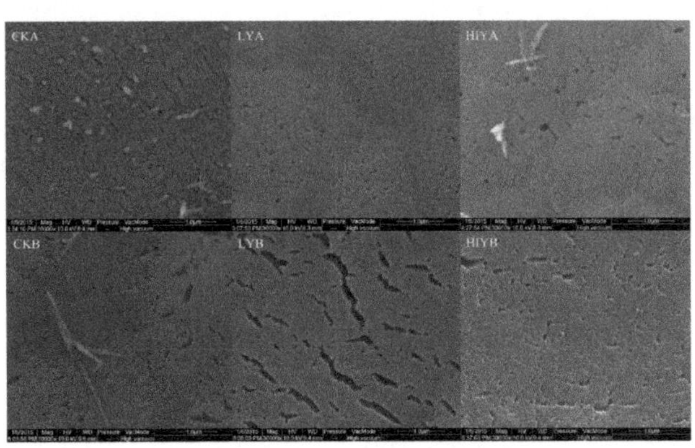

S1 Figure. The original SEM images of the gut microvilli of tilapia fed different diets.

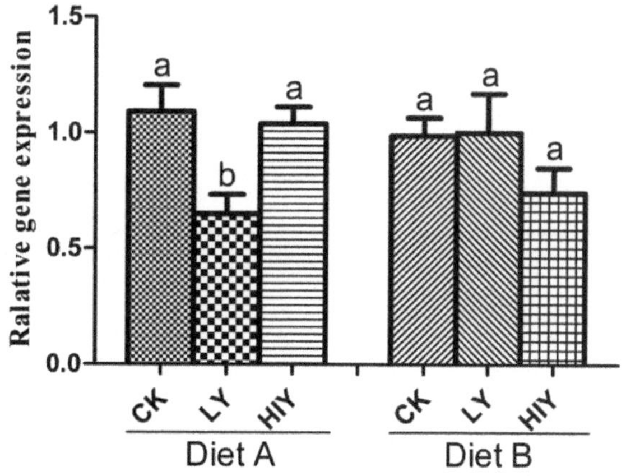

S2 Figure. Intestinal expression of tgf*β* **under different basal diets.** The expression of tgfβ in the intestine was separately analyzed under different basal diets with one way ANOVA. Under each diet, means sharing a same superscript letter were not significantly different.

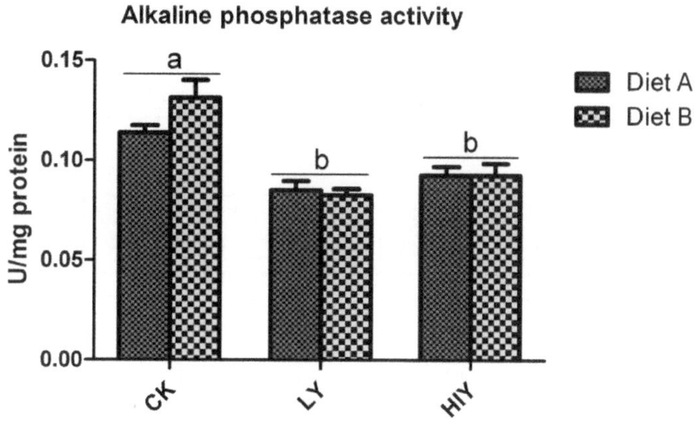

S3 Figure. The gut alkaline phosphatase activity of fish after challenge with Aeromonas hydrophila **NJ-1 (n = 8).** The superscript letters describe the main effect of yeast supplementation from two way ANOVA and Tukey post hoc, with groups sharing the same letter not significantly different (P > 0.05).

S1 Table. Sequences of oligonucleotide primers for qPCR

Target genes	Primers	Sequence (5'-3')	Reference*
Act-β	F	GCTACTCCTTCACCACCACAG	JF957365
	R	CGTCAGGCAGCTCGTAACTC	
HSP70	F	TGCCTTTGTCCAGACCGTAG	JF957370
	R	GTGTCCAACGCTGTCATCAC	
IL-1β	F	TGCACTGTCACTGACAGCCAA	JF957374
	R	ATGTTCAGGTGCACTTTGCGG	
TNF-α	F	CTTCCCATAGACTCTGAGTAGCG	JF957367
	R	GAGGCCAACAAAATCATCATCCC	
TGF-β	F	TGCGGCACCCAATCACACAAC	JF957373
	R	GTTAGCATAGTAACCCGTTGGC	
V3-V4 of 16s rRNA	341F	CCTACGGGAGGCAGCAG	(1)
	806R	GGACTACCVGGGTATCTAAT	(2)

*NCBI accession No.

S2 Table. Effect of yeast and basal diets on the diversity of autochthonous microbiota of Nile tilapia

	Yeast			Basal diet		*P* value	
	CK	LY	HIY	A	B	Yeast	Basal diet
OTU	1491.0	1655.0	1523.5	1496.7	1616.3	0.14	0.10
Shannon	4.4	4.7	4.8	4.5	4.8	0.52	0.30
PD	60.2	66.9	61.3	61.8	63.7	0.15	0.37

Values are means of two or three groups for 'Yeast' and 'Basal diet' unit cells, respectively. Analysis was conducted by two way ANOVA without replication.

S3 Table. Effect of yeast and basal diets on the diversity of allochthonous microbiota of Nile tilapia

	Yeast			Basal diet		*P* value	
	CK	LY	HIY	A	B	Yeast	Basal diet
OTU	1410.5	1399.0	1487.5	1515.7	1349.0	0.25	0.04
Shannon	4.3	4.2	4.5	4.6	4.1	0.42	0.097
PD	60.9	62.0	64.2	65.9	58.9	0.40	0.047

Values are means of two or three groups for 'Yeast' and 'Basal diet' unit cells, respectively. Analysis was conducted by two way ANOVA without replication.

ACKNOWLEDGMENTS

We would like to thank Yang Deng for the fish work. Also, we want to thank Prof.

Jamie Newbold for his helps in improving the writing and structure of the manuscript.

AUTHOR CONTRIBUTIONS

Conceived and designed the experiments: CR ZZ. Performed the experiments: CR LH ZL LX YY. Analyzed the data: CR LH. Contributed reagents/materials/analysis tools: LH PT. Wrote the paper: CR LH PT EA.

REFERENCES

1. FAO 2015: World aquaculture production by inland and marine waters.
2. Teuber M. Veterinary use and antibiotic resistance. Curr Opin Microbiol. 2001; 4: 493–499. pmid:11587923 doi: 10.1016/s1369-5274(00)00241-1
3. Gatesoupe FJ. The use of probiotics in aquaculture. Aquaculture. 1999; 180: 147–165. doi: 10.1016/s0044-8486(99)00187-8
4. Newaj-Fyzul A, Al-Harbi AH, Austin B. Review: Development in the use of probiotics for disease control in aquaculture. Aquaculture. 2013; 431: 1–11. doi: 10.1016/j.aquaculture.2013.08.026
5. Lara-Flores M, Olvera-Novoa MA, Guzman-Méndez BE, López- Madrid W. Use of the bacteria Streptococcus faecium and Lactobacillus acidophilus, and the yeast Saccharomyces cerevisiae as growth promoters in Nile tilapia (Oreochromis niloticus). Aquaculture. 2003; 216: 193–201. doi: 10.1016/s0044-8486(02)00277-6
6. Ortuno J, Cuesta A, Rodríguez A, Esteban MA, Meseguer J. Oral administration of yeast, Saccharomyces cerevisiae, enhances the cellular innate immune response of gilthead seabream (Sparus aurata L.). Vet Immunol Immunopathol. 2002; 85:41–50. pmid:11867166 doi: 10.1016/s0165-2427(01)00406-8
7. Li P, Gatlin DM III. Evaluation of brewer's yeast (Saccharomyces cerevisiae) as a feed supplement for hybrid striped bass (Morone chrysops×M. saxatilis). Aquaculture. 2003; 219: 681–692. doi: 10.1016/s0044-8486(02)00653-1
8. Abdel-Tawwab M, Abdel-Rahman AM, Ismael NEM. Evaluation of commercial live bakers' yeast, Saccharomyces cerevisiae as a growth and immunity promoter for Fry Nile tilapia, Oreochromis niloticus (L.) challenged in situ with Aeromonas hydrophila. Aquaculture. 2008; 280: 185–189. doi: 10.1016/j.

aquaculture.2008.03.055

9. Abdel-Tawwab M, Mousa MAA, Mohammed MA. Use of live baker's yeast, Saccharomyces cerevisiae, in practical diet to enhance the growth performance of Galilee Tilapia, Sarotherodon galilaeus (L.), and its resistance to environmental copper toxicity. J World Aquacult Soc. 2010;41(S2):214–223 doi: 10.1111/j.1749-7345.2010.00361.x
10. Chiu CH, Cheng CH, Gua WR, Guu YK, Cheng W. Dietary administration of the probiotic, Saccharomyces cerevisiae P13, enhanced the growth, innate immune responses, and disease resistance of the grouper, Epinephelus coioides. Fish Shellfish Immunol. 2010; 29: 1053–1059. doi: 10.1016/j.fsi.2010.08.019. pmid:20816806
11. Dimitroglou A, Merrifield DL, Moate R, Davies SJ, Spring P, Sweetman J, et al. Dietary mannan oligosaccharide supplementation modulates intestinal microbial ecology and improves gut morphology of rainbow trout, Oncorhynchus mykiss (Walbaum). J Anim Sci. 2009; 87: 3226–3234. doi: 10.2527/jas.2008-1428. pmid:19617514
12. Kuhlwein H, Merrifield DL, Rawling MD, Foey AD, Davies SJ. Effects of dietary b-(1,3)(1,6)-D-glucan supplementation on growth performance, intestinal morphology and haemato-immunological profile of mirror carp (Cyprinus carpio L.) J Anim Physiol An N. 2014; 98: 279–289. doi: 10.1111/jpn.12078
13. Gao J, Zhang HJ, Yu SH, Wu SG, Yoon I, et al. Effects of yeast culture in broiler diets on performance and immunomodulatory functions. Poult Sci. 2008; 87: 1377–1384. doi: 10.3382/ps.2007-00418. pmid:18577619
14. Calder PC. Branched-chain amino acids and immunity. J Nutr. 2006; 136: 288S–293S. pmid:16365100
15. Vizioli J, Bulet P, Hoffmann JA, Kafatos FC, Muller H- M, Dimopoulos G. Gambicin: a novel immune responsive antimicrobial peptide from the malaria vector Anopheles gambiae. Proc Natl Acad Sci USA 2001; 98: 12630–12635. pmid:11606751 doi: 10.1073/pnas.221466798
16. Eicher SD, McKee CA, Carroll JA, Pajor EA. Supplemental vitamin C and yeast cell wall beta-glucan as growth enhancers in newborn pigs and as immunomodulators after an endotoxin challenge after weaning. J Anim Sci. 2006; 84: 2352–2360. pmid:16908637 doi: 10.2527/jas.2005-770
17. Zanello G, Meurens F, Berri M, Chevaleyre C, Melo S, Auclair E, et al. Saccharomyces cerevisiae decreases inflammatory responses induced by F4+ enterotoxigenic Escherichia coli in porcine intestinal epithelial cells. Vet Immunol Immunop. 2011; 141: 133–138 doi: 10.1016/j.vetimm.2011.01.018
18. Jensen GS, Patterson KM, Yoon I. Yeast culture has anti-inflammatory effects and specifically activates NK cells. Comp Immunol Microbiol Infect Dis. 2008; 31: 487–500. pmid:17915321 doi: 10.1016/j.cimid.2007.08.005
19. Gatesoupe FJ. Live yeasts in the gut: Natural occurrence, dietary introduction, and their effects on fish health and development. Aquaculture. 2007; 267: 20–30. doi: 10.1016/j.aquaculture.2007.01.005
20. Hoseinifar SH, Mirvaghefi A, Merrifield DL. The effects of dietary inactive brewer's yeast Saccharomyces cerevisiae var. ellipsoideus on growth, physiological responses and gut microbiota of juvenile beluga (Huso huso). Aquaculture. 2011; 318: 90–94. doi: 10.1016/j.aquaculture.2011.04.043
21. Li P, Gatlin DM. Dietary brewer's yeast and the prebiotic GroBiotic™ AE influence growth performance, immune responses and resistance of hybrid striped bass (Morone chrysops×M. saxatilis) to Streptococcus iniae infection. Aquaculture. 2004; 231: 445–456. doi: 10.1016/j.aquaculture.2003.08.021
22. Li P, Gatlin DM. Evaluation of the prebiotic GroBioticR-A and brewers yeast as dietary supplements for sub-adult hybrid striped bass (Morone chrysops ×M. saxatilis) challenged in situ with Mycobacterium marinum. Aquaculture. 2005; 248: 197–205. doi: 10.1016/j.aquaculture.2005.03.005
23. Liu W, Yang Y, Zhang J, Gatlin DM, Ringo E, Zhou Z. Effects of dietary microencapsulated sodium butyrate on growth, intestinal mucosal morphology, immune response and adhesive bacteria in juvenile common carp (Cyprinus carpio) pre-fed with or without oxidised oil. Brit J Nutr. 2014; 112: 15–29. doi: 10.1017/S0007114514000610. pmid:24774835
24. Livak KJ, Schmittgen TD. Analysis of relative gene expression data using realtime quantitative PCR and the 2-DDCT method. Methods. 2001; 25: 402e8 doi: 10.1006/meth.2001.1262
25. Qin C, Xu L, Yang Y, He S, Dai Y, Zhao H, Zhou Z. Comparison of fecundity and offspring immunity in zebrafish fed Lactobacillus rhamnosus CICC 6141 and Lactobacillus casei BL23. Reproduction. 2014; 147(1): 53–64. doi: 10.1530/REP-13-0141. pmid:24129154
26. Liu Y, Zhou Z, Yao B, Shi P, He S, Hølvold LB, et al. Effect of intraperitoneal injection of immunostimulatory substances on allochthonous gut microbiota of Atlantic salmon (Salmo salar L.) determined using denaturing gradient gel electrophoresis. Aquacult Res. 2008; 39: 635–646. doi: 10.1111/j.1365-2109.2008.01934.x
27. Hamady M, Walker JJ, Harris JK, Gold NJ, Knight R. Error-correcting barcoded primers for pyrosequencing hundreds of samples in multiplex. Nat Methods. 2008; 5: 235–237. doi: 10.1038/nmeth.1184. pmid:18264105
28. Sogin ML, Morrison HG, Huber JA, Welch DM, Huse SM, Neal PR, et al. Microbial diversity in the deep sea and the underexplored 'rare Biosphere'. Proc Natl Acad Sci USA 2006; 103: 12115–12120. pmid:16880384 doi: 10.1073/pnas.0605127103
29. Andlid T, Vazquez-Juarez R, Gustafsson L.Yeast colonizing the intestine of rainbow trout (Salmo gairdneri) and turbot (Scophtalmus maximus). Microbial Ecol. 1995; 30: 321–334. doi: 10.1007/bf00171938
30. Refstie S, Baeverfjord G, Seim RR, Elvebo O. Effects of dietary yeast cell wall β-glucans and MOS on performance, gut health, and salmon lice resistance in Atlantic salmon (Salmo salar) fed sunflower and soybean meal. Aquaculture. 2010; 305: 109–116. doi: 10.1016/j.

aquaculture.2010.04.005

31. Li P, Gatlin DM III. Nucleotide nutrition in fish: Current knowledge and future applications. Aquaculture. 2006; 251: 141–152. doi: 10.1016/j.aquaculture.2005.01.009

32. Iji PA, Saki AA, Tivey DR. Intestinal structure and function of broiler chickens on diets supplemented with a mannan oligosaccharide. J Sci Food Agric. 2001; 81: 1186–1192. doi: 10.1002/jsfa.925

33. Hooge DM. Meta-analysis of broiler chicken pen trials evaluating dietary mannan oligosaccharide, 1993–2003. Int J Poult Sci. 2004; 3:163–174. doi: 10.3923/ijps.2004.163.174

34. Peet-Schwering CMC, Jansman AJM, Smidt H, Yoon I. Effects of yeast culture on performance, gut integrity, and blood cell composition of weanling pigs. J Anim Sci. 2007; 85: 3099–3109. pmid:17609465 doi: 10.2527/jas.2007-0110

35. Castillo M, Martin-Orue SM, Taylor-Pickard JA, Perez JF, Gasa J. Use of mannanoligosaccharides and zinc chelate as growth promoters and diarrhea preventative in weaning pigs: Effects on microbiota and gut function. J Anim Sci 2008; 86: 94–101. pmid:17911238 doi: 10.2527/jas.2005-686

36. Dimitroglou A, Merrifield DL, Spring P, Sweetman J, Moate R, Davies SJ. Effects of mannan oligosaccharide (MOS) supplementation on growth performance, feed utilisation, intestinal histology and gut microbiota of gilthead sea bream (Sparus aurata). Aquaculture. 2010; 300:182–188. doi: 10.1016/j.aquaculture.2010.01.015

37. Salze G, McLean E, Schwarz MH, Craig SR. Dietary mannan oligosaccharide enhances salinity tolerance and gut development of larval cobia. Aquaculture. 2008; 274: 148–152. doi: 10.1016/j.aquaculture.2007.11.008

38. Reyes-Becerril M, Tovar-Ramírez D, Ascencio-Valle F, Civera-Cerecedo R, Gracia-López V, Barbosa-Solomieu V, et al. Effects of dietary supplementation with probiotic live yeast Debaryomyces hansenii on the immune and antioxidant systems of leopard grouper Mycteroperca rosacea infected with Aeromonas hydrophila. Aquac Res. 2011;42:1676–1686. doi: 10.1111/j.1365-2109.2010.02762.x

39. Rollo A, Sulpizio R, Nardi M, Silvi S, Orpianesi C, Caggiano M, et al. Live microbial feed supplement in aquaculture for improvement of stress tolerance. Fish Physiol Biochem. 2006; 32: 167–177. doi: 10.1007/s10695-006-0009-2

40. Avella MA, Olivotto I, Silvi S, Place AR, Carnevali O. Effect of dietary probiotics on clownfish: a molecular approach to define how lactic acid bacteria modulate development in a marine fish. Am J Physiol-Reg I. 2010; 298: R359–R371. doi: 10.1152/ajpregu.00300.2009

41. Avella MA, Olivotto I, Silvi S, Ribecco C, Cresci A, Palermo F, et al. Use of Enterococcus faecium to improve common sole (Solea solea) larviculture. Aquaculture. 2011; 315: 384–393. doi: 10.1016/j.aquaculture.2011.02.046

42. Avella MA, Place A, Du SJ, Williams E, Silvi S, Zohar Y, et al. Lactobacillus rhamnosus Accelerates Zebrafish Backbone Calcification and Gonadal Differentiation through Effects on the GnRH and IGF Systems. PLoS ONE. 2012; 7: e45572. doi: 10.1371/journal.pone.0045572. pmid:23029107

43. Liu W, Ren P, He S, Xu L, Yang Y, Gu Z, et al. Comparison of adhesive gut bacteria composition, immunity, and disease resistance in juvenile hybrid tilapia fed two different Lactobacillus strains. Fish Shellfish Immunol. 2013; 35: 54–62. doi: 10.1016/j.fsi.2013.04.010. pmid:23608032

44. Ferrucci L, Cherubini A, Bandinelli S, Bartali B, Corsi A, Lauretani F, et al. Relationship of plasma polyunsaturated fatty acids to circulating inflammatory markers. J Clin Endocrinol Metab. 2006; 91: 439–446. pmid:16234304 doi: 10.1210/jc.2005-1303

45. Falco A, Frost P, Miest J, Pionnier N, Irnazarow I, Hoole D. Reduced inflammatory response to Aeromonas salmonicida infection in common carp (Cyprinus carpio L.) fed with β-glucan supplements. Fish Shellfish Immunol. 2012; 32: 1051–1057. doi: 10.1016/j.fsi.2012.02.028. pmid:22406448

46. Pirarat N, Pinpimai K, Endo M, Katagiri T, Ponpornpisit A, Chansue N, et al. Modulation of intestinal morphology and immunity in nile tilapia (Oreochromis niloticus) by Lactobacillus rhamnosus GG. Res Vet Sci. 2011; 91: 92–97. doi: 10.1016/j.rvsc.2011.02.014

47. Panigrahi A, Kiron V, Satoh S, Hirono I, Kobayashi T, Sugita H, et al. Immue modulation and expression of cytokine genes in rainbow trout Oncorhynchus mykiss upon probiotic feeding. Dev Comp Immunol. 2007; 31: 372–382. pmid:17045337 doi: 10.1016/j.dci.2006.07.004

48. Kim DH, Austin B. Cytokine expression in leucocytes and gut cells of rainbow trout, Oncorhynchus mykiss Walbaum, induced by probiotics. 2006; 114: 297–304.

49. Tacchi L, Bickerdike R, Douglas A, Secombes CJ, Martin SAM. Transcriptomic responses to functional feeds in Atlantic salmon (Salmo salar). Fish Shellfish Immunol. 2011; 31: 704–715. doi: 10.1016/j.fsi.2011.02.023. pmid:21377530

50. Lokesh J, Fernandes JMO, Korsnes K, Bergh Ø, Brinchmann MF, Kiron V. Transcriptional regulation of cytokines in the intestine of Atlantic cod fed yeast derived mannan oligosaccharide or b-Glucan and challenged with Vibrio anguillarum. Fish Shellfish Immunol. 2012; 33: 626–631. doi: 10.1016/j.fsi.2012.06.017. pmid:22771966

51. Niewold TA. The nonantibiotic anti-inflammatory effect of antimicrobial growth promoters, the real mode of action? A hypothesis. Poult Sci. 2007; 86(4): 605–609. pmid:17369528 doi: 10.1093/ps/86.4.605

52. Wallace RJ. Ruminal microbiology, biotechnology and ruminant nutrition: progress and problems. J Anim Sci. 1994; 72: 2992–3003. pmid:7730195

53. Newbold CJ, Wallace RJ, Chen XB, Mcintosh FM. Different strains of Saccharomyces cerevisiae differ in their effects on ruminal bacterial numbers in vitro and in sheep. J Anim Sci. 1995; 73:1811–1818. pmid:7673076

54. Shen YB, Piao XS, Kim SW, Wang L, Liu P, Yoon I, et al. Effects of yeast culture supplementation on growth

performance, intestinal health, and immune response of nursery pigs. J Anim Sci. 2009; 87: 2614–2624. doi: 10.2527/jas.2008-1512. pmid:19395514

55. Jung A, Jung-Schroers V. Detection of Deefgea chitinilytica in freshwater ornamental fish. Lett Appl Microbiol. 2011; 52:497–500. doi: 10.1111/j.1472-765X.2011.03030.x. pmid:21323934

56. Chang CI, Liu WY. An evaluation of two probiotic bacterial strains, Enterococcus faecium SF68 and Bacillus toyoi, for reducing edwardsiellosis in cultured European eel, Anguilla anguilla L. J Fish Dis. 2002; 25: 311–315. doi: 10.1046/j.1365-2761.2002.00365.x

57. Panigrahi A, Kiron V, Kobayashi T, Puangkaew J, Satoh S, Sugita H. Immune responses in rainbow trout Oncorhynchus mykiss induced by a potential probiotic bacteria Lactobacillus rhamnosus JCM 1136. Vet Immunol Immunopathol. 2004; 102: 379–388. pmid:15541791 doi: 10.1016/j.vetimm.2004.08.006

58. RingØ E, Olsen RE, Vecino JLG, Wadsworth S, Song SK. Use of immunostimulants and nucleotides in aquaculture: a review. J Marine Sci Res Development. 2012; 1:104. doi: 10.4172/2155-9910.1000104.

CHAPTER 6

EFFECTS OF DIETARY SUPPLEMENTATION OF MAGNESIUM HYDROGEN PHOSPHATE (MGHPO4) AS AN ALTERNATIVE PHOSPHORUS SOURCE ON GROWTH AND FEED UTILIZATION OF JUVENILE FAR EASTERN CATFISH (SILURUS ASOTUS)

Tae-Hyun Yoon, Dong-Hoon Lee, Seung-Gun Won, Chang-Six Ra, Jeong-Dae Kim

Keywords: Silurus asotus, Weight Gain, Feed Efficiency, Phosphorus Availability, Alternative Phosphorus, Plasma Phosphorus

INTRODUCTION

Like other animals, fish have the dietary requirement of phosphorus (P) for growth which was reported to range from 0.3% to 0.6% for channel catfish, *Ictalurus punctatus* (Wilson et al., 1982), rainbow trout, *Oncorhynchus mykiss* (Rodehutscord and Pfeffer, 1995), striped bass, *Morone saxatilis* (Brown et al., 1993; Dougall et al., 1996) and white fish, *Coregonnus lavaretus* (Vilema et al., 2002). However, somewhat higher requirement value of 0.7% to 0.9% was estimated for haddock, *Melanogrammus aeglefinus* (Roy and Lall, 2003), red tilapia, *Tilapia rendalli* (Phromkunthong and Udom, 2008), African catfish, *Clarias gariepinus* (Nwanna et al., 2009), red drum, *Sciaenops ocellatus* (Davis and Robinson, 1987) and yellow croaker, *Pseudosciaena crocea* (Ma et al., 2006). On the other hand, discharged P into water is known to play a vital role in promoting algal growth causing eutrophication (Beveridge, 1984; Auer et al., 1986). Theoretically diet for fish to achieve both maximal growth and minimum P discharge should contain all essential nutrients including P above the needs and maintain total P levels as low as its available requirements. However, the practical diet generally contains an excessive P originating from animal and plant sources. Such sources have low P availability by stomachless species like carp and even by fish with stomach in the early stage of growth due to a lack of gastric juice. Therefore, soluble P sources like monocalcium phosphate (MCP) and dicalcium phosphate (DCP) are being supplemented to the diet to meet the requirement of P for maximum growth. Based on this point of views, some significant discharge of P is inevitable from fish farming (Kim and Ahn, 1993). Kim et al. (1998) reported dietary available P of 0.7% with 2% MCP exerted both maximal growth and minimum P loss in juvenile carp. An adequate combination of low-P protein meals supplemented with 0.5% MCP significantly reduced the P loading from rainbow trout without compromising the growth (Satoh et al., 2003; Hernandez et al., 2004; 2005).

Phosphorus, one of non-renewable elements in the nature, is mostly produced from phosphate rock of which production

Citation: Tae-Hyun Yoon, Dong-Hoon Lee, Seung-Gun Won, Chang-Six Ra, Jeong-Dae Kim, "Effects of Dietary Supplementation of Magnesium Hydrogen Phosphate (MgHPO4) as an Alternative Phosphorus Source on Growth and Feed Utilization of Juvenile Far Eastern Catfish (Silurus asotus). DOI: http://dx.doi.org/10.5713/ajas.2014.14079.

in the world has constantly increased from 198 million metric tons in 2011 to 210 million metric tons in 2012 according to US Geological Survey (USGS, 2013). Shu et al. (2006) expected all reserves of phosphate rock would be depleted by 2090, assuming an annual increase of 1.5% in its demand. Considering this estimation, it is necessary to recover phosphorous from diverse downstream of agricultural and industrial fields. One of resources to retrieve phosphorous could be swine manure which contains high levels of phosphorous and nitrogen. Swine manure becomes a source of pollution on surface waters and induces eutrophication near the site when it is under improper treatment. Thus, the control of wastewater stream must be achieved and struvite precipitation could be an effective way to control P from wastewater stream with the addition of magnesium (Liu et al., 2011). If this magnesium hydrogen phosphate recovered from swine manure could be effectively employed as an available P source for animals including fish, it would not only substitute for import of several phosphates but also protect our environment through recycling of the waste source.

In Korea, far eastern catfish (*Silurus asotus*) is one of major culture species in freshwater and its production reached 4,300 metric tons in 2010 (KOSTAT, 2012). Nevertheless, any nutritional study has not been conducted to investigate P availability and the supplemental effects of various P additives on growth, feed utilization, and hematological and serological parameters of the fish. Furthermore, the potential use of dietary magnesium hydrogen phosphate as an available P source for growth of the fish was not examined until now. This study was therefore carried out to investigate the supplemental effect of various P sources and magnesium hydrogen phosphate (MHP) on growth, feed utilization, hematological and serological parameters of juvenile catfish.

MATERIALS AND METHODS

Preparation of Various Phosphorus Additives

Commercially available P additives, MCP (BIOFOS, Plymouth, MN, USA), DCP (SICHUAN MIANJHUSANJIA FEED Co, Sichuan, China) and tricalcium phosphate (TCP, FOODCHEM, Shandong, China) were obtained from fish feed companies in Korea. The magnesium hydrogen phosphate ($MgHPO_4$) was manufactured from swine manure using a pilot scale reactor with the effective volume of 0.4 m^3 in Kangwon National University. The struvite forming process was performed with hydraulic retention time of 3 h and pH 8 to 9 was maintained by CO_2-stripping (aeration rate of 33 $L/m^3 \cdot min$). Magnesium chloride was added to meet Mg to P ratio of approximately 1.0. Collected precipitate from the reactor was dried and analyzed using X-ray diffractometer (Rigaku, Model D/Max-2500V, Tokyo, Japan) to confirm the formation of struvite. MHP was obtained by removing ammonium-N through incineration at 550°C for 30 min of the recovered struvite. It was finely ground to use as a P additive.

Preparation of Diets

The P sources were incorporated to diet without P source (control) at the level of 2% in lieu of cellulose and the diets were designated as MCP, DCP, TCP, and MHP. Fish meal (25%), soybean meal (40%), wheat flour (27%), fish oil (2%) and soy oil (2%) were employed as major ingredients to formulate control diet containing 42.5% protein and 6.5% lipid (Table 1). Prior to diet formulation, chemical composition of fish meal, soybean meal, wheat flour and 4 different P sources were determined. To make a mixture of 500 kg per diet, weighed ingredients following the formula were ground to 100 mesh size by a hammer mill and thoroughly mixed for 10 min using a V-mixer (Hangjin co., Gwangju, Korea). Then, the mixture was transferred to a twin extruder (Model ATX-2, Fesco Precision Co., Daegu, Korea) and manufactured to the sinking pellets with two sizes of 1.5 and 3.5 mm. Extrusion conditions were as follows: feeder speed, 16 to 18 rpm; conditioner temperature, 80°C to 90°C; main screw speed, 250 to 320 rpm; temperature of the 2nd and the 3rd barrel compartment, 105°C to 135°C; steam heater pressure, 4 to 6 kgf/cm^2 and temperature of the 4th barrel compartment, 80°C to 90°C. Extruded pellets were oven-dried at 60°C for 6 h to maintain the moderate moisture content of 5% to 8%.

Table 1: Ingredient and chemical composition of the experimental diets

Diet	Control	MCP	DCP	TCP	MHP
Ingredient (%)					
Fish meal	25.00	25.00	25.00	25.00	25.00
Soybean meal	40.00	40.00	40.00	40.00	40.00
Wheat flour	27.18	27.18	27.18	27.18	27.18
Soya oil	2.00	2.00	2.00	2.00	2.00
Fish oil	2.00	2.00	2.00	2.00	2.00
Vitamin mix[1]	0.70	0.70	0.70	0.70	0.70
Mineral mix[2]	0.30	0.30	0.30	0.30	0.30
Lysine-HCl	0.30	0.30	0.30	0.30	0.30
DL-methionine	0.20	0.20	0.20	0.20	0.20
Choline-HCl	0.30	0.30	0.30	0.30	0.30
Antioxidant	0.02	0.02	0.02	0.02	0.02
P source	–	2.00	2.00	2.00	2.00

Cellulose	2.00	–	–	–	–
Total	100.00	100.00	100.00	100.00	100.00
Composition (%, DM)[3]					
Crude protein	42.27	42.85	42.20	42.70	42.52
Crude lipid	6.90	6.67	6.62	6.65	6.65
Crude ash	8.41	9.38	9.77	10.33	9.63
Ca	1.74	1.93	2.16	2.30	1.75
P	1.20	1.57	1.53	1.51	1.53
Available P[4]	0.46	0.78	0.70	0.48	0.76

MCP, monocalcium phosphate; DCP, dicalcium phosphate; TCP, tricalcium phosphate; MHP, magnesium hydrogen phosphate; DM, dry matter; BHT, butylated hydroxytoluene; BHA, butylated hydroxyanisol.

[1] Vitamin added to supply the following (per kg diet): vitamin A, 4,000 IU; vitamin D_3, 800 IU; vitamin E, 150 IU; vitamin K_3, 20 mg; thiamine HCl, 25 mg; riboflavin, 50 mg; D-Ca pantothenate, 100 mg; biotin, 1 mg; folic acid, 20 mg; vitamin B_{12}, 0.2 mg; niacin, 200 mg; pyridoxine HCl, 20 mg; ascorbic acid, 500 mg; inositol, 200 mg; BHT, 15 mg; BHA, 15 mg.

[2] Mineral added to supply the following (per kg diet): copper sulfate (25.4% Cu), 30.5 mg; zinc sulfate (22.7% Zn), 230 mg; manganous sulfate (32.5% Mn), 100 mg; cobalt chloride (24.8% Co), 20 mg; potassium iodide (76.4% I), 6.5 mg; sodium selenite (45.6% Se), 2.2 mg; sodium fluoride (45.2% F), 8 mg.

[3] Values are means of 2 determinations.

[4] Available P calculated based on apparent digestibility coefficient values.

An aliquot of 10 kg of each extruded diet were fully ground and mixed with 1.0% chromic oxide for P digestibility measurement. Then, 20% distilled water were added to each diet mixture and the mixture was pelletized using a meat chopper and dried for 6 hours in a ventilated oven at 60°C. The diets were stored in a freezer at −20°C for P digestibility measurement following the growth trial.

Growth Trial

Far eastern catfish fry of 5,000 with around body weight of 6 g were purchased from a private hatchery and acclimated to the experimental conditions for 3 weeks. During this period, they were fed a control diet. Following a 24 h fasting, 5 groups (three replicates/group) of 450 fish of a mean body weight of 11.3 g were randomly allotted to each of 15 tanks (0.4×0.6×0.36 cm, water level of 66 L). The feeding experiment lasted 8 weeks during which each diet was hand-fed to apparent satiety twice a day (08:30 and 16:30) at the 4% of body weight during 6 days per week. A recirculation freshwater system, where water temperature and dissolved oxygen were maintained at 26±1.2°C and 5.5 to 6.4 mg O_2/L, respectively, was employed. The flow rate was held at 5 L/min. The extruded pellet of 1.5 and 3.5 mm sizes were fed for 1st and 2nd feeding of 4 weeks, respectively. Fish were bulk-weighed at the beginning of the experiment and every 4 weeks. Mortality was daily recorded and correction of dead fish was made based on the specific growth rate (SGR) using the equation described by Hardy and Barrows (2002). Daily feed intake (DFI, %/av. body weight/d)), weight gain (WG, %), feed efficiency (FE, %), protein efficiency ratio (PER), SGR (%), survival rate (SR, %) and relative efficiencies (RE, %) of MCP, DCP and TCP to MHP in terms of WG and FE were calculated as follows:

DFI (%/av. body wt/d)
= feed intake (g, dry matter)
/((initial wt+final wt)/2)/experimental days×100

WG (%) = (final weight (g) initial weight (g))
× 100/initial weight (g)

FE (%) = wet weight gain (g)×100/feed intake (g, DM)

PER = wet weight gain (g)/protein intake

SGR (%) = (Ln final weight (g)–Ln initial weight (g))
/experimental days×100

SR (%) = final fish number/initial fish number×100

RE (%) = WG or FE of fish groups fed MCP, DCP, or TCP containing diets/WG or FE of fish group fed MHP containing diet×100

Digestibility Trial

At the end of the growth trial, digestibility measurement was conducted to calculate the available P of the experimental diets. Following a 24 h fasting, 250 fish (mean body weight, 52.2 g) were randomly distributed into each of five 130 L capacity tanks (50 fish/tank) with a fecal collection column. Following one week of feeding, fecal collections were made for 3 consecutive weeks as described by Kim et al. (2006). Each diet was fed by hand to apparent satiety twice a day (08:30 and 16:30). One hour after final feeding of the day, the drain pipes and fecal collection columns were thoroughly cleaned with a brush to remove feed residues and feces from the system. The settled feces and surrounding water were carefully collected into 250 mL centrifuge bottles each morning (08:00). Apparent digestibility coefficient (ADC) of P in the experimental diets was calculated according to the following equation (Maynard and Loosli, 1969):

$$ADC(\%) = \left(1 - \frac{ID \times PF}{IF \times PD}\right) \times 100$$

where *ID* is % indicator in the diet, *PF* represents % P in the feces, *IF* indicates % indicator in the feces, and *PD* is % P in the diet.

P availability of the phosphorus additives was calculated according to the following equation:

$$P\ availability\ (\%) = \frac{APDP - APCD}{TPDP - TPCD} \times 100$$

where APDP indicates % available P in the diet containing P source, APCD is % available P in control diet, TPDP shows % total P in the diet containing P source, and TPCD is % total P in control diet.

Sample Collection and Analysis

At the end of the experimental period, fish were anesthetized with AQUI-S (New Zealand Ltd., Lower Hutt, NZ) and bulk-weighed and counted for calculation of WG, FE, SGR, PER, and SR. Blood samples were obtained from the caudal vessels with a heparinized syringe from two fish of each tank after fish were starved for 24 h and anesthetized with AQUI-S. Feces collected in the same bottle from each tank for 6 days a week were used as one replicate for the treatment. After collection of three replicate samples of each diet during 3 weeks, fecal samples were lyophilized, finely ground and frozen at −20°C until analysis.

Chemical analyses of feed ingredient, diets and feces were performed by the standard procedure of AOAC (1990) for moisture, crude protein, crude fat and crude ash. Moisture content was obtained after drying in an oven at 105°C for 24 h. Crude protein (N×6.25) was determined by Kjeldahl method after acid digestion. Crude fat was determined by the soxhlet extraction method by using Soxtec system 1046 (Foss, Hoganas, Sweden) and crude ash from incineration in a muffle furnace at 550°C for 12 h. Chromium in diets and feces for P digestibility measurement was analyzed using a spectrophotometer (Shimadzu, UV-120-12) at a wavelength of 440 nm after perchloric acid digestion (Bolin et al., 1952). Ca and P were measured using inductively coupled plasma mass spectrometer (ICP-MS) (Perkin-Elmer, NexION 300D, Waltham, MA, USA) after the pretreatment of test materials following the method from US Environmental protection agency (USEPA, 1996). Hematocrit (PCV, %) and hemoglobin (Hb, g/dL) were measured with the same fish by the microhematocrit method (Brown, 1980) and the cyan-methemoglobin procedure using Drabkins solution, respectively. Hb standard prepared from human blood (Sigma Chemical, St. Louis, MO, USA) was employed. Blood plasma was obtained after blood centrifugation (3,500×g, 5 min, 4°C) and stored at −80°C until AST (aspartate aminotransferase), ALT (alanine aminotransferase), TP (total protein), ALB (albumin), GLU (glucose), TCHO (total cholesterol) and inorganic P were analyzed. The plasma parameters were measured using a blood chemical analyzer (HITACHI 7600-210, Hitachi High-Technologies co. Ltd., Tokyo, Japan) with commercial clinical investigation reagent (Pureauto S AST, Pureauto S ALT, Clinimate TP, Clinimate ALB, Pureauto S GLU, Pureauto S CHO-N and Clinimate IP, Sekisui medical co. Ltd., Tokyo, Japan).

Statistical Analysis

Data of growth trial (DFI, WG, FE, PER, SGR, SR, and RE) and hematological and serological parameters were analyzed using one-way analysis of variance (ANOVA) and significant differences among treatment means were compared using Duncan's multiple range test (Duncan, 1955). Prior to the analysis, homogeneity of variance of all data was verified using Cochran's test (Sokal and Rohlf, 1994). All statistical analyses were carried out using the SPSS Version 10 (SPSS, 1999). Statistical significance of the differences was determined by a significant level of 5% ($p<0.05$).

RESULTS

Growth Performance and P Digestibility

At the end of the 8-week growth trial, DFI (%) ranged from 2.38% (MCP and MHP) to 2.89% (control). Fish fed MCP showed the highest WG, which was not significantly different ($p>0.05$) from those of fish fed DCP and MHP, while fish fed control and TCP showed lower WG than the other groups ($p<0.05$). The FE ranged from 107.4% (MCP) to 82.4% (control). The PER of fish fed MCP, DCP, and MHP were not significantly different ($p>0.05$), while fish fed control (2.06) and TCP (2.12) showed lower PER than the other groups ($p<0.05$). The SGR of fish fed MCP was the highest (3.15%), while that of fish fed control was the lowest (2.79%). The SR were not significantly different ($p>0.05$) among fish groups. Mortality of fish was caused by leaping from the tanks only during 1st 4 week growth trial and any of dead fish did not show pathogenic symptoms. Relative efficiencies of MCP, DCP, and TCP to MHP were found to be 100.5% and 101.35, 92.0% and 91.6%, and 79.1% and 80.9% for WG and FE, respectively (Table 2). Apparent availability of P in the experimental diets varied from 32.0% to 49.8% for control and MCP, respectively. The values of P were found to be 88.1%, 75.2%, 8.7%, and 90.9% for MCP, DCP, TCP, and MHP, respectively (Table 3).

Table 2: Growth performance of catfish fed diets containing various phosphorus sources for 8 weeks[1]

Parameters	Diet				
	Control	**MCP**	**DCP**	**TCP**	**MHP**
DFI[2]	2.89±0.08[a]	2.38±0.02[c]	2.60±0.03[b]	2.78±0.06[a]	2.38±0.06[c]
WG[3]	376.77±28.40[b]	484.39±12.09[a]	443.37±13.56[a]	381.09±31.72[b]	481.96±16.98[a]

FE[4]	82.38±4.45[c]	107.42±0.69[a]	97.14±1.87[b]	85.72±4.20[c]	106.02±3.61[a]
PER[5]	2.06±0.11[b]	2.62±0.02[a]	2.45±0.05[a]	2.12±0.10[b]	2.59±0.09[a]
SGR[6]	2.79±0.10[b]	3.15±0.04[a]	3.02±0.04[a]	2.80±0.12[b]	3.13±0.05[a]
SR[7]	97.78±3.14 (2)[ns]	96.67±2.72 (3)	90.00±4.71 (9)	94.44±5.67 (5)	98.89±1.57 (1)

MCP, monocalcium phosphate; DCP, dicalcium phosphate; TCP, tricalcium phosphate; MHP, magnesium hydrogen phosphate; ns, nonsignificant; SE, standard error; DM, dry matter.

[1] Values (means±SE of triplicates) with different superscripts in the same row are significantly different ($p<0.05$).

[2] Daily feed intake (%/av. wt/d) = dry feed intake (g/fish)/((initial wt+final wt)/2)/experimental days×100.

[3] Weight gain (%) = (final weight (g)– initial weight (g))×100/initial weight (g).

[4] Feed efficiency (%) = wet weight gain (g)×100/feed intake (g, DM).

[5] Protein efficiency ratio = wet weight gain (g)/protein intake.

[6] Specific growth rate (%) = (Ln final weight (g)–Ln initial weight (g))/experimental days×100.

[7] Survival rate (%) = final fish number/initial fish number×100; total numbers of dead fish in each fish group are shown in parenthesis.

Table 3: Apparent availability (%) of phosphorus in diets and various phosphorus sources[1]

	P availability				
	Control	**MCP**	**DCP**	**TCP**	**MHP**
Diet	38.0±1.6 [b]	49.8±1.8 [a]	46.0±4.8 [a]	32.0±3.8 [c]	49.4±2.7 [a]
Phosphorus sources		88.1±3.7 [a]	75.2±4.9 [b]	8.7±2.5 [c]	90.9±3.3 [a]

MCP, monocalcium phosphate; DCP, dicalcium phosphate; TCP, tricalcium phosphate; MHP, magnesium hydrogen phosphate; SD, standard deviation.

[1] Means±SD of three replicates of each group with different superscript letter in the same row are significantly different ($p<0.05$).

Hematological and Serological Characteristics

Hematological and serological characteristics of fish fed the experimental diets are shown in Table 3. Hematocrit (%) of fish fed control (33.3) showed a significant difference ($p<0.05$) from that of fish fed MHP (39.8). Hb (g/dL) of fish ranged from 9.4 (MCP) to 10.3 (TCP and MHP), which were not significantly different among fish groups ($p>0.05$). The highest AST (IU/L) were obtained in fish fed TCP (115.5), which were not significantly different ($p>0.05$) from that of fish fed control (110.3). However, fish fed MCP (72.7), DCP (79.8), and MHP (83.2) showed relatively lower AST than the two groups. Alanine aminotransferase (IU/L) of fish ranged from 7.3 (DCP) to 11.5 (TCP), the latter was not significantly different ($p>0.05$) from those of fish fed control (10.0) and MCP (9.8). The total protein (g/dL) was lowest in fish fed MCP (3.4), which was significantly different ($p>0.05$) from those (3.7) of fish fed control and MHP. Albumin (g/dL) ranged from 1.0 to 1.2, which did not show significant differences ($p>0.05$) among fish groups. Glucose (mg/dL) of fish fed MCP showed the highest (143.3), while the lowest (103.8) was found in fish fed TCP. Total cholesterol (mg/dL) ranged from 115.0 (MCP) to 143.3 (control), which were not significantly different ($p>0.05$). Fish fed MCP (15.4) and MHP (15.8) showed a significantly higher P (mg/dL) in plasma among fish groups, while the P level in fish fed the other diets was kept at 13.2 (DCP) to 13.9 (TCP) (Table 4).

Table 4: Hematological and serological characteristics of catfish fed diets with various phosphorus sources for 8 weeks1

Parameters	**Diet**				
	Control	**MCP**	**DCP**	**TCP**	**MHP**
PCV (%)[2]	33.3±5.2[b]	38.0±4.0[ab]	38.3±2.7[ab]	34.8±5.4[ab]	39.8±1.8[a]
Hb (g/dL)3	9.7±0.7[ns]	9.4±0.5	9.8±0.4	10.3±1.6	10.3±1.6
AST (IU/L)[4]	110.3±48.5[a]	72.7±18.4[b]	79.8±16.0[ab]	115.5±34.2[a]	83.2±7.8[ab]
ALT (IU/L)[5]	10.0±2.9[a]	9.8±1.7[ab]	7.3±2.5[b]	11.5±4.2[a]	7.8±1.6[b]
TP (g/dL)[6]	3.7±0.2[a]	3.4±0.2[b]	3.6±0.2[ab]	3.5±0.1[ab]	3.7±0.2[a]
ALB (g/dL)[7]	1.1±0.1[ns]	1.0±0.1	1.1±0.1	1.1±0.1	1.2±0.1
GLU (mg/dL)[8]	108.8±15.2[b]	143.3±37.5[a]	126.7±16.5[ab]	103.8±25.3[b]	112.2±18.3[b]

TCHO (mg/dL)[9]	143.3±32.5^{ns}	115.0±7.4	128.2±16.1	140.3±21.3	133.5±24.0
Phosphorus (mg/dL)	13.3±1.3^{b}	15.4±0.9^{a}	13.2±0.6^{b}	13.9±1.8^{b}	15.8±0.8^{a}

MCP, monocalcium phosphate; DCP, dicalcium phosphate; TCP, tricalcium phosphate; MHP, magnesium hydrogen phosphate; PCV, hematocrit; Hb, hemoglobin; AST, aspartate aminotransferase; ALT, alanine aminotransferase; TP, total protein; ALB, albumin; ns, nonsignificant; GLU, glucose; TCHO, total cholesterol; SD, standard deviation.

[1] Means±SD of each group (n = 6) with different superscript letter in the same row are significantly different (p>0.05).

DISCUSSION

Magnesium hydrogen phosphate (MHP) was newly developed to recycle P from swine manure. It is first attempt to investigate the effect of the MHP as dietary P source on growth and feed utilization of fish. Fish fed MCP showed the best WG at the end of growth trial, although it was not significantly different from those of fish fed MHP and DCP. Same tendencies were found in PER and SGR (Table 2). The result suggests that far eastern catfish could utilize P from MHP and DCP as effectively as MCP. On the other hand, fish group fed DCP had FE lower than fish groups fed MCP and MHP, while the group had FE higher than control and TCP groups. From the results, it was evident that MHP was a good P source competitive with MCP in terms of WG and FE in juvenile far eastern catfish. As shown inFigure 1, RE of MCP, DCP, and TCP to MHP at the level of 2% in each diet were 100.5%, 92.0%, and 79.1% on WG and 101.3%, 91.6%, and 80.9% on FE for MCP, DCP, TCP, respectively. Such relative differences among various P sources might be due to the difference in availability of P, by which the requirement could be met or not. Available P requirement is known to be 0.45% (Lovell, 1978) and 0.8% (Andrews et al., 1973) for channel catfish and 0.67% to 0.82% (Nwanna, 2009) for African catfish. Although the requirement for far eastern catfish has not been evaluated to date, it may be anticipated to be higher than 0.7% based on WG and FE from the present study.

Ogino et al. (1979) extensively studied the P availability of inorganic P sources and various feed ingredients by fish. They reported that the availability of MCP, DCP, and TCP was 94%, 46% and 13%, and 94%, 71% and 64% by carp and rainbow trout, respectively, using egg ALB based diet. Pimentel-Rodrigues and Oliva-Teles (2007) reported the availability of MCP, DCP, and TCP using corn gluten based diet by European sea bass was found to be 65%, 66%, and 42%, respectively at higher inclusion level (1.6%, 2.2%, and 2.0% in each diet) of the phosphate. However, the values were found to be 47%, 71%, and 60%, respectively at lower inclusion level (0.8%, 1.1%, and 1.0% in each diet). Even though the P availabilities were not affected by dietary P levels in their study, the values showed great variation by the inclusion levels as well as the additive sources. Sarker et al. (2009) determined P availability of different P sources by yellowtail using ALB based diet. They incorporated MCP (2.7%), DCP (3.7%) and TCP (3.6%) into the respective diets and obtained P availabilities of 92.4%, 59.2%, and 48.8% for MCP, DCP and TCP, respectively. Lovell (1978) reported that the availability was found to be 94% and 65% for reagent grade MCP and DCP, respectively in channel catfish. On the other hand, Eya and Lovell (1997) determined the net absorption of P from various P sources using all-plant basal diet in channel catfish. They obtained the values of 81.2%, 74.8%, and 54.8% for MCP, DCP, and TCP, respectively.

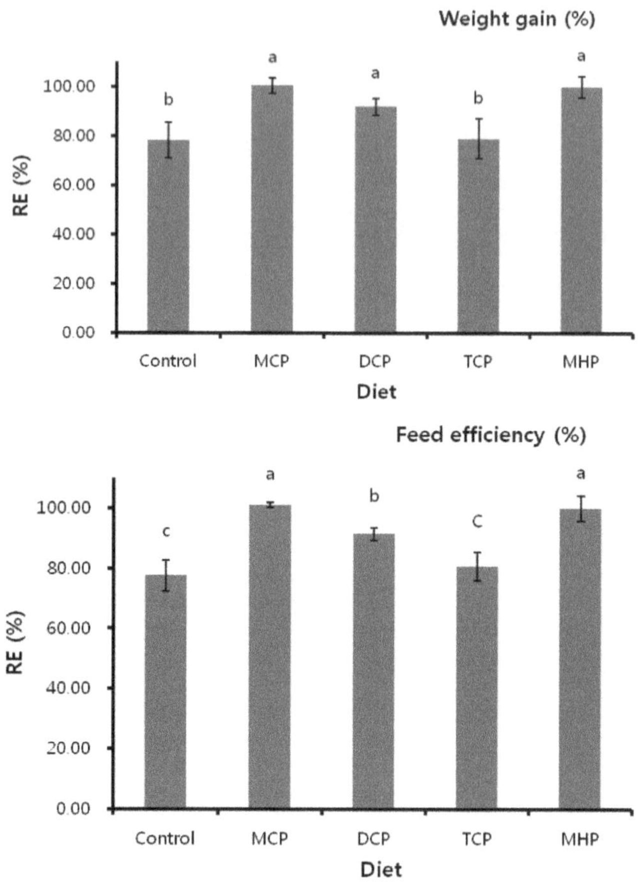

Figure 1: Relative efficiency of various phosphorus sources on weight gain (%) and feed efficiency (%) of juvenile catfish. RE, relative efficiencies; MCP, monocalcium phosphate; DCP, dicalcium phosphate; TCP, tricalcium phosphate; MHP, magnesium hydrogen phosphate.

To our knowledge, this is the first time P availability of feed grade MCP, DCP, and TCP as well as MHP was determined for far eastern catfish. In the present study, P availabilities of both MCP and DCP were found to be high while that of TCP was very low (Table 3). P availability of MCP seems to be lower than those reported by Ogino et al. (1979), Sarker et al. (2009) and Lovell (1978). However, it was comparable to that reported by Kim et al. (1997) and higher than those reported by Pimentel-Rodrigues and Oliva-Teles (2007) and Eya and Lovell (1997). The availability of DCP was comparable to those obtained in rainbow trout (Ogino et al., 1979) and channel catfish (Eya and Lovell, 1997), while it was higher than those obtained in carp (Ogino et al., 1979), European sea bass (Pimentel-Rodrigues and Oliva-Teles, 2007), Yellowtail (Sarker et al., 2009) and channel catfish (Lovell, 1978). On the other hand, reported P availability of TCP by stomach fish ranged from 42% (Pimentel-Rodrigues and Oliva-Teles, 2007) to 64% (Ogino et al., 1979). In contrast, the value from the present study was as low as that obtained by carp, stomachless species (Ogino et al., 1979). Bioavailability of dietary P is influenced by several factors including chemical form, digestibility of diet, particle size, interaction with other nutrients, feed processing and water chemistry (Lall, 1991). It remains to be explained whether such discrepancies in P availability of various P sources, especially TCP, are due to the differences in species and diet composition employed for the digestibility measurement as well as methodological approach in fecal collection (Kim et al., 1996). On the other hand, P availability of MHP was as high as those of MCP obtained from carp and rainbow trout by Ogino et al. (1979) and from yellowtail by Sarker et al. (2009), suggesting that MHP could be a potential P source for fish.

Hematological and serological parameters of blood plasma are useful in monitoring the physiological status of fish and as indicators of the health of the aquatic environment, although they are not routinely used in fish disease diagnosis (El-Sayed et al., 2007). Hematocrit (%) provides an indirect measurement of the body's oxygen carrying ability, while Hb (g/dL) a direct measurement of the oxygen carrying capacity of the blood (McClatchey, 1994). It was reported that hematological parameters could be influenced by nutritional status (Spannhof et al., 1979), infectious disease (Barham et al., 1980; Iwama et al., 1986), environmental changes (Giles et al., 1984) and stress (Ellsaesser and Clem, 1986). The normal ranges of healthy adult Atlantic salmon were reported 44 to 49 and 8.9 to 10.4 for PCV and Hb, respectively (Sandnes and Waagbo, 1988). Somewhat lower values for cichlid fish were reported by Vazquez and Guerrero (2007), which were 22.5 to 39.2 for PCV and 5.2 to 8.3 for Hb. Recently, Rahimnejad and Lee (2013) reported 30.7 to 34.3 for PCV and 4.4 to 5.4 for Hb of red sea bream fed various dietary valine levels. Our findings for the parameters are in good agreement with those obtained from tilapia (Hrubec et al., 2000) and striped bass (Hrubec et al., 2001), although there is no available information on the effect of dietary available P on the parameters of fish. Alanine aminotransferase and AST are two of the most useful measures of liver cell injury, although the AST is less liver specific than the ALT level (Suman and Carey, 2006). In addition to the parameters, TP, ALB, GLU, and TCHO are generally employed as valuable diagnostic means in nutritional studies for fish (Lee et al., 2003; Yan et al., 2007; Cho et al., 2007; Lee et al., 2012; Rahimnejad and Lee, 2013). However, standard ranges of such parameters in fish species as wells as effects of dietary available P on them were not reported to date. Therefore, it remains to be explained why fish fed diet containing low available P (control and TCP) showed higher AST and ALT as well as TP values than fish fed the other diets in the present study. On the other hand, fish fed MCP showed GLU significantly higher than fish fed control and TCP as well as MHP, while ALB and TCHO were not significantly different among fish groups. Due to lack of data related to serological parameters of fish fed diet containing various levels of dietary P, more researches need to be conducted to explain the discrepancies in the present results. An increase in available P in diet resulted in an increase in plasma Pi (Vielma and Lall, 1998; Bureau and Cho, 1999; Avila et al., 2000). The present study clearly showed that plasma P increased in fish fed MCP and MHP with higher available P, resulting in the significant improvement in WG and FE by meeting dietary P need (Table 4). From the present results, the potential use of MHP recovered from swine wastewater was proven sufficiently to replace MCP as an alternative P source with respect to WG and FE as well as P availability, although it remains to be elucidated the optimal dietary level to maximize growth and feed utilization of far eastern catfish. Further, such a re-use of P from wastewater stream could positively influence on the environmental protection and resource security.

Kim JD., & Ahn KH. 1993. Effects of monocalcium phosphate supplementation on phosphorus discharge and growth of carp (Cyprinus carpio) grower. Asian Australas J Anim Sci. 6:521–526.

ACKNOWLEDGMENTS

We are grateful to the funding by Rural Development Administration, Korea. Also, this study was partially supported by 2013 Research Grant from Kangwon National University (No. 120131248).

REFERENCES

1. Andrews JW., Murai T., & Campbell C. 1973. Effects of dietary calcium and phosphorus on growth, food conversion, bone ash and hematocrit levels of catfish. J Nutr. 103:766–771.
2. AOAC. 1990. Official Methods of Analysis. Association of Official Analytical Chemists;Arlington, VA, USA: p. 1298
3. Auer MT., Kiesser MS., & Canale RP. 1986. Identification of critical nutrient levels through field verification of models for phosphorus and phytoplankton growth. Can J Fish Aquat Sci. 43:379–388.
4. Avila EM., Tu H., Basantes S., & Ferraris RP. 2000. Dietary phosphate regulates intestinal transport and plasma concentrations of phosphate in rainbow trout. J Comp Physiol. 170B:210–209.
5. Barham WT., Smit GL., & Schoonbee HJ. 1980. The haematological assessment of bacterial infection in rainbow trout, *Salmo gairdneri* Richardson. J Fish Biol. 17:275–281.
6. Beveridge MCM. 1984. Cage and pen fish farming: carrying capacity models and environmental impact. FAO Fish Tech Pap. 255:131
7. Bolin DW., King RP., & Klosrerman EW. 1952. A simplified method for the determination of chromic oxide (Cr_2O_3) when used as an inert substance. Science. 116:3023634–635.
8. Brown BA. 1980. Routine hematology procedures. Hematology: Principles and Procedures. Lea and Febiger;Philadelphia, PA, USA: p. 71–112.
9. Brown ML., Jaramillo, Jr F., & Gatlin DM. 1993. Dietary phosphorus requirement of juvenile sunshine bass, *Morone chrysops* × *M. saxatilis*. Aquaculture. 113:355–363.
10. Bureau DP., & Cho CY. 1999. Phosphorus utilization by rainbow trout (*Oncorhynchus mykiss*): estimation of dissolved phosphorus waste output. Aquaculture. 179:127–140.
11. Cho SH., Lee SM., Park BH., Ji SC., Choi CY., Lee JH.,

Kim CY., Lee JH., & Oh SY. 2007. Effect of daily feeding ratio on growth and body composition of subadult olive flounder, *Paralichthys olivaceus*, fed an extruded diet during summer season. J World Aquacult Soc. 38:68–73.

12. Davis DA., & Robinson EH. 1987. Dietary phosphorus requirement of juvenile red drum *Sciaenops ocellatus*. J World Aqucult Soc. 18:129–136.
13. Dougall DS., Wood LC., Douglass LW., & Soares JH. 1996. Dietary phosphorus requirement of juvenile striped bass *Morone saxatilis*. J World Aquacult Soc. 27:82–91.
14. Duncan DB. 1955. Multiple range and multiple 'F'tests. Biometrics. 11:1–42.
15. Ellsaesser CF., & Clem LW. 1986. Haematological and immunological changes in channel catfish stressed by handling and transport. J Fish Biol. 28:511–521.
16. El-Sayed YS., Saad TT., & El-Bahr SM. 2007. Acute intoxication of deltamethrin in monosex Nile tilapia, *Oreochromis niloticus* with special reference to the clinical, biochemical and haematological effects. Environ Toxicol Phamacol. 24:212–217.
17. Eya JC., & Lovell RT. 1997. Net absorption of dietary phosphorus from various inorganic sources and effect of fungal phytase on net absorption of plant phosphorus by channel catfish *Ictalurus punctatus*. J World Aquacult Soc. 28:386–391.
18. Giles MA., Majewski HS., & Hobden B. 1984. Osmoregulatory and haematological responses of rainbow trout (*Salmo gairdneri*) to extended environmental acidification. Can J Fish Aquat Sci. 41:1686–1694.
19. Hardy RW., & Barrows FT. 2002. Diet formulation and manufacture. Fish Nutrition. Halver JE, Hardy RW, editors3rd edAcademic Press;London, UK: p. 505–600.
20. Hernandez A., Satoh S., Kiron V., & Watanabe T. 2004. Phosphorus retention efficiency in rainbow trout fed diets with low fish meal and alternative protein ingredients. Fish Sci. 70:580–586.
21. Hernandez A., Satoh S., & Kiron V. 2005. Effect of monocalcium phosphate supplementation in a low fish meal diet for rainbow trout based on growth, feed utilization, and total phosphorus loading. Fish Sci. 71:817–822.
22. Hrubec TC., Cardinale JL., & Smith SA. 2000. Hematology and plasma chemistry reference intervals for cultured tilapia (*Oreochromis* hybrid). Vet Clin Pathol. 29:7–12.
23. Hrubec TC., Smith SA., & Robertson JL. 2001. Age related changes in haematology and chemistry values of hybrid striped bass chrysops*Morone saxatilis*. Vet Clin Pathol. 30:8–15.
24. Iwama GK., Greer GL., & Randall DJ. 1986. Changes in selected haematological parameters in juvenile Chinook salmon subjected to a bacterial challenge and a toxicant. J Fish Biol. 28:563–572.
25. Kim JD., & Ahn KH. 1993. Effects of monocalcium phosphate supplementation on phosphorus discharge and growth of carp (*Cyprinus carpio*) grower. Asian Australas J Anim Sci. 6:521–526.
26. Kim JD., Kim KS., Song JS., Lee JY., & Jeong KS. 1998. Optimum level of dietary monocalcium phosphate based on growth and phosphorus excretion of mirror carp, *Cyprinus carpio*. Aquaculture. 161:337–344.
27. Kim JD., Kim KS., Song JS., Jeong KS., Woo YB., Choi NJ., & Lee JY. 1996. Comparison of feces collection methods for determining apparent phosphorus digestibility of feed ingredients in growing mirror carp (*Cyprinus carpio*). Korean J Anim Nutr Feed. 20:201–206.
28. Kim JD., Kim KS., Lee SB., & Jeong KS. 1997. Nutrient and energy digestibilities of various feedstuffs fed to Israeli strain of growing common carp (*Cyprinus carpio*). J Aquacult. 10:327–334.
29. Kim JD., Tibbets SM., Milley JE., & Lall SP. 2006. Effect of the incorporation level of herring meal into test diet on apparent digestibility coefficients for protein and energy by juvenile haddock, *Melanogrammus aeglefinus* L. Aquaculture. 258:479–486.
30. KOSTAT (Statistics Korea). 2012. Fishery production survey report. http://kostat.go.krAccessed February 16, 2012
31. Lall SP. 1991. Digestibility, metabolism and excretion of dietary phosphorus in fish. Nutritional Strategies and Aquaculture Wastes. Cowey CB, Cho CY, editorsUniversity of Guelph;Ontario, Canada: p. 21–36.
32. Lee DH., Lim SR., Ra CS., & Kim JD. 2012. Effects of dietary garlic extracts on whole body amino acid and fatty acid composition, muscle free amino acid profiles and blood plasma changes in juvenile sterlet sturgeon, *Acipenser ruthenus*. Asian Australas J Anim Sci. 25:1419–1429.
33. Lee SM., Lee JH., & Kim KD. 2003. Effect of dietary essential fatty acids on growth, body composition and blood chemistry of juvenile starry flounder (*Platichthys stellatus*). Aquaculture. 225:269–281.
34. Liu YH., Kumar S., Kwag JH., Kim JH., Kim JD., & Ra CS. 2011. Recycle of electronically dissolved struvite as an alternative to enhance phosphate and nitrogen recovery from swine wasterwater. J Hazard Mater. 195:175–181.
35. Lovell RT. 1978. Dietary phosphorus requirements of channel catfish. Trans Am Fish Soc. 107:617–621.
36. Ma K., Zhang C., Ai Q., Duan Q., Xu W., Zhang L., Luifu Z., & Tan B. 2006. Dietary phosphorus requirement of large yellow croaker, *Pseudosciaena crocea* R. Aquaculture. 251:346–353.
37. Maynard LA., & Loosli JK. 1969. Animal Nutrition. 6th edMcGraw-Hill;New York, USA: p. 613
38. McClatchey KD. 2002. Clinical Laboratory Medicine. Williams & Wilkins;Baltimore, MD, USA: p. 833–839.
39. Nwanna LC., Adebayo IA., & Omitoyin BO. 2009. Phosphorus requirements of African catfish, *Clarias gariepinus*, based on broken-line regression analysis methods. ScienceAsia. 35:227–233.
40. Ogino C., Takeuchi L., Takeda H., & Watanabe T. 1979. Availability of dietary phosphorus in carp and rainbow trout. Bull Jap Soc Sci Fish. 45:1527–1532.

41. Phromkunthong W., & Udom U. 2008. Available phosphorus requirement of sex-reversed red tilapia fed all-plant diets. Songklanakarin. J Sci Technol. 30:7–16.

42. Pimentel-Rodrigues A., & Oliva-Teles A. 2007. Phosphorus availability of inorganic phosphates and fish meal in European sea bass (*Dicentrarchus labrax* L.) juveniles. Aquaculture. 267:300–307.

43. Rahimnejad S., & Lee KJ. 2013. Dietary valine requirement of juvenile red sea bream *Pagrus major*. Aquaculture. 416–417. 212–218.

44. Rodehutscord M., & Pfeffer E. 1995. Requirement for phosphorus in rainbow trout (*Oncorhynchus mykiss*) growing from 50 to 200 g. Water Sci Technol. 31:137–141.

45. Roy PK., & Lall SP. 2003. Dietary phosphorus requirement of juvenile haddock (*Melanogrammus aeglefinus* L.). Aquaculture. 221:451–468.

46. Sandnes K., Lie O., & Waagbo R. 1988. Normal ranges of some blood chemistry parameters in adult farmed Atlantic salmon, *Salmo salar*. J Fish Biol. 32:129–136.

47. Sarker PK., Fukada H., & Masumoto T. 2009. Phosphorus availability from inorganic phosphorus sources in yellowtail (*Seriola quinqueradiata* Termminck and Schlegel). Aquaculture. 289:113–117.

48. Satoh S., Hernandez A., Tokoro T., Morishita Y., Kiron V., & Watanabe T. 2003. Comparison of phosphorus retention efficiency between rainbow trout (*Oncorhynchus mykiss*) fed a commercial diet and a low fish meal based diet. Aquaculture. 224:271–282.

49. Shu L., Schneider P., Jegatheesan V., & Johson J. 2006. An economical evaluation of phosphorus recovery as struvite from digester supernatant. Bioresour Technol. 97:2211–2216.

50. Sokal RR., & Rohlf FJ. 1994. Nested analysis of variance. Biometry. 3rd edFreeman W. H.;New York: p. 272–342.

51. Spannhof L., Nasev D., & Kreutzmann HL. 1979. Early recognition of metabolic disturbances in trout (*Salmo gairdneri* Rich.) stocks. Aquaculture. 18:317–323.

52. SPSS Inc. 1999. SPSS Base 100 for Windows User's Guide. SPSS Inc;Chicago IL, USA:

53. Suman A., & Carey WD. 2006. Assessing the risk of surgery in patients with liver disease. Cleve Clin J Med. 73:398–404.

54. USEPA (US Environmental protection agency). 1996. Acid digestion of sediments, sludge, and solids, Method 3050B, Revision 2. http://www.epa.gov/epaoswer/hazwaste/test/3050b.pdfAccessed December, 1996

55. USGS (US Geological survey). 2013. Mineral commodity summaries. Digital Data Serieshttp://minerals.usgs.gov/minerals/pubs/mcs/2013/mcs2013.pdfAccessed January, 2013

56. V'azquez GR., & Guerrero GA. 2007. Characterization of blood cells and hematological parameters in *Cichlasoma dimerus* (Teleostei, Perciformes). Tissue Cell. 39:151–160.

57. Vilema J., Koskela J., & Ruohonen K. 2002. Growth, bone mineralization, and heat and low oxygen tolerance in European whitefish (*Coregonnus lavaretus* L.) fed with graded levels of phosphorus. Aquaculture. 212:321–333.

58. Vielma J., & Lall SP. 1998. Control ofphosphorus homeostasis of Atlantic salmon (*Salmo sala*) in fresh water. Fish Phosiol Biochem. 19:83–93.

59. Wilson RP., Robinson EH., Gatlin DM., & Poe WE. 1982. Dietary phosphorus requirement of channel catfish. J Nutr. 112:1197–1202.

60. Yan Q., Xie S., Zhu X., Lei W., & Yang Y. 2007. Dietary methionine requirement for juvenile rockfish, *Sebastes schlegeli*. Aquacult Nutr. 13:163–169.

SECTION III: REPRODUCTION AND PHYSIOLOGY

CHAPTER 7

MELATONIN INDUCED CHANGES IN SPECIFIC GROWTH RATE, GONADAL MATURITY, LIPID AND PROTEIN PRODUCTION IN NILE TILAPIA OREOCHROMIS NILOTICUS (LINNAEUS 1758)

Ruchi Singh, A. K. Singh, Madhu Tripathi

Keywords: Melatonin, Gonad Structure, GSI, Condition Factor, SGR%, Biochemical Changes, Oreochromis niloticus

INTRODUCTION

Melatonin, the hormone synthesized by the pineal organ in fish mainly, was first isolated from beef pineal organs (Lemer et al., 1958) and is chemically known as N-acetyl-5-methoxytryptamine. It is a naturally occurring compound found in animals, plants, and microbes, (Caniato et al., 2003) and it has been described to influence feeding and locomotory rhythms (Pinillos et al., 2001; Zhdanova et al., 2001), as well as glucose homeostasis in fish (Delahunty and Tomlinson, 1984). In addition, it has also been found as a potent antioxidant, in recent years (Reiter et al., 2000; Boutin et al., 2005). In vertebrates, circulating levels of melatonin vary in a diurnal cycle, thereby regulating the circadian rhythms of several biological functions (Altun and Ugur-Altun, 2007; Paredes, 2009). In mammals, the photoperiodic message is reported to be mediated via melatonin, but it is unclear whether this also applies to fish as well (Bromage et al., 1995). Administration of melatonin has been reported to mimic a short non-stimulatory photoperiod cycle and to influence maturation in male three-spined sticklebacks *Gasterosteus aculeatus*, kept under stimulatory long photoperiods (Bornestaf et al., 2001). The fact that lighting influences fish reproduction, has been exploited in aquaculture, where artificial lights have been used to submit fishes to shorter or longer day lengths, and thus managed to advance or delay gonad maturation or spawning time (Bromage et al., 1995). Experimental evidences indicate that the pineal organ and/or melatonin control seasonal maturation of gonads in fish (Bromage et al., 2001; Maitra et al., 2001; Bayarri et al., 2004). Melatonin-treated *Rana ridibunda* maintained on a long photoperiod (18 L:6 D) produced significantly smaller gonads than control animals. Frogs injected daily with melatonin exhibited a substantial decrease in ovarian weight when compared to saline-injected animals (Delgado et al., 1983).

Earlier workers have demonstrated the influence of melatonin on the action of maturation inducing hormone (MIH) thereby affecting the maturation of oocytes in carps (Chattoraj et al., 2005). Ghosh and Nath (2005) have shown that melatonin is effective in inhibiting ovarian vitello-genesis, plasma GtH II levels, and gonado-somatic-index (GSI) in

Citation: Ruchi Singh, A. K. Singh, Madhu Tripathi, "Melatonin Induced Changes in Specific Growth Rate, Gonadal Maturity, Lipid and Protein Production in Nile Tilapia Oreochromis niloticus (Linnaeus 1758) DOI: http://dx.doi.org/10.5713/ajas.2011.11139

a catfish, *Clarias batrachus*. Melatonin also, influences the inhibition of hypothalamic dopamine secretion in common carp (Popek et al., 2006). The fish pineal, through the cyclical synthesis and release of melatonin, seems to be involved in the control of reproduction (Joy and Agha, 1991; Vanecek, 1998; Bromage et al., 2001). It is reported that rhythmical secretion of melatonin by the pineal gland subsequently induces physiological, metabolic and gonadal changes in fish (Koger et al., 1999; Porter et al., 1999; Bromage et al., 2001; Oppedal et al., 2001, 2003). Experimental evidences indicate that the pineal organ and/or melatonin control maturation of gonads in fish (Bromage et al., 2001; Maitra et al., 2001; Bayarri et al., 2004) and also in sex change in tilapia (Singh, 2009) and rice field eel (Shi Qiong, 2005). In view of the above facts, the present study was undertaken to assess the effect of melatonin on fish growth, gonado-somatic index, maturity of gonad, total gonadal protein and lipid in Nile tilapia *Oreochromis niloticus* (Linnaeus, 1758). The result of this study will be useful to enhance knowledge on the role of melatonin in fish reproduction.

MATERIALS AND METHODS

300 live fish specimens of Nile tilapia *Oreochromis niloticus* weighing 2.8±1.58 g and measuring 3.14±1.22 cm was procured and acclimatized to laboratory conditions (12: 12 h L:D condition) in glass aquaria (300 L capacity) for two weeks prior to their use in the experimental study. Fishes were fed twice a day with commercially available granulated food (Taiyo Pet Productions Pvt. Ltd.) which contained fish meal, wheat flour, soybean meal, corn meal, yeast, vitamins and minerals. The nutritional content of the feed was 32% crude protein, 4% crude fat, 5% crude fiber and 10% moisture. In addition to feeding with granules, fishes were also fed with planktons collected from the fish farm of the institute, once daily at noon to keep them healthy and to enable fishes to attain maturity. All the fish were divided into two groups and maintained them in glass aquaria in replicates. The first group received melatonin (Sigma Chemicals Co., USA) in the dose of 25 µg/L through rearing water under normal photoperiod for three weeks. The control group of fishes was held in normal day night condition on similar diet but without MLT treatment to rearing water. The water parameter of keeping tank during experiment was maintained at 25.5±2°C temperature, 7.2 pH and 4.8–5.2 mg/L dissolved oxygen.

Somatic and Gonadal Growth

Length and weight (F_L and F_W) of control and MLT treated fish was registered with the help of a digital caliper meter scale and a digital weighing machine (Denver Instrument) at the end of 21 day. Condition factor (k) was calculated as (weight in g)/(length in cm) $^3\times100$ and specific growth rate (SGR % day^{-1}) was calculated as SGR (% day^{-1}) = 100 (log final weight-log initial weight)/n. Where, n = days of experiment. The GSI was determined as Gonad weight/body weight×100; The GSI was calculated with a precision of 0.01 g to show any changes in fish maturity.

Biochemical Analysis

Both control as well as MLT treated fish was gently netted out the aquarium and five fishes from each experimental group was anesthetized by putting them in water containing 100 mg tricaine-methane-sulfonate (MS-222, Sigma Chemical Co., USA) per liter buffered with sodium bicarbonate1:1. Blood was drawn from the caudal fin of the anaesthetized fish and collected in heparinized tubes. About 2 ml of the blood was centrifuged at 5,000 rpm for 5 minute and the plasma so obtained was used for analysis of protein (Lowry et al., 1951) using kit (Bangalore GENEI/Qualigens). Similarly, the gonadal protein was also determined using 2 mg gonadal tissue. Absorbance reading for protein was taken with Spectrophotometer (Tecan, USA) at an optical density of 660nm. For measurement of serum lipid, blood was collected into tubes containing 1 mg EDTA/ml. Lipid extraction was done following the method of Bligh and Dyer (1959) by using a mixture of Chloroform and methanol (1:2) followed by a mixture of chloroform and water (1:1). Reading of lipid concentration was taken using a spectrophotometer (Tecan, USA) on 440 nm. Similarly, gonads collected from the control as well as MLT treated fishes were also subjected to total lipid estimation by the same method. Triglycerides and cholesterol were estimated by the method described by Stein (1986).

Histological Examination

Anesthetized fish used in biochemical studies were dissected after collecting blood and gonads were removed immediately afterwards and fixed in 10% natural buffer formalin solution, dehydrated in different grades (30%, 70%, 90% and absolute) ethanol and embedded in paraffin wax. Sections (5–7 µm thick) were cut, dewaxed dehydrated with different grades of ethanol, stained with Haematoxylin and Eosin, dehydrated and mounted in Canada balsam (Merk, Darmstadt, Germany). Sections were examined under a compound microscope (Olympus Co., Japan).

Statistical Analysis

All data were expressed as, mean±standard error mean (X±SEM). Data of all parameters were analyzed by one way analysis of variance (ANOVA) using a computer programmed software, Statistical Package for Social Science Version (SPSS)-10.0.

RESULTS

The experimental fishes were found healthy throughout the observation period and there was hardly any mortality in both the groups i.e., control and MLT treated fishes. The length and weight of fish of control group was 9.6±0.23 cm and 11.0±0.39 g respectively after three weeks. However, the weight of MLT treated fish reduced to 8.6±0.47 g which was 78.18% low as compared to control. The length of MLT treated fish also declined to 7.9±0.421 cm after the same period and there was 82.29% reduction (Figure 1). The specific growth rate (SGR % day^{-1}) when calculated was found to decline after MLT treatment and it was 0.9±0.04 as compared to 1.23±0.02 observed in control fishes (Figure 1). The calculated, condition factor (k) was found to be 0.99±0.026 in control group while it was 0.80±0.04 in MLT treated fishes. The Ganado-somatic index (GSI) of control fish and MLT treated fish were found negatively correlated with each other, and the correlation coefficient (r) was −0.51. The GSI value for control group of fishes was 2.56±0.25 and 1.77±0.25 in the MLT treated group after three weeks and it was significantly ($p<0.05$) lower in case of MLT treatment (Figure 2).

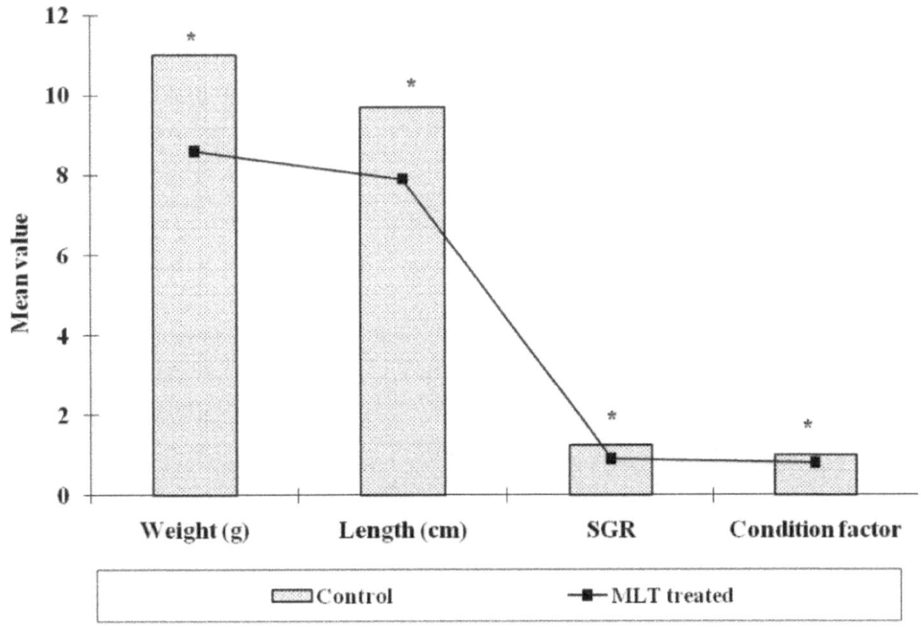

Figure 1: Effect of melatonin (MLT) treatment on physiological parameters (weight, length, SGR% day^{-1} and condition factor) of *O. niloticus* (Significance level, * $p<0.01$).

Total gonadal protein was 14.67±1.45 mg/g in control group and it significantly ($p<0.01$) increased to 22.67±0.66 mg/g after MLT treatment. Serum protein level however, increased after MLT treatment and it was 9.33±2.90 mg/ml in control group while in MLT treated fish it was 11.67±1.45 mg/ml and the difference was insignificant (Figure 3). The serum cholesterol level in control group of fishes was 132.0±1.65 mg/dl which declined to 126.66±0.88 mg/dl after MLT treatment and was significant ($p<0.01$). Similarly, serum triglyceride level also declined significantly ($p<0.001$) from 123.0±1.23 mg/dl in the control group to 86.16±1.078 mg/dl in fishes exposed to MLT (Figure 4). The gonadal total lipid was found 1.05±0.01 μg/g in the control group of fishes which reduced significantly ($p<0.05$) to almost half and it was 0.50±0.20 μg/g after MLT treatment.

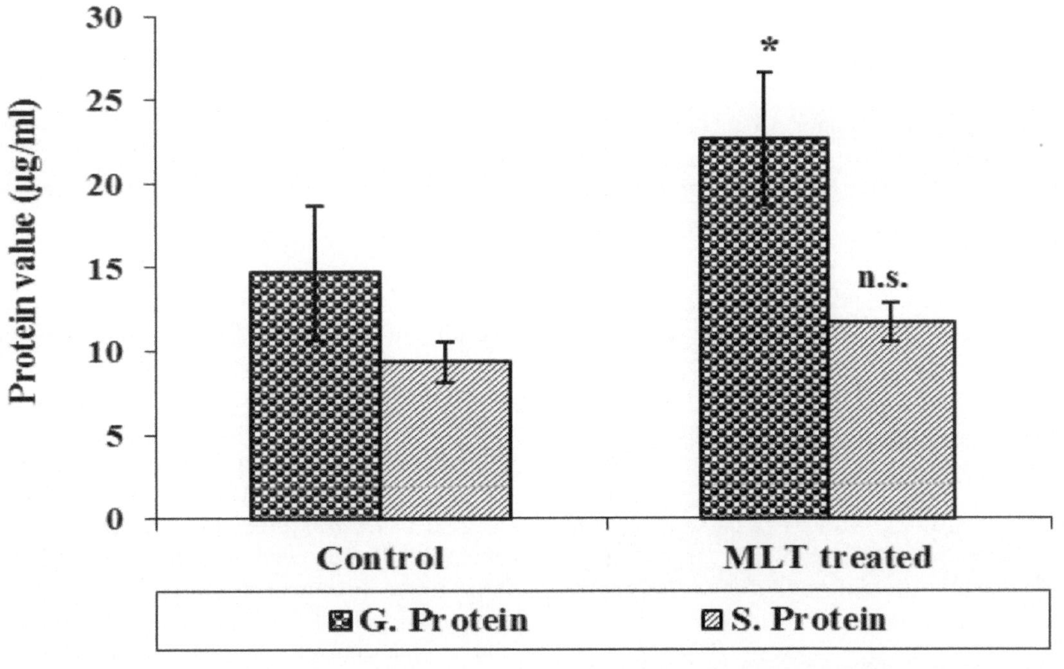

Figure 3: Serum protein and gonadal protein after melatonin (MLT) treatment to normal *O. niloticus* (n.s. = statistically not significant and * statistically significant $p<0.01$).

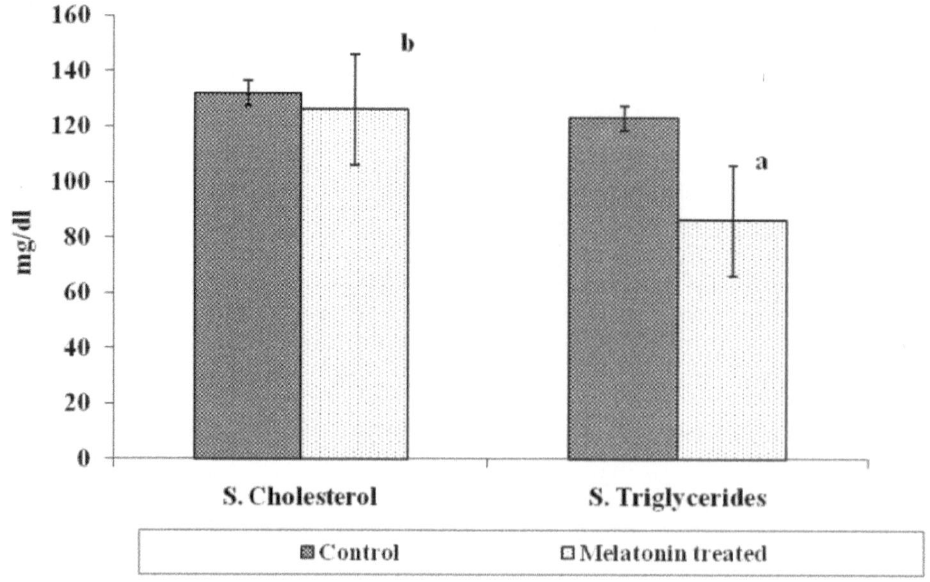

Figure 4: Serum cholesterol and serum triglycerides of *O. niloticus* after melatonin (MLT) treatment (significance level: 'a' $p<0.001$, 'b' $p<0.01$).

The gonadal development of *O. niloticus* was found asynchronous. In female, the oogonia contained a nucleus with a conspicuous nucleolus and a weak staining cytoplasm. The cortical alveloli were pushed in the periphery of the oocytes and most of the cytoplasm was filled with yolk globules (Microphotograph 1A). The histological observations of the ovary of MLT treated fish revealed suppressed developing stage of ovary. Small eggs were found in early perinucleolus stage. MLT treated fish did not show maturing oocytes unlike control fish. In the MLT treated fish, ovary of *O. niloticus* showed deeply basophilic cytoplasm in follicular cells having nucleolus in nucleus (Microphotograph 1B). In male, each spermatogonium contained a large nucleus, a distinct nucleolus and a network of chromatin. MLT treatment to fish did not show any conspicuous changes in the testicular structure. Testicular development of controlled fish and MLT treated fish was more or less similar.

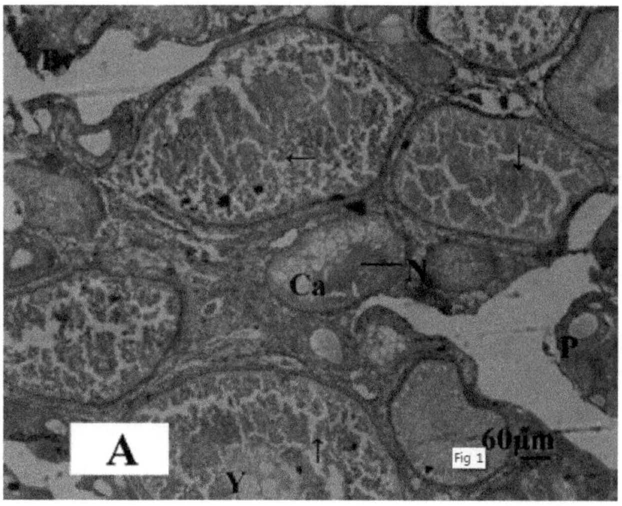

Microphotograph 1A: Histological section of ovary of control *Oreochromis niloticus* showing oocyte in the first growth phase (→), cortical alveolar cells (Ca), Perinucleaur cells (P), Blood vessels (Bv), Yolk bodies (Y), and Nucleolus of oocytes. Haematoxylin and Eosin, X-bar 60 µm.

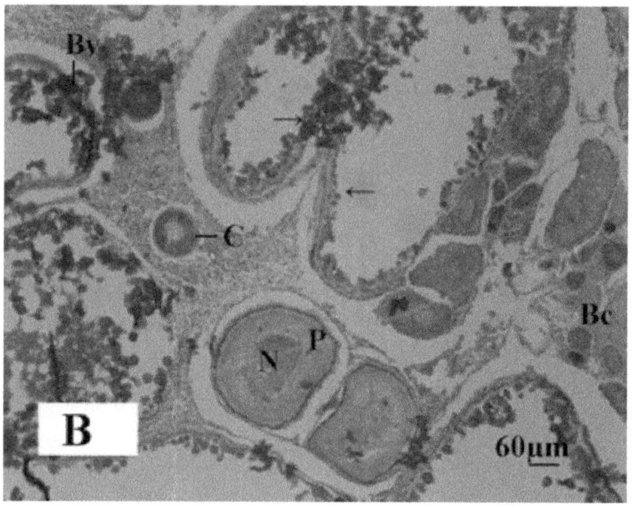

Microphotograph 1B: Histological section of the melatonin treated ovary of *Oreochromis niloticus* showing distribution of early perinuclear stage oocytes (→). Perinuclear cells (P), Blood capillaries (Bc), Blood Vessels (Bv) and Cortex of ovary (C). Haematoxylin and Eosin, X-bar 60 µm.

DISCUSSION

Results of this study indicated that continuous treatment of melatonin to fish inhibited somatic and gonadal growth in *O. niloticus*. The finding of this study revealed that MLT treatment suppressed length and weight of fish which supported the earlier report where melatonin in the dose of 10 µg/g body weight has been shown to reduce body weight of gold fish *Carassius auratus* (Lopez et al., 2006;De Pedro et al., 2008). The condition factor of MLT treated fish was a little lower in comparison to control fish in our study which suggested that MLT did not affect the performance of the fish. Our findings corroborated with earlier report where melatonin was found to affect the body mass and condition factor of Arctic charr, *Salvelinus alpines* (Aarseth et al., 2010).

Melatonin normally has been reported to decline protein level in blood and gonadal tissue in fish (Falcon et al., 2007). However, our findings indicated that protein level increased after MLT treatment. Information concerning the lipid and protein value in *O. niloticus* after MLT treatment is limited. Lipid was recognized as a major form of energy stored in teleosts during gonadogenesis. Shulman (1974) reported that, lipid stored in the body and viscera of teleost supplied the energy necessary for gonadal maturation. Total lipid level of fish was reduced by the administration of melatonin in Nile tilapia suggesting that low level of lipid might have negatively affected the gonadal development. Lipid peroxides, uric acid, blood glucose, total lipid, triglycerides and cholesterol have been reported to decline in fish as well as in rats after the oral administration of melatonin (Mamdouh and Abdel-Raheim, 2003). The results of this study delineated that melatonin regulated metabolic activity of protein and lipid in *O. niloticus* (Lopez et al., 2006).

The effect of MLT on gonadal development of *O. niloticus* suggested that melatonin exerted potential inhibitory effect on GSI and gonadal development. In many animals, melatonin has been reported to possess anti-gonadal effect (Sreenivasan et al., 2009). Numerous data suggest involvement of melatonin in reproductive physiology (Tamura et al., 2009). However, information about the role of melatonin in fish reproduction is still fragmentary (Guererro et al., 2007; Chattoraj et al., 2009a, b; Falcon et al., 2009a, b; Renuka and Joshi, 2010).Renuka and Joshi (2010) immersed *Channa punctatus* in water containing melatonin (100 µg/L) for 24 and 17 h daily and found that the GSI increased, but when the same dose was given by way of injection, GSI decreased. Melatonin has been reported to possess inhibitory effects in catfish *Clarias batrachus* (Ghosh and Nath, 2005) while MLT implants (18 mg) in fish suppressed the gonadal maturation of turbot females three months after implantation (Alvaro et al., 2001). The result of this study provided evidences of inhibitory influences of exogenous melatonin on gonadal maturity in tilapia. The antigonadal effects of melatonin has also been reported in *Oryzias latipes* (Uraski, 1972), and *Heteropneustes fossilis* (Sundararaj and Keshavanath, 1976). The effect of MLT on the structure of ovary and ovarian cells was similar as reported earlier (Borg and Ekstrom, 1981; Fatemet et al., 2007; Wang et al., 2008). The minimum employed dose of melatonin was 25 µg/100 g body weight/day, and this is frequently reported dose to study the effect of melatonin on the gonads (Maitra et al., 2005). Effect of melatonin on reproductive processes in fish, most often, points to its stimulatory or inhibitory effect depending on the species, season of the year, time of the day, and degree of sexual maturity, but fails to show how and where melatonin affects the reproductive system (Sokolowaska et al., 2004).

CONCLUSION

It is concluded that exogenous melatonin can change the biometrics, physiology and gonadal development of fish. The results obtained on the biometric parameters, biochemical data, GSI and gonadal structures demonstrated that melatonin is involved in a wide complex system that regulates growth, lipid and protein metabolism and reproduction in *O. niloticus*. However, the role of melatonin in the control of reproduction is still debatable as very little is known about the mechanism by which melatonin contributes to this process.

ACKNOWLEDGEMENTS

The authors are grateful to Dr J. K. Jena, the Director, NBFGR for his kind support, encouragement and providing all necessary facilities. Authors are thankful to Mr. Abubakar Ansari, Senior Research Fellow at NBFGR for his help in day to day monitoring of the experimental fishes and data collection.

REFERENCES

1. Aarseth JJ., Froiland E., & Jorgensen EH. 2010. Melatonin implantation during spring and summer does not affect the seasonal rhythm of feeding in anadromous Arctic charr (*Salvelinus alpines*). Polar Biol. 33:379–388.
2. Altun A., & Ugur-Altun B. 2007. Melatonin: therapeutic and clinical utilization. Int J Clin Pract. 61:835–845.
3. Alvaro JMR., Rebollar PG., Olmedo M., Alvarez-Blazquez B., Ubilla E., & Peleterio BPD. 2001. Effects of melatonin implants on reproduction and growth of turbot brood stock. Aquacult Int. 9:477–487.
4. Bayarri MJ., Rodriguez L., Zanuy S., Madrid JA., Sanchez-Vazquez FJ., Kagawa H., Okuzawa K., & Carrillo M. 2004. Effects of photoperiod manipulation on the daily rhythms of melatonin and reproductive hormones in caged European sea bass *Dicentrachus labrax*. Gen Comp Endocrinol. 136:72–81.
5. Bligh EG., & Dyer WJ. 1959. A rapid method of total lipid extraction and purification. Can J Biochem Physiol. 37:911–917.
6. Borg B., & Ekstrom P. 1981. Gonadal effects of Melatonin in the three-spined stickeleback, *Gasterosteus aculeatus* L., during different season and photoperiods. Reprod Nutr. 21:919–927.
7. Bornestaf C., Mayer I., & Borg B. 2001. Melatonin and maturation pace in female three-spined stickleback, *Gasterosteus aculeatus*. Gen Comp Endocrinol. 122:341–348.
8. Boutin JA., Audinot V., Ferry G., & Delagrange P. 2005. Molecular tools to study melatonin pathways and actions. Trends Pharmacol Sci. 26:412–419.
9. Bromage N., Porter M., & Randall C. 2001. The environmental regulation of maturation in farmed finfish with special reference to the role of photoperiod and melatonin. Aquaculture. 197:63–98.
10. Bromage NR., Randall CF., Porter MJR., & Davies B. 1995. How do photoperiod, the pineal, melatonin and circannual rhythms interact to coordinate seasonal reproduction in salmonid fish? Reprod Physiol. 95:164–166.
11. Caniato R., Filippini R., Piovan A., Puricelli L., Borsarini A., & Cappelletti EM. 2003. Melatonin in plants. Adv Exp Med Biol. 527:593–597.
12. Chattoraj A., Bhattacharyya S., Basu D., Bhattacharya S., Bhattacharya S., & Maitra K. 2005. Melatonin accelerates maturation inducing hormone (MIH) induced oocyte maturation in carps. Gen Comp Endocrinol. 140:145–155.
13. Chattoraj A., Seth M., & Maitra SK. 2009a. Localization

and dynamics of melatonin receptor in the ovary of carp, *Catla catla* in relation to serum melatonin levels. Com. Biochem. Physiol. A, Mol Integr Physiol. 152:327–333.

14. Chattoraj A., Seth M., Basu A., Shrivastava TG., Porta S., & Maitra SK. 2009b. Temporal relationship between the circulating profiles of melatonin and ovarian steroids under natural photothermal conditions in an annual reproductive cycle in carp *Catla catla*. Biol Rhythm Res. 40:347–359.
15. Delahunty G., & Tomlinson M. 1984. Hypoglycemic effects of melatonin in the goldfish, *Carassius auratus*. Comp Biochem Physiol. 78:871–875.
16. Delgado MJ., Gutierrez P., & Alonso-Bedate M. 1983. Effects of daily melatonin injection on the photoperiodic gonadal response of the female frog *Rana ridibunda*. Comp Biochem Physiol. 76:289–292.
17. De Pedro N., Martinez-Alvarez R., & Delgado MJ. 2008. Melatonin reduces body weight in goldfish (*Carassius auratus*): effects on metabolic resources and some feeding regulators. J Pineal Res. 45:32–39.
18. Falcón J., Laurence B., Sandrine S., & Gilles B. 2007. Melatonin effects on the hypothalamo-pituitary axis in fish. Trends Endocrinol Metabol. 18:281–88.
19. Falcon J., Besseau L., Fuentes M., Sauzet M., Magnanou E., & Boeuf G. 2009a. Structural and functional evaluation in pineal melatonin system in vertebrates. Ann NY Acad Sci. 1163:101–111.
20. Falcon J., Migaud H., Munoz-Cueto JA., & Carrillo M. 2009b. Current knowledge on the melatonin system in teleost fish. Gen Comp Endocrinol. 165:469–482.
21. Fatemeh M., Naderil M., & Adibi S. 2007. Effects of melatonin on histopathological changes after experimental ovarian torsion-detorsion in cat. Iran J Reprod Med. 5:177–181.
22. Ghosh J., & Nath P. 2005. Seasonal effects of melatonin on ovary and plasma gonadotropin and vitellogenin levels in intact and pinealectomized catfish, *Clarias batrachus*(Linn). Ind J Exp Biol. 43:224–232.
23. Guerrero HY., Cardillo E., Poleo G., & Macrcano D. 2007. Daily and annual variations of plasma melatonin in the annual gonadal maturation. In : Proceedings of the 8th international symposium on reproductive physiology of fish; Saint-Malo. 32:p. 22–24.
24. Joy KP., & Agha AK. 1991. Seasonal effect of administration of melatonin and 5-methoxy-trytophol on ovarian activity in the catfish *Heteropneustes fossilis*(Bloch). J Pineal Res. 10:65–70.
25. Koger CS., Teh SJ., & Hinton DE. 1999. Variations of light and temperature regimes and resulting effects on reproductive parameters in medaka (*Oryzias latipes*). Biol Reprod. 61:1287–1293.
26. Lemer AB., Case JD., Takahashi Y., Lee Y., & Mori W. 1958. Isolation of melatonin, the pineal gland factor that lightens melanocytes. J Am Chem Soc. 80:2587
27. Lopez LF., Olmeda JA., Madrid FJ., & Sanchez-Vazquez . 2006. Melatonin effects on food intake and activity rhythms in two fish species with different activity patterns: Diurnal (goldfish) and nocturnal (tench). Comp Biochem Physiol. 144:180–187.
28. Lowry OH., Rosebrough NJ., Farr AL., & Randall RJ. 1951. Protein measurement with folin's phenol reagents. J Biol Chem. 193:265–275.
29. Maitra SK., Chattoraj A., & Bhattacharyya S. 2005. Implication of melatonin in oocyte maturation in Indian major carp *Catla catla*. Fish Physiol Biochem. 31:201–207.
30. Maitra SK., Dey R., & Bhattacharya S. 2001. Seasonal reproduction in fish: a functional interplay between the pineal organ and photoperiods. J Endocrinol Reprod. 5:1–12.
31. Mamdouh MA., & Abdel-Raheim MA. 2003. Oxidative stress in streptozotocin - induced diabetic rats: Effects of garlic oil and melatonin. Comp. Biochem. Physiol Part A. 135:539–547.
32. Oppedal F., Juell JE., Taranger GL., & Hansen T. 2001. Artificial light and season affect vertical distribution and swimming behaviors of post-smolt Atlantic salmon in sea cages. J Fish Biol. 58:1570–1584.
33. Oppedal F., Taranger GL., & Hansen T. 2003. Growth performance and sexual maturation in diploid and triploid Atlantic salmon (*Salmo salar* L.) in seawater tanks exposed to continuous light or simulated natural photoperiod. Aquaculture. 215:145–162.
34. Paredes SD., Korkmaz A., Manchester LC., Tan DX., & Reiter RJ. 2009. Phytomelatonin: a review. J Exp Bot. 60:57–69.
35. Pinillos ML., Pedro ND., Alonso-Gomez AL., Alonso-Bedate M., & Delgado MJ. 2001. Food intake inhibition by melatonin in goldfish (*Carassius auratus*). Physiol Behav. 72:629–634.
36. Popek W., Luszczek-Trojnar E., Drog-Kozak E., Rzsa J., & Epler P. 2006. Effect of melatonin on dopamine secretion in the hypothalamus of mature female common carp, *Cyprinus carpio* L. Acta Ichthyol Piscat. 36:134–141.
37. Porter MJR., Duncan NJ., Mitchell D., & Bromage NR. 1999. The use of cage lighting to reduce plasma melatonin in Atlantic salmon (*Salmo salar*) and its effects on the inhibition of grilling. Aquaculture. 176:237–244.
38. Reiter RJ., Tan DX., & Acuna-Castroviejo D. 2000. Melatonin: mechanisms and actions as an antioxidant. Curr Top Biophys. 24:171–183.
39. Renuka K., & Josh BN. 2010. Melatonin-induced changes in ovarian function in the freshwater fish *Channa punctatus*(Bloch) held in long days and continuous light. Gen Comp Endocrinol. 165:42–46.
40. Qiong Shi. 2005. Melatonin is involved in sex change of the rice field eel, *Monopterus albus* Zuiew. Rev Fish Biol Fish. 15:23–36.
41. Shulman GE. 1974. Life cycles of fish physiology and biochemistry. New York: Hulsted Press, John Wiley and sons.
42. Singh AK. 2009. Role of pineal melatonin during environmental sex determination in fishes. Recent

advances in Hormonal Physiology of fish and Shell fish Reproduction. Singh BN, Pandey AK, editorsNarendra Publishing House;New Delhi: p. 265–271.

43. Sokołowska E., Kalamarz H., & Kulczykowska X. 2004. Seasonal changes in brain melatonin concentration in the three-spined stickleback (*Gasterosteus aculeatus*): towards an endocrine calendar. Comp. Biochem. Physiol. Part A: Mol Integr Physiol. 13:365–369.

44. Sreenivasan V., Spence WD., Pandi-Perumal SR., Zakharia R., Bhatnagar KP., & Brzezinski A. 2009. Melatonin and human reproduction: shedding light on the darkness hormone. Gynecol Endocrinol. 25:779–785.

45. Stein EA. 1986. Text book of clinical chemistry. Tietz NW, editorW B Saunders Co;Philadelphia: p. 1818–1829.

46. Sundararaj BI., & Keshavanath P. 1976. Effects of melatonin and prolactin treatment on the hypophysial-ovarian system in the catfish,*Heteropneustes fossilis*(Bloch). Gen Comp Endocrinol. 129:84–96.

47. Tamura H., Nakamura Y., Korkmaz A., Manchester LC., & Tan DX. 2009. Melatonin and the ovary: physiological and pathophysiological implications. Fertil Steril. 92:328–343.

48. Uraski H. 1972. Effect of restricted photoperiods and melatonin: administration on gonadal weights in Japanese Kilifish. J Endocrinol. 55:619–620.

49. Vanecek J. 1998. Cellular mechanism of melatonin action. Physiol Rev. 78:687–711.

50. Wang Y., Peng KM., Li J., Song H., Heli S., Wei L., & Wang JX. 2008. Ultrastructure and melatonin 1a receptor distribution in the ovaries of African ostrich chick. Cytotechnology. 56:187–195.

51. Zhdanova IV., Wang SY., Leclair OU., & Danilova NP. 2001. Melatonin promotes sleep-like state in zebra fish. Brain Res. 903:263–268.

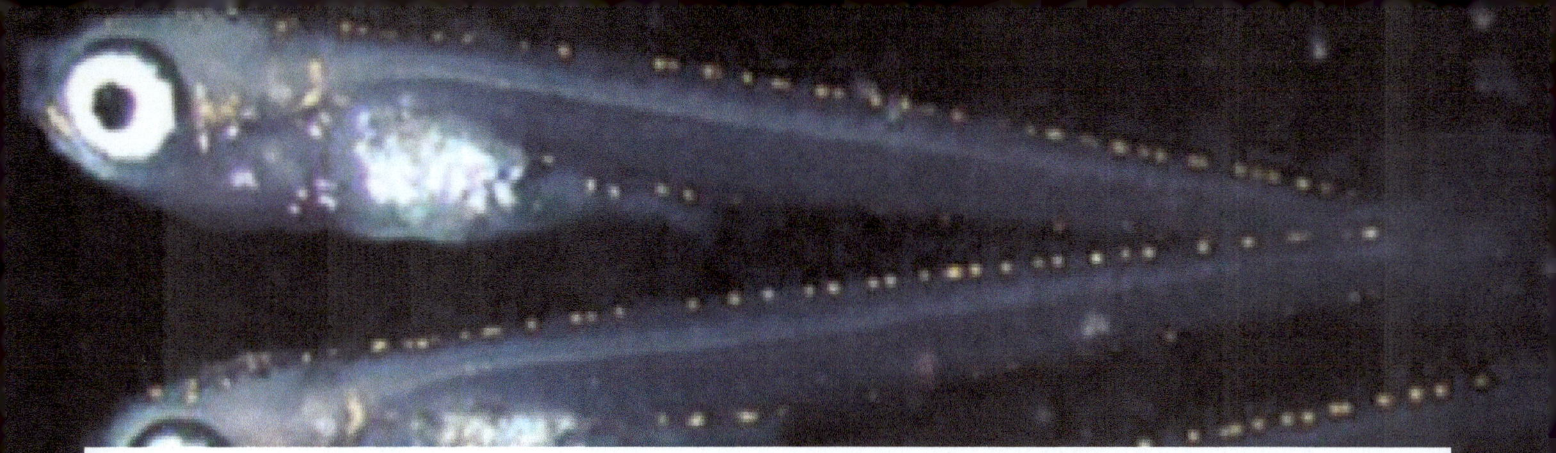

CHAPTER 8

GENETIC ARCHITECTURE OF SEX DETERMINATION IN FISH: APPLICATIONS TO SEX RATIO CONTROL IN AQUACULTURE.

Paulino Martínez, Ana M. Viñas, Laura Sánchez, Noelia Díaz, Laia Ribas and Francesc Piferrer

Keywords: Sex determination, fish, genetic architecture, sex ratio, aquaculture

INTRODUCTION

Fish represent the most diverse group of vertebrates including more than 28,000 species (Nelson, 2006). This diversity is a reflection of their high capacity for adaptation to a broad spectrum of environmental conditions. As a result, fish show amazing morphological, physiological, and behavioral adaptations to live in the highly diverse aquatic environment. Fish also show all types of reproductive strategies, including gonochorism, proterandrous, protogynous, and simultaneous hermaphroditism, and unisexuality (Devlin and Nagahama, 2002). These reproductive strategies emerged independently in different lineages during evolution demonstrating a polyphyletic origin (Avise and Mank, 2009). Domestication of fish for production is an ancient practice and again shows the high adaptation capacity of this group, especially considering that more than 354 fish species are cultivated all over the world (Food and Agriculture Organization [FAO], 2014). Production of domestic fish largely relies on reproduction, and a vast amount of information has been gathered for its control. Reproduction techniques include production of monosex populations because the existence of sexual growth dimorphism, either in favor of males or females depending on the species, and also because sometimes the most valuable trait is associated with one sex (e.g., color, shape, secondary sexual ornaments).

Here, we review the available data on the genetic architecture of sex determination (SD) in fish and how sex ratio is controlled in aquaculture production. We contrast this information with models emerging from the classical studies in *Drosophila*, mammals, or birds with highly conserved mechanisms associated to marked sex chromosome heteromorphisms. We show the huge intra and interspecific diversity of SD systems in fish associated to a high evolutionary turnover. So, its genetic architecture, although commonly supported by major genes, is also influenced by minor genes, and environmental factors approaching it to a complex trait. We illustrate this diversity and its influence on the strategies aimed at the production of the most desired sex in the last section of the paper by analyzing the genetic basis of SD in two important

Citation: Martínez P, Viñas AM, Sánchez L, Díaz N, Ribas L and Piferrer F (2014) Genetic architecture of sex determination in fish: applications to sex ratio control in aquaculture.. doi: 10.3389/fgene.2014.00340

species of the European aquaculture, turbot (*Scophthalmus maximus*), and European sea bass (*Dicentrarchus labrax*). We also consider two of the main fish groups with established breeding programs, salmonids, and tilapias, all of them with remarkable differences in sex determination.

GENETIC ARCHITECTURE OF SEX DETERMINATION IN FISH

The genetic architecture of a complex trait refers to the genes involved in that trait, their influence and interactions to establish the phenotype. It takes also into account the influence of environmental factors and their interactions with the genotype on the final phenotype (McKay, 2001). Although getting all this information is a very ambitious enterprise, the more we approximate to this goal the better we shall understand key questions related to the genetic variation for its exploitation in genetic breeding programs.

Main Features of Gonad Development

Gonad development includes all developmental processes aimed to transform an undifferentiated primordium into a mature gonad, either ovary of testis. It is the result of two concatenated development processes controlled by a hierarchical genetic network: SD and gonad differentiation (GD; Figure 1). The sex of gonads is essentially determined by processes, either genetic or environmental, operating at the beginning of development, where a binary decision is taken related to the fate of the undifferentiated primordium (Kobayashi and Nagahama, 2009; Siegfried, 2010; Uller and Helanterä, 2011). Once the future of the gonad has been established, morphogenetic GD processes work until maturation is completed.

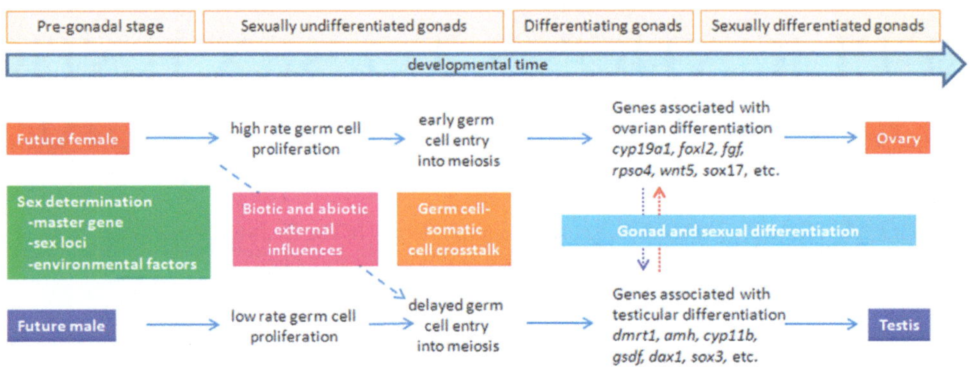

Figure 1. Major events leading to ovarian vs. testicular differentiation in fish. The first event, sex determination – the establishment of gender – can be triggered by the action of a major sex determining master gene, several sex-associated loci, an environmental factor (e.g., temperature) in an ecologically relevant context (i.e., occurring normally in the habitat of the species) or a combination of them, typically when the gonads are still sexually undifferentiated or even before they are formed at the most rudimentary stage (pre-gonadal stage). Successive events are connected by horizontal arrows and include differences in the proliferative rate of germ cells (females > males), which can be one of the first effects of the sex determining factor, whether genetic or environmental. During this period, the germ cell-somatic cell crosstalk is very important, but still largely uncharacterized. Also, during these early events, biotic and abiotic factors (e.g., stress, abnormally high temperature, etc.) can change the course of subsequent sex differentiation, usually in the female → male direction (diagonal dashed arrow). Finally, also at the beginning of gonad differentiation – the transformation of an undifferentiated gonad into a testis or ovary – the accidental (i.e., contamination) or deliberate (e.g., sex control treatment) incorporation of sex steroids, androgens or estrogens can result in female → male (vertical blue dotted arrow) or male → female (vertical red dotted arrow) sex-reversal, i.e., in that genotypic females and males develop into phenotypic males and females, respectively.

As a consequence of the hierarchical nature of gonad development, a single gene or environmental cue operating at the beginning of development can drive the gonad pathway towards one direction or another. Thus, in contrast to other characters influenced by many genes operating in different routes with important additive effects, here gene interactions likely represent an important genetic component. Particularly, epistatic effects may be relevant because a single gene acting upstream or even downstream of a preexisting SD gene (SDg) may take the control of gonad fate, thus, masking extant genetic variation at other involved loci. Epistatic interactions have been reported between major SD loci in different fish groups (Cnaani et al., 2008; Ser et al., 2010; Parnell and Streelman, 2013), and also epistatic allelic variants have been reported segregating in populations of species with a well-known SD genetic system like medaka (*Oryzias latipes*; Shinomiya et al., 2010). In fact, notable interactions occur between gene products at the initial stages of gonad development such as the suppression of *wnt4* and β-*catenin,* key genes for ovarian development, by *sox9* and *fgf9* (Nef and Vassalli, 2009); the modulation of gonadal aromatase (cypb19a), responsible of the balance between androgens and estrogens, by the action of other genes or environmental factors such as temperature (Navarro-Martín et al., 2011); or the interaction between the anti-müllerian hormone (amh1) and its receptor (amhr2), which triggers an essential signaling pathway for testis development (Kamiya et al., 2012).

Gonad development of fish is unusual in the sense that the sexually undifferentiated period can last from weeks until years (Saito and Tanaka, 2009; Berbejillo et al., 2012) opening a large developmental window in which the sexual fate can be influenced by abiotic or biotic environmental factors (Penman and Piferrer, 2008; Baroiller et al., 2009). In such a long period, it is tempting to speculate that the brain may be involved through the hypothalamic–pituitary–gonadal axis (Baroiller et al., 2009). However, despite the fact that the brain certainly integrates environmental stimuli and, in particular, social interactions, which have been shown to be implicated in the process of sex-change in hermaphrodites, currently there is no convincing proof that the brain plays any role in the SD process in gonochoristic fishes.

Homoeothermic vertebrates show a conserved morphogenetic development supported by a strongly canalized genetic sex determining system (Charlesworth et al., 2005). However, fish show inter-specific differences in the morphogenetic events occurring along gonad development. Variation exists both regarding the general pattern of differentiation, the interaction between somatic and germ cells, and in the time of occurrence and relative weight of the different steps. The amount of primordial germ cells have been reported to be the first development difference between males and females in species such as medaka (*O. latipes*; Kondo et al., 2009) and stickleback (*Gasterosteus aculeatus*; Lewis et al., 2008), and this has been related to the possible influence of growth-related factors on SD (Schlueter et al., 2007; Baroiller et al., 2009). Also, gonad development has been classified as undifferentiated or differentiated type, respectively, depending on the existence of an initial transitory female stage that can subsequently revert to testis like in zebrafish (*Danio rerio*; Orban et al., 2009), or the lack of that stage like in medaka or tilapia (*Oreochromis niloticus*; Saito and Tanaka, 2009). Interaction between somatic and germ cells is also recognized as an important feature for gonad development. In fact, several key genes related to the initial steps of differentiation like *dmrt1*, *amh*, or *sox9* in males (Lee et al., 2009; Wu et al., 2010) and *cyp19a1a* or *foxl2* in females (Nakamura et al., 2009) are expressed in Sertoli or granulosa/theca cells, respectively, and thus, communication between somatic and germ cells is essential for GD. In this communication there are species-specific differences and, for example, the ablation of the female germ cells determines the reversal of development towards a testis in zebrafish (*D. rerio*) and tilapia (Siegfried and Nusslein-Volhard, 2008; Fujimoto et al., 2010), while in goldfish (*Carassius auratus*) the female pathway is maintained (Goto et al., 2012).

Sex Determination: Origin and Evolution

A consensus existed until recently on the high conservation of the gene network controlling gonad development among vertebrates, differences being mainly related to changes in the switching mechanism. This hierarchical development controlled process would facilitate the control of sex ratio by a single-gene mechanism, but at the same time it would open the opportunity for changing the SD factor in response to new evolutionary scenarios (Marín and Baker, 1998).

Theories on the evolution and genetic architecture of SD in animals have been largely influenced by studies on *Drosophila*, mammals, and birds, all of them showing convergent patterns, with a heteromorphic sex chromosome pair and, as a consequence, a particular sex-linked inheritance model (Bachtrog et al., 2014). A generalized theory on the origin and evolution of SD systems emerged from these data, which assumed a sexual conflict between antagonistic alleles at specific loci favorable to one sex but detrimental to the other (Figure 2; Charlesworth et al., 2005). To maintain the beneficial association between the antagonistic allele and the SD locus, recombination would be restricted, giving rise to the permanent heterozygous state of that portion of the sexual pair (Bergero and Charlesworth, 2009). That circumstance would promote the accumulation of repetitive elements and deleterious variants in the SDg-bearing chromosome, contributing to its progressive degeneration and the typical heteromorphic shape of the sexual pair (Charlesworth et al., 2005). Mathematical models based on this theory suggested that only one gene should underlie the SD system, and if more than one gene were segregating, this should represent an unstable equilibrium towards a new SD mechanism (Rice, 1986).

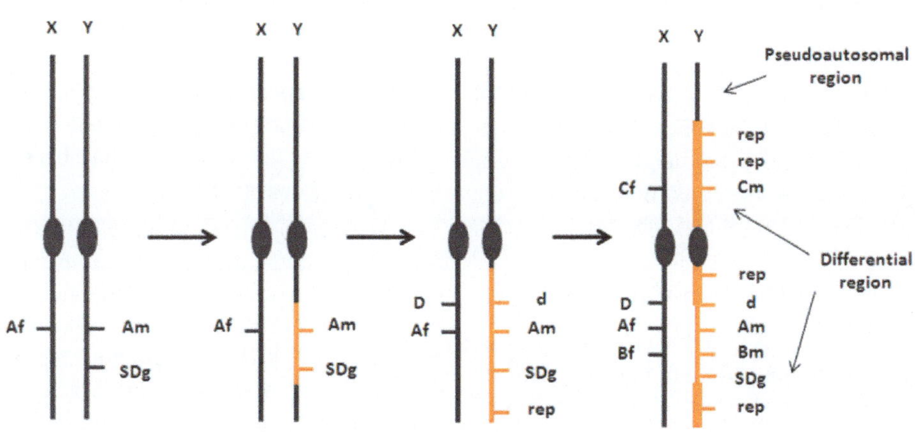

Figure 2. Model on the origin and evolution of the SD region-bearing chromosome from studies in mammals, birds and *Drosophila*. This model, although has been demonstrated in some fish, shows a large variation on its progression, which is reflected on the degree of differentiation between the chromosomes of the sexual pair. The origin of a new sexual pair is related to the origin of genes (A,B) with antagonistic alleles favorable to females (Af, Bf) or to males (Am, Bm) associated with a new SDg. Subsequent steps involve accumulation of repetitive elements (rep) and the degeneration of the Y chromosome because of its permanent heterozygotic state at the differential region (d: recessive nonfunctional variant of a sex-linked gene).

Initial data in ectothermic vertebrates, particularly fish, demonstrated a sharply different picture (Devlin and Nagahama, 2002). More than one sex-associated gene has been reported in many fish species (Cnaani et al., 2008; Ser et al., 2010). In addition, the conservation of the GD cascade has been to some degree questioned, and notable genetic differences have been observed not only at the top, but also downstream of the GD network (Böhne et al., 2013; Herpin et al., 2013). For example, the well-known female-associated aromatase genes (*cyp19a1a* and *cyp19a1b*) have shown a new role in testis differentiation in African cichlids (Böhne et al., 2013). Although antagonistic alleles demonstrated to be associated with the SD region in fish (Roberts et al., 2009; Parnell and Streelman, 2013), only 7% species showed heteromorphic sex chromosomes (Penman and Piferrer, 2008; Oliveira et al., 2009). This has been related to the huge evolutionary turnover of SD mechanism which limits chromosome differentiation and, as a consequence, different SD systems have been reported between closely related species and even between populations of the same species. However, degeneration and differentiation of sex chromosomes can evolve very quickly (Charlesworth et al., 2005) and a broad heteromorphism degree has been observed among species of the same genus in Neotropical fish of the genera *Eigenmannia* (Henning et al., 2011), *Characidium* (Vicari et al., 2008), and *Leporinus* (Parise-Maltempi et al., 2013). Thus, sex heteromorphism can involve the whole chromosome and be detectable with the usual cytogenetic techniques, as reported in Neotropical fish (Henning et al., 2011; Parise-Maltempi et al., 2013); be cryptic at cytogenetic level but involving an important degeneration of the SDg-bearing chromosome, as in stickleback (Ross and Peichel, 2008; Shikano et al., 2011); embrace no more than a few kilobases as in medaka (*O. latipes*; Matsuda et al., 2002); or show a very tiny differentiation as the single SNP observed in the *amhr2*receptor, the only detectable difference between X and Y chromosomes in fugu (*Takifugu rubripes*; Kamiya et al., 2012).

To understand the origin of SD regions both external pressures (i.e., sexual selection) and the internal context (available genes and genomic structure) should be considered. Several genes with different functions have been recruited along evolution as SDg in fish, which shows the opportunistic nature of selection to face new evolutionary pressures. In this regard, the specific genome duplication occurred within teleosts may have provided a suitable raw material for new sex determinants (Mank and Avise, 2009). A remarkable case which exemplifies the many options available to change SD mechanisms is the association of sex to B chromosomes, usually considered as junk DNA, in species of the genus *Astyanax* and in two species of pufferfish (Vicente et al., 1996; Noleto et al., 2012). The high turnover of the SD region in fish has led to suggest that changes in the SD mechanism may be associated with speciation (Ser et al., 2010; Böhne et al., 2013; Parnell and Streelman, 2013).

The Genetic Basis of Sex Determination in Fish

High genetic variation has also been described between fish species regarding the gene responsible for SD, the number of genes involved in such decision and the relationships between them. Currently, five different master genes have been documented in fish: *dmY*, *gsdf*, *amhy*, *amhr2*, and *sdY*. *dmY* (DM-domain gene on the Y chromosome), the SDg of medaka and the first one described in fish (Matsuda et al., 2002), is a transcription factor expressed in the somatic cells surrounding germ cells before sex differentiation and in the testis thereafter and it is involved in germ cell proliferation and development of pre-Sertoli cells into Sertoli cells. It originated from a segmental duplication of a small autosomal region containing the precursor *dmrt1*, followed by an insertion of the duplicated region on the proto-Y chromosome (Matsuda et al., 2002). *gsdf*, *amhy*, and*amhr2* are members of the TGF-ß superfamily involved in cell signaling controlling cell proliferation (Heule et al., 2014). *gsdf* (gonadal soma derived growth factor) is a downstream gene of *dmY* in the SD cascade that has taken the role of master SDg in *Oryzias curvinotus* (Myosho et al., 2012). *amhy* (Y chromosome-specific anti-müllerian hormone) is expressed in the presumptive Sertoli cells of XY males of *Odonthestes hatcheri* during the onset and subsequent GD. This gene has been inserted upstream of *amh* in the cascade of male development, becoming a male SDg (Hattori et al., 2012). In *T. rubipres*, *amhr2* (anti-müllerian hormone receptor type 2) is expressed in somatic cells surrounding germ cells and it is thought to be the SDg in this species. This gene contains a specific SNP variant in the kinase domain of *amhr2* in the X chromosome which determines lower affinity for the amh hormone, thus fating the female pathway when homozygous (XX;Kamiya et al., 2012). Finally, *sdY* (sexually dimorphic on the Y chromosome) is linked to the SEX locus of salmonids and is necessary and sufficient to induce testicular differentiation. It has evolved through neofunctionalization from *irf9* by losing its role in *ifn* signaling pathway and acquiring a new role in SD (Yano et al., 2012).

From an evolutionary perspective different SDgs have been reported either in closely related species like *O. latipes (dmY)* and *Oryzias luzonensis* (*gsdf1*; Myosho et al., 2012) or in divergent ones such as rainbow trout (*Oncorhynchus mykiss*; *sdY;* Yano et al., 2012), fugu (*amhr2*; Kamiya et al., 2012), or pejerrey (*O. hatcheri*;*amhY*; Hattori et al., 2012). Also, the same SDg has been identified in closely related species like *O. latipes* and*O. curvinotus* (*dmY*; Matsuda et al., 2002); *T. rubripes, Takifugu pardalis*, and *Takifugu poecilonotus*(*amhr2*; Kamiya et al., 2012); and in most salmonid species studied so far (*sdY*; Yano et al., 2013). Very recently, *dmrt1* has been suggested as a strong candidate in the half-smooth tongue sole (*Cynoglossus semilaevis*) based on its association with sex and its pseudogenization in the W chromosome (Chen et al., 2014). This would constitute the first SDg reported in a ZW system species and within flatfish, a group of species of great commercial value and with a particular metamorphosis to adapt to demersal life. However, no functional demonstration has been reported to date thus requiring further investigation.

Additional information exists from marker and QTL sex-associated studies which are usually unraveling variation on major genes controlling the fate of the undifferentiated primordium, and thus, basically related to SD. The relationships between the different genomic regions identified through this approach has been established through comparative mapping using model fish as a bridge, taking advantage of the conserved macrosynteny pattern observed in teleosts (Kai et al., 2011; Bouza et al., 2012). In most fish groups analyzed, these SD regions demonstrated to be non-homologous (Piferrer et al., 2012), so in the *Oryzias* (medaka) genus up to five different genes/genomic regions seem to be involved in sex determination, and only two species among the six analyzed show the same SDg (Tanaka et al., 2007); in the Gasterosteidae (stickleback) family four different genomic regions have been identified in the five species analyzed, and in the fifth, a fusion of two previously reported SD chromosomes gave rise to the sexual chromosome (Ross et al., 2009); in the tilapiine (tilapia) cichlid tribe two chromosomes have been identified [linkage group 1 (LG1) and LG3], two species associated with LG1, other two with LG3, and the remaining two showing both linkage groups, but other studies also demonstrated association with LG23 (Cnaani, 2013); in the Salmoniformes (salmonids) order most species showed different non-homologous sex-associated genomic regions (Phillips et al., 2001); and finally, in the Poeciliidae (guppy and platyfish) family, up to four different chromosomes are involved in SD (Tripathi et al., 2009). Very likely these non-homologous regions include different SDgs, although in salmonid species showing non-homologous SD genomic regions, the SDg appears to be the same (Yano et al., 2013).

The diversity of SDg in fish highlights the many options available at the undifferentiated stage of gonads to switch and drive gonad fate, although some genes have been recurrently used because of their prominent position in the development cascade (Graves and Peichel, 2010; Heule et al., 2014). Among them, *dmrt1* and related genes found in medaka and likely in half-smooth tongue sole have also been reported in different vertebrates, including birds and amphibians (Smith and Sinclair, 2004; Yoshimoto et al., 2008), which illustrate processes of convergent evolution related to the suitability of some genes acting at the beginning of development. Furthermore, a great plasticity has been shown, since it can be found in XY or ZW systems and acting on a presence/absence model (medaka and frog) or in a dose-dependent manner (birds and likely tongue sole; Koopman, 2009; Chen et al., 2014). Three other genes, *gsdf1*, *amh1*, and *amhr2*, have also been reported to be activated at the beginning of development in the male pathway, and thus their SD role fits to the top of the GD invoked by theory (Heule et al., 2014). Contrary to previous findings in vertebrates (*sry* and*dmrt*-derived genes), *gsdf1*, *amh1*, and *amhr2* are not transcription factors being involved in cell signaling controlling cell proliferation (Heule et al., 2014). Finally, the *sdY* gene, the major SD factor in rainbow trout, represents an unexpected SDg of unknown function, whose carboxi-terminal extreme is homologous to an interferon-related gene, thus exemplifying the vast source of genes available for leading the SD process (Yano et al., 2012).

In addition to species with a single major SDg, many fish have shown more than one gene of big effect involved in sex determination. Indeed, in the most investigated fish groups at least two major genes (or genomic regions) related to SD have been reported in the same species: within tilapiinid, two major male and female determinant genes on LG1 and LG3, respectively, (Cnaani et al., 2008); in the platyfish, a poecilid species, a multifactorial SD system with X,Y, and W sex chromosomes (Schultheis et al., 2009); in the sticklebacks, *Gasterosteus weathlandi* a multiple chromosome system due to the fusion of two major SD chromosomes presented in other species of Gasterosteidae (Ross et al., 2009); and in cichlids from lake Malawi, two main SD systems on LG5 and LG7, the first one representing a female epistatically dominant factor (Ser et al., 2010). Finally, a polygenic SD system has also been documented in other species like European sea bass (Vandeputte et al., 2007) and zebrafish (Liew et al., 2012)

The lowering cost of new generation sequencing (NGS) methodologies will allow obtaining much more information in the near future to get a more accurate picture of SD in fish. Restriction-site associated DNA (RAD) sequencing, a technique which combines the powerful of NGS with the simplification of genomes through restriction enzyme digestion (Baird et al., 2008; Davey et al., 2011), is enabling to perform dense genomic screening to study the genetic architecture of multigene traits (Hohenlohe et al., 2010). This methodology is being applied for the study of SD in zebrafish (Anderson et al., 2012) and to identify sex-associated genomic regions in Nile tilapia (*O. niloticus*; Palaiokostas et al., 2013a) and Atlantic halibut (*Hippoglossus hippoglossus*; Palaiokostas et al., 2013b) for its application in aquaculture production. In addition, the high capacity of RAD sequencing for SNP discovery and constructing genetic maps will aid to get dense maps at candidate regions to narrow them and facilitate the identification SDgs (Taboada et al., 2014).

Gene expression studies linked to NGS methodologies will also be essential to understand the relationship between morphogenetic effects and the underlying genetic network. Microarrays have been used as a powerful tool for assessing gene expression profiles along GD (Gardner et al., 2012; Sreenivasan et al., 2014), but more recently, RNA-Seq is being applied due to its higher sensitivity, accuracy, and also because it provides additional information on genetic variants linked to expression differences (Sun et al., 2013; Tao et al., 2013). Finally, the evaluation of the pattern of methylation through bisulfite sequencing and other methodologies is providing quick genomic evaluation of the epigenomic maps along development (Cokus et al., 2008) constituting a valuable tool for understanding the regulation of SD and gonadogenesis (Piferrer, 2013).

Environmental Factors on Sex Determination

In environmental sex determination (ESD), the first difference between the two sexes is established by differences in the value of an environmental factor. ESD has been well studied in reptiles like crocodiles and turtles (Valenzuela and Lance, 2004). In *Menidia menidia* (Conover and Kynard, 1981), the first fish species where temperature-dependent sex determination (TSD), a form of ESD, was first described, some populations have shown a genetic component underlying SD (Conover and Heins, 1987). Much of the literature in this field demonstrated the influence of environmental factors in laboratory conditions, extreme in some cases, but not necessarily reflecting the conditions that species experience in nature (Ospina-Álvarez and Piferrer, 2008). Nevertheless, the presence of TSD in fish demonstrates the plasticity of gonad development (Baroiller et al., 2009).

Social interactions represent important environmental cues for SD in hermaphroditic species (Godwin, 2009). In these cases, the brain has shown to be a major player translating social cues into a physiological signal (Kobayashi et al., 2010). Although this is not a mechanism driving the gonad fate at the beginning of development, it is a good example of the plasticity of gonad development in fish and also an evidence on the existence of bipotential primordium cells in the differentiated gonads of adult fish (Zhou and Gui, 2010).

Temperature is the environmental factor with highest influence in SD in fish (Ospina-Álvarez and Piferrer, 2008; Baroiller et al., 2009). Temperature influence on sex shifting can be exerted at several points of the differentiation cascade. High temperatures usually tend to produce more males and low temperatures have no effects or produce more females in some cases (Ospina-Álvarez and Piferrer, 2008). The ultimate mechanism (if there were a single one) connecting temperature and sex ratio is not known, and several have been proposed. The influence of temperature on SD has been related to a higher stress, giving rise to changes in circulating cortisol levels. In fact, the administration of cortisol in the diet has demonstrated a significant influence on sex ratio (Mankiewicz et al., 2013). However, it is far from clear the ultimate molecular mechanism connecting cortisol levels and masculinization. On one hand, Hayashi et al. (2010) proposed a direct up-regulation of the follicule stimulant hormone (FSH) receptor, which would be connected to germ cell proliferation. On the other, Fernandino et al. (2013) suggested an up-regulation of hsd11b2, a steroidogenic enzyme implicated both in the metabolism of cortisol into cortisone, and in the synthesis of biologically active androgens such as 11-ketotestosterone. Epigenetic regulation of aromatase expression mediated by temperature has also been proposed. Thus, Navarro-Martín et al. (2011) demonstrated that hypermethylation of aromatase promoter is correlated with high temperature during the thermosensitive period in the European sea bass, strongly suggesting that sex differentiation is under epigenetic control in this species.

Genetic Variation within Species: Sex Determination as a Complex Trait

Most studies in SD in fish have been focused on identifying the master SDg expected according to previous SD models. However, new data are consistently showing that other minor genetic and environmental factors are also involved in SD, even in species with well-studied master genes like *dmY* of medaka (Matsuda et al., 2002), and thus, more effort should be devoted to investigate this variation to get the closest picture as possible on SD in fish.

The information exposed so far shows that: (i) although environmental factors may influence sex ratio, usually the genetic component represent the main SD factor in most studied species; (ii) major genes, those which explain a high proportion of the trait phenotypic variance, are on the basis of SD in many species, likely an expected fact, because of the hierarchical nature of the sex development pathway; and (iii) the interaction of other genes involved in SD with the major factors and the environment. Although major genes are involved in SD of many fish species, available data suggest that SD studies and their applications for fish aquaculture should emphasize the complex nature of this trait and thus, using appropriate quantitative genetics tools for its study.

In fact, consistent variation among families has been reported on sex ratio in European sea bass (Vandeputte et al., 2007), turbot (Haffray et al., 2009; Martínez et al., 2009), Nile tilapia (Lozano et al., 2013) and zebrafish (Liew et al., 2012). In some species, the additive genetic component underlying SD or sex ratio was even estimated (Vandeputte et al., 2007; Lozano et al., 2013). In zebrafish, a polygenic SD system was suggested based on inter-family and inter-strain variation, and consequently several QTL at different genomic locations were identified, some of them associated to the different strains studied (Bradley et al., 2011; Anderson et al., 2012; Howe et al., 2013). In other species where major loci are involved, the application of QTL screening or genomic association analysis to look for the SD region demonstrated to be efficient and new SD-related genomic regions were identified (Martínez et al., 2009; Hermida et al., 2013), sometimes denoting important intraspecific variation such as in *Eigenmannia* or in cichlid species complex (Ser et al., 2010; Henning et al., 2011; Parnell and Streelman, 2013). As a consequence, selection has demonstrated to be efficient to change sex ratio in progenies of several species and in other related traits like the sensitivity to temperature on sex ratio (Baroiller et al., 2009; Liew et al., 2012; Lühmann et al., 2012; Lozano et al., 2013).

SEX-ASSOCIATED TRAITS IN FINFISH AQUACULTURE: SOME RELEVANT EXAMPLES

In many fish species one sex grows faster or matures earlier than the other and these differences may be accentuated under aquaculture conditions (Breder and Rosen, 1966; Parker, 1992; Piferrer et al., 2012). Sex-associated growth differences generate size dispersion, and therefore, classification must be performed for feeding and to avoid cannibalism or size hierarchies affecting social relations (Dou et al., 2004). This represents more work in animal husbandry, and a higher number of production units to adjust different growth groups (Piferrer et al., 2012). Sexual growth dimorphism can favor males (e.g., tilapias) or, more frequently, females (flatfish, sea bass, among others). In some cases, as in the turbot, females can be 50% larger than males (Imsland et al., 1997). In other cases, as in the European sea bass, the rearing conditions result in highly male-biased stocks (Piferrer et al., 2005). A great deal of research towards the development of sex control methods has been carried out in fish (Piferrer, 2001; Cnaani and Levavi-Sivan, 2009).

Sex-associated markers are very useful in this context for precocious sex identification, especially in those species lacking morphological sexual dimorphism. This can aid to identify the sex of potential broods in genetic breeding programs and to avoid sex bias in the selected population. However, the most relevant application of sex-associated markers is to identify the genetic sex of sex-reversed individuals after hormonal treatment to accelerate the processes for establishing monosex populations (Penman and Piferrer, 2008). The availability of sex-associated markers or even better the SD master gene makes it possible to shorten this process using marker assisted selection (MAS) or gene assisted selection (GAS), respectively.

Turbot

The strong sexual growth dimorphism of turbot has promoted the interest of industry for all-female populations. No sex-associated karyotype heteromorphism have been detected in turbot, either after analyzing the mitotic or the 11-fold longer and higher resolution meiotic chromosomes (Bouza et al., 1994; Cuñado et al., 2002). This suggests that the SD region in turbot is small or not large enough to be detected with these cytogenetic techniques. A QTL screening performed with 100 homogeneously distributed microsatellites identified a major SD region in the proximal region of LG5 between two markers separated by 17.4 cM (Martínez et al., 2009). Assuming a single SD region with full penetrance, the SD master gene (SDg) was located at 2.6 cM of the Sma-USC30 microsatellite locus, representing 1.4 Mb according to the general relationship between genetic and physical maps in turbot (Bouza et al., 2007). Also, the analysis of segregation of Sma-USC30 in all families demonstrated that the mother is responsible for sex, supporting a ZZ/ZW system (Martínez et al., 2009) in accordance with the sex ratios observed in progenies from hormonal sex reversed parents (Haffray et al., 2009). However, sex association of Sma-USC30 showed variation among families (between 84 and 100%) and, in addition, other minor QTL were detected at LG6, LG8, and LG21. Temperature also showed some influence on sex ratios (Haffray et al., 2009), although without the general trend reported in most species where the proportion of males increased with temperature (Ospina-Álvarez and Piferrer, 2008).

Using the Sma-USC30 marker, it was possible to classify correctly 98.4% of the individuals in four out of five families analyzed. This information was essential to develop a molecular tool for precocious sex identification in turbot, currently under a Spanish patent (Ref. number: 2 354 343). Since sex cannot be identified in turbot until fish maturation, precocious sex identification is relevant in breeding programs to estimate sex ratio in selected progenies. This molecular tool is also essential to facilitate the achievement of all-female populations. Because turbot displays a ZW mechanism, getting all-female populations requires a three-generation pedigree starting from hormonal sex-reversed ZW neomales until obtaining WW superfemales in the progeny of the second cross (Figure 3). These superfemales would produce all-female offspring after being crossed with normal ZZ males. However, the chromosome constitution of ZW neomales or WW superfemales require individual progeny testing of hormone-treated larvae (ZZ or ZW) and of female offspring in cross II (ZW or WW).

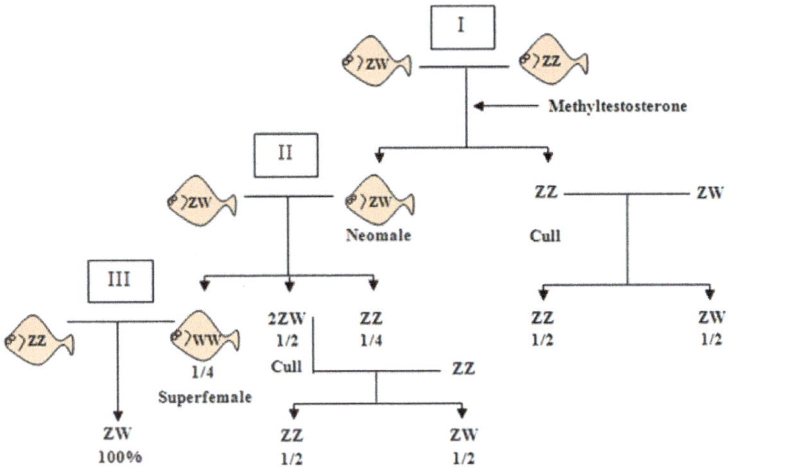

Figure 3. Scheme of crosses aimed at obtaining all-female populations of turbot. I, II, and III represent the three generations required for the full process. Neomales (ZW) are obtained in generation I by applying methyltestosterone in the diet in undifferentiated fry. Identification of neomales (I) and superfemales (II) is usually done by individual progeny testing, so parents producing 50:50 f/m in cross I and II are culled because they are not neomales and superfemales, respectively. Crossing normal males (ZZ) with superfemales (WW) would produce 100% females (ZW) assuming a single full penetrant SDg.

Progeny testing is long and involves at least 2–3 years until fish are mature for checking sex ratio in their progenies. Additionally, sex can only be visually identified 4–6 months after hatching in sacrificed offspring, which globally represents a minimum of 2.5 years for progeny testing per generation. The availability of a genetic marker closely linked to the SDg has enabled assessment of the genotypic sex of fish just after 4–6 months when a fin clip can be obtained without damage, thus saving a minimum of 5 years for the production of all-female progenies. However, some limitations still remain, so this technology should be further refined. First, production of 100% females is not often fulfilled because turbot sex also depends on other minor genetic and environmental factors. In addition, because Sma-USC30 is a linked marker to the SDg, we need to establish association of SmaUSC-E30 to sex at family level because this is not a sex-specific marker. Finally, because crossovers can take place between the marker and the SDg, some of the selected ZW neomales or WW females would not show the expected genetic constitution. Nevertheless, this tool is being used by turbot companies with encouraging results.

The availability of a denser genetic map with around 600 markers (Bouza et al., 2012; Hermida et al., 2013), the enriched database with reproduction and immune genes (Pereiro et al., 2012; Ribas et al., 2013), and the recently assembled turbot genome (Figueras et al., in preparation) is enabling a more refined analysis of the SD region to identify the SDg, to analyze its relationship with the previous suggestive QTL, and to study the evolution of sex chromosomes (Taboada et al., 2014). The fine mapping performed has narrowed the genomic position of the SDg to a few kilobases and now much more genetic markers closely associated to the SDg are available, thus facilitating the precocious evaluation of sex. However, although several strong candidates were identified in that region (*sox2*, *dnajc19*, *and fxr1*), none of them were associated with sex at the species level, which illustrates the difficulty of such enterprise (Hermida et al., 2013). In addition, this work has provided additional support for the existence of minor factors at LG6, LG8, and LG21 in new families and demonstrated an interactive rather than an additive component between minor and the major SD QTL.

European Sea Bass

The European sea bass is a gonochoristic marine teleost of the family Moronidae that is present in the NE Atlantic, from Norway to NW Africa and in the Mediterranean up to the Black Sea. As in most teleosts, sea bass does not show sex chromosomes (Devlin and Nagahama, 2002). Female homogamety (XX) has been ruled out since sex ratios of normal diploid and gynogenetic offspring are equivalent (Felip et al., 2002), and offspring from masculinized females is not female-biased (Blázquez et al., 1999). Other data from hormone-treated fish suggested that male homogamety with environmentally male-biased sex ratio would still be a possibility (Vandeputte et al., 2007). However, a polygenic sex determining system (Vandeputte et al., 2007) influenced by temperature (Piferrer et al., 2005) seems to fit better to data. Sex ratio shows interfamily differences with specific and measurable paternal and maternal components and at least two genes would be necessary to explain the variation pattern observed (Vandeputte et al., 2007), although no sex-associated QTL screening was carried out. The temperature can greatly influence European sea bass sex ratios (Piferrer et al., 2005) and a genetic component determining differential sensitivity to the masculinizing power of temperature has been reported (Saillant et al., 2002). Variation found on female percent due to thermal treatments show that genetic and environmental components are of comparable magnitude, supporting the notion of a continuum between GSD and ESD components of SD in the European sea bass.

In the European sea bass, gonads remain undifferentiated during post-larval stages until fish attain about 8 cm standard length, usually at about 5–6 months of age (Bruslé and Roblin, 1984; Blázquez et al., 1999; Navarro-Martín et al., 2009a; Díaz et al., 2013). European sea bass females differentiate earlier, are bigger, and mature later than males (Blázquez et al., 1999; Navarro-Martín et al., 2009a; Díaz et al., 2013). While still sexually undifferentiated, European sea bass gonads can be influenced by environmental abiotic factors or external factors such as sex steroids (Navarro-Martín et al., 2009a), but once sex is determined remains throughout life (Zanuy et al., 2001).

One of the main genes involved in fish GD as outlined above is the aromatase (cytochrome P450). This gene is present in two paralogous copies as a result of the teleost genome duplication, one predominantly expressed in the ovary (*cyp19a1a*; Dalla Valle et al., 2002) and the other in the brain (*cyp19a1b*; Blázquez and Piferrer, 2004). Interestingly, the *cyp19a1a* promoter exhibits important conserved binding sites for several genes of the GD network such as *sf1*, *sox*, *foxl2* or *ar*.

Recent work by Navarro-Martín et al. (2011) showed how temperature during early development is linked to the production of male-biased populations through differences in the methylation levels specifically on the gonadal aromatase promoter at one year. Different CpGs loci within the *cyp19a1a* promoter showed different sensitivities to temperature, suggesting a different role on regulation of aromatase expression. Methylation of gonadal aromatase promoter is thought to be the cause of the lower expression of aromatase in the temperature-masculinized fish linked to the suppressed ability of *sf1* and *foxl2* transcription factors to stimulate gonadal aromatase expression (Navarro-Martín et al., 2011). While 15°C has been proposed as the optimal temperature for larval rearing in European sea bass (Koumoundouros et al., 2001), optimal growth in juveniles is found at 26°C and 13°C is considered as detrimental (Person-Le Ruyet et al., 2004). European sea bass hatcheries usually apply high temperatures (>20°C) to speed up growth rates, but male-bias progenies determine a loss of biomass. In the European sea bass the thermosensitive period includes from half-epiboly to mid-metamorphosis (~17–18 mm total length; ~70 dph; Koumoundouros et al., 2002), and treating fish with high temperatures masculinize a high proportion of the population (Navarro-Martín et al., 2009b), while temperatures never surpassing 17°C until metamorphosis yielded the maximum female proportion (Pavlidis et al., 2000). However, raising fish at low temperatures (15°C) for a long period also masculinize the population even more than a high temperature thermal treatment (Saillant et al., 2002). A thermal protocol to maximize the number of females was developed (Navarro-Martín et al., 2009a) and recently patented (patent no N200802927). Such protocol consisted on maintaining 17°C water temperature until the end of the thermosensitive period, and then increasing the temperature as a ratio of 0.5°C/day until 21°C to allow high growth rates. It should be stated that there is no temperature regime that will increase the proportion of females in the European sea bass. Instead, proper management of temperature during early development will avoid induced masculinization by applying high water temperatures before the thermosensitive period is over. What a proper thermal regime does is allowing the production of as many females as the polygenic system will allow, bearing in mind the sex ratio of a given brood, even reared at the optimal thermal regime, will ultimately depend on the genetic constitution of both parents.

With this information at hand, different strategies have been devised by European sea bass industry to improve growth rate. On one hand, classical breeding programs may incorporate sex ratio as additional phenotypic information to get female-biased progenies. This strategy would be much more efficient if sex-related QTL screening were performed and sex-associated genetic markers explaining an important proportion of trait variance were incorporated following MAS selection programs. On the other hand, controlling sex-ratio through larval rearing temperature protocols could increase female proportions by avoiding masculinization due to elevated water temperature. Currently, efforts are underway aimed at selecting broodstock that will produce the highest number of females and investigating the possibility that, among these fish, those that are more resistant to the masculinizing temperature can also be selected. An additional step would be further investigation of the epigenetic mechanisms responsible for the inheritance of the high-temperature masculinization, as already done in the half-smoth tongue sole (Chen et al., 2014). In the case of the European sea bass, the goal would be the opposite, i.e., to select as future broodstock those fish that despite being reared at high (masculinizing temperature) they do not become masculinized. In this way, the production of the highest number of females across different generations perhaps could be achieved.

Salmonids

The Salmonidae family (11 genera, 70 species; salmon, trout, char, whitefishes, and graylings) includes several of the most economically important species for aquaculture and fisheries industry (the third largest world fish production). Salmonids are worldwide distributed and some species have played a vital role in the life and culture of the North hemisphere societies for thousands of years.

Atlantic salmon (*Salmo salar*) is the leading intensively farmed marine fish. In 1998 global farmed production exceeded the world's total wild salmon captures, and in 2010 around 1.2 million tons were produced worldwide (Food and Agriculture Organization [FAO], 2014). Atlantic salmon breeding companies have achieved more than 100% growth increase in around six generations of selection, and significant improvement in disease resistance and delay at the onset of sexual maturation. The vast majority of farmed Atlantic salmon eggs and smolts are now sourced by such breeding companies (Bostock et al., 2010). Another important salmonid species, the rainbow-trout (*O. mykiss*), is the most-widely cultivated cold fresh water fish species in the world.

In salmonids, early maturation occurs differentially in males and females and is responsible for some problems related to intensive culture (Felip et al., 2006) including reducing growth, increasing disease susceptibility and changing of organoleptic properties. Thus, in this group, females are preferred for production. In all salmonid species investigated so far, SD is strictly genetic with a male heterogametic sex-determination system (Davidson et al., 2009), although no heteromorphisms or only slight morphological differentiation have been reported associated to sex chromosomes in some species. In rainbow trout, all-female production is generalized in Europe since females are still immature at harvest. Neomales (hormone sex-reversed genotypic females) are used to produce all-female progenies from XX neomale crossed to XX females. Triploids are also produced using chromosome set manipulation techniques to avoid sexual maturation for the production of individuals of bigger size (more than 3 kg; Piferrer et al., 2009).

Both, endocrine and genetic technologies have been implemented for sex control on a production scale in salmonids. Nevertheless, the phenotypic differentiation of males and females is still problematic until the fish become sexually mature and, thus, in some species it is still necessary a genetic/molecular test for sexing. Sex-specific markers, including a linked sequence to the growth hormone pseudogene, have been developed in the last decade for Pacific species of the *Oncorhynchus* genus, rainbow trout (*O. mykiss*), Chinook salmon (*O.tshawystscha*), coho salmon (*O. kisutch*), chum salmon (*O. keta*), and pink salmon (*O. gorbuscha*; Brunelli and Thorgaard, 2004).

Recently, the *sdY* (sexually dimorphic on the Y chromosome) was identified as a male-specific linked gene on the Y chromosome in most salmonids (Salmoninae, Coregoninae and Thymallinae subfamilies), strongly suggesting that *sdY* may be the conserved master sex-determining gene of this group (Yano et al., 2012, 2013). However, because this gene is not located at homologous genomic positions among the different salmonid species, it has been suggested its jumping associated to mobile elements (Yano et al., 2013). Irrespective of its location, sequences of this gene may represent a useful tool for sexing. However, some exceptions were observed to this general rule, and in the Coregoninae subfamily, while *Stenodus leucichthys* showed *sdY* as a male specific gene, *Coregonus lavaretus* and *Coregonus clupeaformis*, both males and females contain the *sdY*gene, so a different SD mechanism appears to be involved.

The analysis of sex-associated markers in several families of the SALTAS Tasmanian Atlantic salmon program from different male lineages allowed the discovery of three sex-associated markers (Ssa03, Ssa06, and Ssa2) mapping at three different linkage groups. Ssa2 is the same sex-associated marker previously reported in Atlantic European populations (Eisbrenner et al., 2014), but Ssa03 and Ssa06 represent new genomic positions. These three loci showed positive amplification for *sdY* gene and most individuals analyzed showed a good concordance between phenotypic sex and *sdY* PCR amplification, suggesting the movement of *sdY* to new positions within this species. However, some inconsistencies were detected among the sex marker associated genotypes, the presence of *sdY* gene and the phenotypic sex. Based on these findings, and in the fact that the sdY protein lacks a DNA binding domain, these authors suggested the existence of another sex determining gene in Atlantic salmon upstream to *sdY*. Also, they emphasized the importance of using many families to identify sex-associated markers in salmonid species (Eisbrenner et al., 2014). Furthermore, the recent characterization of the male-specific region on the Y chromosome of rainbow trout which contains the *sdY* gene and the male specific marker (OmY1), revealed several male specific SNPs associated with 12 single-copy protein coding sequences whose role in the SD should be further analyzed (Phillips, 2013). Data show that even in a species with an apparent well-established SD mechanism variation is observed. Thus, caution should be taken when applying sex-associated markers for establishing associations and also to for their application to precocious sex determination.

Tilapia

The tribe Tilapiini includes more than 80 species of cichlid fish (family Cichlidae, order Perciformes). Tilapias are endemic to Africa and the Middle East, but they have been introduced into most tropical and subtropical countries for aquatic weed control and aquaculture. Tilapia culture was considered a resource to improve protein supply in developing countries, but nowadays there is also an important market for tilapias in Japan, United States, European Union as well as in other developed countries. Tilapia is one of the fastest growing fish farming sector, being China the production leader, and it constitutes the second most important group of farmed fish after carps and the most widely grown of any farmed fish (85 producer countries). The main aquaculture species is the Nile tilapia (*O. niloticus*) with a production exceeding 3.2 million metric tons in 2012 (Food and Agriculture Organization [FAO], 2014).

Tilapias exhibit an important sexual growth dimorphism in favor of males. Additionally, tilapias show early maturation (e. g., 4–5 months old in Nile tilapia), which determines successive spawning during the growing period, leading to stunting of growth (Beardmore et al., 2001). All these circumstances make it difficult to establish a uniform product, and all male fry are preferred.

Monosex culture can be obtained by different approaches: manual separation of males and females; hybridization between species to produce all-male offspring; or artificial sex reversal using hormones. The most frequent method for producing all-male populations in tilapia was the treatment with 17α-methyltestosterone included in the diet of sexually undifferentiated fry. If properly applied, farms can produce male populations with 98 to 100% effectiveness. However, marketing of hormonally treated fish can also be a problem for health and the direct use of hormones is usually forbidden by food safety regulations in European Community, although in other countries it may be allowed. One way to get through this problem is to combine a method for sex reversal with a breeding scheme aimed at obtaining broodstock that produces monosex fry following the reverse procedure as outlined before for turbot (Figure 3). In the case of Nile tilapia, it is necessary to produce YY supermales by crossing neofemales (XY) with regular males (XY). As in turbot, the use of sex-linked DNA markers could shorten the process by distinguishing XX, XY, or YY individuals, thus avoiding the identification of individual genotypes by progeny testing.

Sex chromosomes are not identifiable in tilapias using standard cytogenetic techniques. Most species show 22 chromosome pairs but there is no a heteromorphic sexual chromosome pair. Association studies suggested that sex is determined in tilapias by the existence of major genes located at linkage groups 1, 3, and 23 in the different species (Cnaani et al., 2008; Cnaani, 2013). Additionally, the heterogametic sex can be either the male or the female, depending upon the species, and ZZ/ZW (LG3) and XX/XY (LG1) systems have been reported within the same genus, i.e., in *O. niloticus and Tilapia zillii* (XX/XY), and in *Tilapia mariae, Oreochromis aureus, Oreochromis karongae, Oreochromis tanganicae* (ZZ/ZW). Some families of blue tilapia have been found segregating for both loci, and in these cases the ZW locus appears to be epistatic over the XY (ZW/XY individuals are female; Lee et al., 2004). In Mozambique tilapia (*Oreochromis mossambicus*), the SD locus was found at LG1 (Liu et al., 2013). However, Cnaani et al. (2008) found sex-associated markers on LG1 and LG3 in three families of this species. These discrepancies may be determined by the different genetic background of families and strains used in these studies (Liu et al., 2013). The differences in SD within species show that minor genetic factors segregate and interact with major genes in addition to the influence of environment factors (Baroiller et al., 2009), suggesting that SD should be treated as a quantitative trait and its dissection approached using QTL screening (Eshel et al., 2012).

The main cultured species is the Nile tilapia (GIFT project; Lozano et al., 2013). Most data pointed to LG1 as the sex chromosome in this species (Cnaani et al., 2008) and recently a small candidate region was narrowed by RAD (restriction site associated DNA) sequencing on a 1–2 Mb region on LG1 (Palaiokostas et al., 2013a). However, other sex-linked markers have been identified in *O. niloticus* and its hybrids (*O. niloticus x O. aureus*) mapping on LG3 and LG23 (Eshel et al., 2012). Also, it has been shown by linkage analysis that genetic factors are involved in the sensitivity of SD to temperature and that these factors are located depending on families on the same chromosomal regions as the major QTL at LG1, LG3, and LG23 (Lühmann et al., 2012).

The closely sex-associated genomic region identified on LG1 includes 10 genes not previously related to SD in other species, and two SNPs probed to be very useful for sexing individuals, thus being worthy for production all male stocks by industry (Palaiokostas et al., 2013a). However, some of these markers were not associated to sex in different strains or families, so checking several markers on different linkage groups should be done before its application. Most studies on SD in *O. niloticus* have been carried out on fish derived from Lake Manzala in Egypt population, but today this species show a worldwide production and new information is arising which probably will confirm the necessity for checking a set of markers previous to sexing specific strains.

CONCLUSION

Major genetic factors can explain a high proportion of the SD variance in fish in accordance with the hierarchical gonad development of vertebrates and with the models proposed to explain its origin and evolution. However, several other minor genetic and environmental factors also influence sex following a complex interactive pattern. Thus, the currently available information supports the idea that sex can be regarded as a complex trait in fish, with the influence of one or more genetic factors in addition to possible environmental influences, depending upon the species. The presence of genetic factors regardless of whether SD is under the control of a master gene, a polygenic system or driven by an environmental factor enables their application in MAS programs to exploit the benefits of a particular sex.

ACKNOWLEDGMENTS

This research work was supported by the Spanish Government (Consolider Ingenio Aquagenomics: CSD2007-00002 project) and Spanish Ministerio de Ciencia e Innovación (AGL2009-13273 and AGL2010-15939) projects to Paulino Martínez and Francesc Piferrer.

REFERENCES

1. Anderson, J. L., Mari, A. R., Braasch, I., Amores, A., Hohenlohe, P., Batzel, P.,et al. (2012). Multiple sex-associated regions and a putative sex chromosome in zebrafish revealed by RAD mapping and population genomics. *PLoS ONE* 7:e40701. doi: 10.1371/journal.pone.0040701
2. Avise, J. C., and Mank, J. E. (2009). Evolutionary perspectives on hermaphroditism in fishes. *Sex. Dev.* 3, 152–163. doi: 10.1159/000223079
3. Bachtrog, D., Mank, J. E., Peichel, C. L., Kirkpatrick, M., Otto, S. P., Ashman, T. L.et al. (2014). Sex determination: why so many ways of doing it?.*PLoS Biol.* 12:e1001899. doi: 10.1371/journal.pbio.1001899
4. Baird, N. A., Etter, P. D., Atwood, T. S., Currey, M. C., Shiver, A. L., Lewis, Z. A.,et al. (2008). Rapid SNP discovery and genetic mapping using sequenced RAD markers. *PLoS ONE* 3:e3376. doi: 10.1371/journal.pone.0003376
5. Baroiller, J. F., D'Cotta, H., and Saillant, E. (2009). Environmental effects on fish sex determination and differentiation. *Sex. Dev.* 3, 118–135. doi: 10.1159/000223077
6. Beardmore, J. A., Mair, G. C., and Lewis, R. I. (2001). Monosex male production in finfish as exemplified by tilapia: applications, problems, and prospects. *Aquaculture* 197, 283–301. doi: 10.1016/

S0044-8486(01)00590-7

7. Berbejillo, J., Martinez-Bengochea, A., Bedo, G., Brunet, F., Volff, J. N., and Vizziano-Cantonnet, D. (2012). Expression and phylogeny of candidate genes for sex differentiation in a primitive fish species, the Siberian sturgeon, *Acipenser baerii*. *Mol. Reprod. Dev.* 79, 504–516. doi: 10.1002/mrd.22053

8. Bergero, R., and Charlesworth, D. (2009). The evolution of restricted recombination in sex chromosomes. *Trends Ecol. Evol.* 24, 94–102. doi: 10.1016/j.tree.2008.09.010

9. Blázquez, M., Carrillo, M., Zanuy, S., and Piferrer, F. (1999). Sex ratios in offspring of sex-reversed sea bass and the relationship between growth and phenotypic sex differentiation. *J. Fish Biol.* 55, 916–930. doi: 10.1111/j.1095-8649.1999.tb00730.x

10. Blázquez, M., and Piferrer, F. (2004). Cloning, sequence analysis, tissue distribution, and sex-specific expression of the neural form of P450 aromatase in juvenile sea bass (*Dicentrarchus labrax*). *Mol. Cell Endocrinol.* 219, 83–94. doi: 10.1016/j.mce.2004.01.006

11. Böhne, A., Heule, C., Boileau, N., and Salzburger, W. (2013). Expression and sequence evolution of aromatase cyp19a1 and other sexual development genes in East African cichlid fishes. *Mol. Biol. Evol.* 30, 2268–2285. doi: 10.1093/molbev/mst124

12. Bostock, J., McAndrew, B., Richards, R., Jauncey, K., Telfer, T., Lorenzen, K.,et al. (2010). Aquaculture: global status and trends. *Philos. Trans. R. Soc. Lond. B Biol. Sci.* 365, 2897–2912. doi: 10.1098/rstb.2010.0170

13. Bouza, C., Hermida, M., Pardo, B. G., Fernández, C., Fortes, G., Castro, J.,et al. (2007). A microsatellite genetic map of the turbot (*Scophthalmus maximus*). *Genetics* 177, 2457–2467. doi: 10.1534/genetics.107.075416

14. Bouza, C., Hermida, M., Pardo, B. G., Vera, M., Fernández, C., de la Herrán, R.,et al. (2012). An Expressed Sequence Tag (EST)-enriched genetic map of turbot (*Scophthalmus maximus*): a useful framework for comparative genomics across model and farmed teleosts. *BMC Genet.* 13:54. doi: 10.1186/1471-2156-13-54

15. Bouza, C., Sánchez, L., and Martínez, P. (1994). Karyotypic characterization of turbot (*Scophthalmus maximus*) with conventional, fluorochrome, and restriction endonuclease banding techniques. *Mar. Biol.* 120, 609–613. doi: 10.1007/BF00350082

16. Bradley, K. M., Breyer, J. P., Melville, D. B., Broman, K. W., Knapik, E. W., and Smith, J. R. (2011). An SNP-based linkage map for zebrafish reveals sex determination loci. *G3 (Bethesda)* 1, 3–9. doi: 10.1534/g3.111.000190

17. Breder, C. M., and Rosen, D. E. (1966). *Modes of Reproduction in Fishes.* New York: Natural History Press.

18. Brunelli, J. P., and Thorgaard, G. H. (2004). A new Y-chromosome-specific marker for Pacific salmon. *Trans. Am. Fish Soc.* 133, 1247–1253. doi: 10.1577/T03-049.1

19. Bruslé, J., and Roblin, C. (1984). *Sexualité du loup Dicentrarchus Labrax en Condition d'élevage Controlé.* Paris: INRA.

20. Charlesworth, D., Charlesworth, B., and Marais, G. (2005). Steps in the evolution of heteromorphic sex chromosomes. *Heredity (Edinb)* 95, 118–128. doi: 10.1038/sj.hdy.6800697

21. Chen, S., Zhang, G., Shao, C., Huang, Q., Liu, G., Zhang, P.,et al. (2014). Whole-genome sequence of a flatfish provides insights into ZW sex chromosome evolution and adaptation to a benthic lifestyle. *Nat. Genet.* 46, 253–260. doi: 10.1038/ng.2890

22. Cnaani, A. (2013). The tilapias' chromosomes influencing sex determination. *Cytogenet. Genome Res.* 141, 195–205. doi: 10.1159/000355304

23. Cnaani, A., Lee, B. Y., Zilberman, N., Ozouf-Costaz, C., Hulata, G., Ron, M.,et al. (2008). Genetics of sex determination in tilapia species. *Sex. Dev.*2, 43–54. doi: 10.1159/000117718

24. Cnaani, A., and Levavi-Sivan, B. (2009). Sexual development in fish, practical applications for aquaculture. *Sex. Dev.* 3, 164–175. doi: 10.1159/000223080

25. Cokus, S. J., Feng, S., Zhang, X., Chen, Z., Merriman, B., Haudenschild, C. D.,et al. (2008). Shotgun bisulphite sequencing of the *Arabidopsis*genome reveals DNA methylation patterning. *Nature* 452, 215–219. doi: 10.1038/nature06745

26. Conover, D. O., and Heins, W. W. (1987). Adaptive variation in environmental and genetic sex determination in a fish. *Nature* 326, 496–498. doi: 10.1038/326496a0

27. Conover, D. O., and Kynard, B. E. (1981). Environmental sex determination, interaction of temperature and genotype in a fish. *Science* 213, 577–579. doi: 10.1126/science.213.4507.577

28. Cuñado, N., Terrones, J., Sánchez, L., Martínez, P., and Santos, J. L. (2002). Sex-dependent synaptic behaviour in triploid turbot, *Scophthalmus maximus* (Pisces, *Scophthalmidae*). *Heredity* 89, 460–464. doi: 10.1038/sj.hdy.6800165

29. Dalla Valle, L., Lunardi, L., Colombo, L., and Belvedere, P. (2002). European sea bass (*Dicentrarchus labrax* L.) cytochrome P450arom: cDNA cloning, expression and genomic organization. *J. Steroid Biochem. Mol. Biol.* 80, 25–34. doi: 10.1016/S0960-0760(01)00170-4

30. Davey, J. W., Hohenlohe, P. A., Etterm, P. D., Boone, J. Q., Catchen, J. M., and Blaxter, M. L. (2011). Genome-wide genetic marker discovery and genotyping using next-generation sequencing. *Nat. Rev. Genet.* 12, 499–510. doi: 10.1038/nrg3012

31. Davidson, W. S., Kuang, T. K., Fujiki, K., von Schalburg, K. R., and Koop, B. F. (2009). The sex-determining loci and sex chromosomes in the family *salmonidae*. *Sex. Dev.* 3, 78–87. doi: 10.1159/000223073

32. Devlin, R. H., and Nagahama, Y. (2002). Sex determination and sex differentiation in fish: an overview of genetic, physiological and environmental influences. *Aquaculture* 208, 191–364. doi: 10.1016/S0044-8486(02)00057-1

33. Díaz, N., Ribas, L., and Piferrer, F. (2013). The relationship

between growth and sex differentiation in the European sea bass (*Dicentrarchus labrax*). *Aquaculture* 408–409, 191–202. doi: 10.1016/j.aquaculture.2013.05.032

34. Dou, S. Z., Masuda, R., Tanaka, M., and Tsukamoto, K. (2004). Size hierarchies affecting the social interactions and growth of juvenile japanese flounder, *Paralichthys olivaceus*. *Aquaculture* 233, 237–249. doi: 10.1016/j.aquaculture.2003.09.054
35. Eisbrenner, W. D., Botwright, N., Cook, M., Davidson, E. A., Dominik, S., Elliott, N. G.,et al. (2014). Evidence for multiple sex-determining loci in Tasmanian Atlantic salmon (*Salmo salar*). *Heredity (Edinb)* 113, 86–92. doi: 10.1038/hdy.2013.55
36. Eshel, O., Shirak, A., Weller, J. I., Hulata, G., and Ron, M. (2012). Linkage and physical mapping of sex region on LG23 of Nile tilapia (*Oreochromis niloticus*). *G3 (Bethesda)* 2, 35–42. doi: 10.1534/g3.111.001545
37. Food and Agriculture Organization [FAO]. (2014). *Cultured Aquatic Species Information Programme. FAO Fisheries and Aquaculture Department.* Available at: http://www.fao.org/fishery/culturedspecies
38. Felip, A., Piferrer, F., Carrillo, M., and Zanuy, S. (2002). Growth, gonadal development and sex ratios of meiogynogenetic diploid sea bass. *J. Fish Biol.* 61, 347–359. doi: 10.1111/j.1095-8649.2002.tb01570.x
39. Felip, A., Zanuy, S., and Carrillo, M. (2006). Comparative analysis of growth performance and sperm motility between precocious and non-precocious males in the European sea bass, *Dicentrarchus labrax*, L. *Aquaculture* 256, 570–578. doi: 10.1016/j.aquaculture.2006.02.014
40. Fernandino, J. I., Hattori, R. S., Moreno Acosta, O. D., Strüssmann, C. A., and Somoza, G. M. (2013). Environmental stress-induced testis differentiation: androgen as a by-product of cortisol inactivation. *Gen. Comp. Endocrinol.* 192, 36–44. doi: 10.1016/j.ygcen.2013.05.024
41. Fujimoto, T., Nishimura, T., Goto-Kazeto, R., Kawakami, Y., Yamaha, E., and Arai, K. (2010). Sexual dimorphism of gonadal structure and gene expression in germ cell-deficient loach, a teleost fish. *Proc. Natl. Acad. Sci. U.S.A.* 107, 17211–17216. doi: 10.1073/pnas.1007032107
42. Gardner, L. D., Jayasundara, N., Castilho, P. C., and Block, B. (2012). Microarray gene expression profiles from mature gonad tissues of Atlantic bluefin tuna, *Thunnus thynnus* in the Gulf of Mexico. *BMC Genomics* 13:530. doi: 10.1186/1471-2164-13-530
43. Godwin, J. (2009). Social determination of sex in reef fishes. *Semin. Cell Dev. Biol.* 20, 264–270. doi: 10.1016/j.semcdb.2008.12.003
44. Goto, R., Saito, T., Takeda, T., Fujimoto, T., Takagi, M., Arai, K.,et al. (2012). Germ cells are not the primary factor for sexual fate determination in goldfish. *Dev. Biol.* 370, 98–109. doi: 10.1016/j.ydbio.2012.07.010
45. Graves, J. A. M., and Peichel, C. L. (2010). Are homologies in vertebrate sex determination due to shared ancestry or to limited options? *Gen. Biol.* 11:205. doi: 10.1186/gb-2010-11-4-205
46. Haffray, P., Lebègue, E., Jeu, S., Guennoc, M., Guiguen, Y., Baroiller, J. F.,et al. (2009). Genetic determination and temperature effects on turbot*Scophthalmus maximus* sex differentiation: an investigation using steroid sex-inverted males and females. *Aquaculture* 294, 30–36. doi: 10.1016/j.aquaculture.2009.05.004
47. Hattori, R. S., Murai, Y., Oura, M., Masuda, S., Majhi, S. K., Sakamoto, T.,et al. (2012). A Y-linked anti-Müllerian hormone duplication takes over a critical role in sex determination. *Proc. Natl. Acad. Sci. U.S.A.* 109, 2955–2959. doi: 10.1073/pnas.1018392109
48. Hayashi, Y., Kobira, H., Yamaguchi, T., Shiraishi, E., Yazawa, T., Hirai, T.,et al. (2010). High temperature causes masculinization of genetically female medaka by elevation of cortisol. *Mol. Reprod. Dev.* 77, 679–686. doi: 10.1002/mrd.21203
49. Henning, F., Moysés, C. B., Calcagnotto, D., Meyer, A., and de Almeida-Toledo, L. F. (2011). Independent fusions and recent origins of sex chromosomes in the evolution and diversification of glass knife fishes (*Eigenmannia*). *Heredity (Edinb)* 106, 391–400. doi: 10.1038/hdy.2010.82
50. Hermida, M., Bouza, C., Fernández, C., Sciara, A. A., Rodríguez-Ramilo, S. T., Fernández, J.,et al. (2013). Compilation of mapping resources in turbot (*Scophthalmus maximus*): a new integrated consensus genetic map. *Aquaculture* 414–415, 19–25. doi: 10.1016/j.aquaculture.2013.07.040
51. Herpin, A., Adolfi, M. C., Nicol, B., Hinzmann, M., Schmidt, C., Klughammer, J.,et al. (2013). Divergent expression regulation of gonad development genes in medaka shows incomplete conservation of the downstream regulatory network of vertebrate sex determination. *Mol. Biol. Evol.* 30, 2328–2346. doi: 10.1093/molbev/mst130
52. Heule, C., Salzburger, W., and Böhne, A. (2014). Genetics of sexual development: an evolutionary playground for fish. *Genetics* 196, 579–591. doi: 10.1534/genetics.114.161158
53. Hohenlohe, P. A., Bassham, S., Etter, P. D., Stiffler, N., Johnson, E. A., and Cresko, W. A. (2010). Population genomics of parallel adaptation in threespine stickleback using sequenced RAD tags. *PLoS Genet.* 6:e1000862. doi: 10.1371/journal.pgen.1000862
54. Howe, K., Clark, M. D., Torroja, C. F., Torrance, J., Berthelot, C., Muffato, M.,et al. (2013). The zebrafish reference genome sequence and its relationship to the human genome. *Nature* 496, 498–503. doi: 10.1038/nature12111
55. Imsland, A., Folkvord, A., Stefansson, O., and Taranger, L. (1997). Sexual dimorphism in growth and maturation of turbot, *Scophthalmus maximus* (Rafinesque, 1810). *Aquacult. Res.* 28, 101–114. doi: 10.1111/j.1365-2109.1997.tb01022.x
56. Kai, W., Kikuchi, K., Tohari, S., Chew, A. K., Tai, A., Fujiwara, A.,et al. (2011). Integration of the genetic map and genome assembly of fugu facilitates insights into distinct features of genome evolution in Teleosts and

mammals. *Genome Biol. Evol.* 3, 424–442. doi: 10.1093/gbe/evr041

57. Kamiya, T., Kai, W., Tasumi, S., Oka, A., Matsunaga, T., Mizuno, N.,et al. (2012). A trans-species missense SNP in Amhr2 is associated with sex determination in the tiger pufferfish, *Takifugu rubripes* (fugu). *PLoS Genet.* 8:e1002798. doi: 10.1371/journal.pgen.1002798
58. Kobayashi, T., and Nagahama, Y. (2009). Molecular aspects of GD in a teleost fish, the Nile tilapia. *Sex. Dev.* 3, 108–117. doi: 10.1159/000223076
59. Kobayashi, Y., Alam, M. A., Horiguchi, R., Shimizu, A., and Nakamura, M. (2010). Sexually dimorphic expression of gonadotropin subunits in the pituitary of protogynous honeycomb grouper (*Epinephelus merra*): evidence that follicle-stimulating hormone (FSH) induces gonadal sex change. *Biol. Reprod.* 82, 1030–1036. doi: 10.1095/biolreprod.109.080986
60. Kondo, M., Nanda, I., Schmid, M., and Schartl, M. (2009). Sex determination and sex chromosome evolution: insights from medaka. *Sex. Dev.* 3, 88–98. doi: 10.1159/000223074
61. Koopman, P. (2009). Sex determination: the power of DMRT1. *Trends Genet.* 25, 479–481. doi: 10.1016/j.tig.2009.09.009
62. Koumoundouros, G., Divanach, P., Anezaki, L., and Kentouri, M. (2001). Temperature-inducedontogenetic plasticity in sea bass (*Dicentrarchus labrax*). *Mar. Biol.* 139, 817–830. doi: 10.1007/s002270100635
63. Koumoundouros, G., Pavlidis, M., Anezaki, L., Kokkari, C., Sterioti, K., Divanach, P.,et al. (2002). Temperature sex determination in the European sea bass, *Dicentrarchus labrax* (L., 1758) (*Teleostei, Perciformes, Moronidae*): critical sensitive ontogenetic phase. *J. Exp. Zool.* 292, 573–579. doi: 10.1002/jez.10095
64. Lee, B. Y., Hulata, G., and Kocher, T. D. (2004). Two unlinked loci controlling the sex of blue tilapia (*Oreochromis aureus*). *Heredity (Edinb)*92, 543–549. doi: 10.1038/sj.hdy.6800453
65. Lee, K. H., Yamaguchi, A., Rashid, H., Kadomura, K., Yasumoto, S., and Matsuyama, M. (2009). Germ cell degeneration in high-temperature treated pufferfish, *Takifugu rubripes*. *Sex. Dev.* 3, 225–232. doi: 10.1159/000228723
66. Lewis, Z. R., McClellan, M. C., Postlethwait, J. H., Cresko, W., and Kaplan, R. (2008). Female-specific increase in primordial germ cells marks sex differentiation in threespine stickleback, *Gasterosteus aculeatus*. *J. Morphol.* 269, 909–921. doi: 10.1002/jmor.10608
67. Liew, W. C., Bartfai, R., Lim, Z. J., Sreenivasan, R., Siegfried, K. R., and Orban, L. (2012). Polygenic sex determination system in zebrafish. *PLoS ONE* 7:e34397. doi: 10.1371/journal.pone.0034397
68. Liu, F., Sun, F., Li, J., Xia, J. H., Lin, G., Tu, R. J.,et al. (2013). A microsatellite-based linkage map of salt tolerant tilapia (*Oreochromis mossambicus* x *Oreochromis* spp.) and mapping of sex-determining loci. *BMC Genomics* 14:58. doi: 10.1186/1471-2164-14-58
69. Lozano, C. A., Gjerde, B., Odegard, J., and Bentsen, H. B. (2013). Heritability estimates for male proportion in the GIFT Nile tilapia (*Oreochromis niloticus* L.). *Aquaculture* 372, 137–148. doi: 10.1016/j.aquaculture.2012.10.029
70. Lühmann, L. M., Knorr, C., Hörstgen-Schwark, G., and Wessels, S. (2012). First evidence for family-specific QTL for temperature-dependent sex reversal in Nile tilapia (*Oreochromis niloticus*). *Sex. Dev.* 6, 247–256. doi: 10.1159/000339705
71. Mank, J. E., and Avise, J. C. (2009). Evolutionary diversity and turn-over of sex determination in teleost fishes. *Sex. Dev.* 3, 60–67. doi: 10.1159/000223071
72. Mankiewicz, J. L., Godwin, J., Holler, B. L., Turner, P. M., Murashige, R., Shamey, R.,et al. (2013). Masculinizing effect of background color and cortisol in a flatfish with environmental sex-determination. *Int. Comp. Biol.* 53, 755–765. doi: 10.1093/icb/ict093
73. Marín, I., and Baker, B. S. (1998). The evolutionary dynamics of sex determination. *Science* 281, 1990–1994. doi: 10.1126/science.281.5385.1990
74. Martínez, P., Bouza, C., Hermida, M., Fernández, J., Toro, M. A., Vera, M.,et al. (2009). Identification of the major sex-determining region of turbot (*Scophthalmus maximus*). *Genetics* 183, 1443–1452. doi: 10.1534/genetics.109.107979
75. Matsuda, M., Nagahama, Y., Shinomiya, A., Sato, T., Matsuda, C., Kobayashi, T.,et al. (2002). DMY is a Y-specific DM-domain gene required for male development in the medaka fish. *Nature* 30, 559–563. doi: 10.1038/nature751
76. McKay, T. F. (2001). The genetic architecture of quantitative traits. *Annu. Rev. Genet.* 35, 303–339. doi: 10.1146/annurev.genet.35.102401.090633
77. Myosho, T., Otake, H., Masuyama, H., Matsuda, M., Kuroki, Y., Fujiyama, A.,et al. (2012). Tracing the emergence of a novel sex-determining gene in medaka, *Oryzias luzonensis*. *Genetics* 191, 163–170. doi: 10.1534/genetics.111.137497
78. Nakamura, S., Kurokawa, H., Asakawa, S., Shimizu, N., and Tanaka, M. (2009). Two distinct types of theca cells in the medaka gonad: germ cell-dependent maintenance of cyp19a1-expressing theca cells. *Dev. Dyn.* 238, 2652–2657. doi: 10.1002/dvdy.22068
79. Navarro-Martín, L., Galay-Burgos, M., Sweeney, G., and Piferrer, F. (2009a). Different sox17 transcripts during sex differentiation in sea bass,*Dicentrarchus labrax*. *Mol. Cell. Endocrinol.* 299, 240–251. doi: 10.1016/j.mce.2008.11.013
80. Navarro-Martín, L., Blzquez, M., Viñas, J., and Piferrer, F. (2009b). Balancing the effects of rearing at low temperature during early development on sex ratios, growth and maturation in the European sea bass (*Dicentrarchus labrax*): limitations and opportunities for the production of highly female-biased stocks. *Aquaculture* 296, 347–358. doi: 10.1016/j.aquaculture.2009.07.022
81. Navarro-Martín, L., Viñas, J., Ribas, L., Díaz, N., Gutiérrez, A., Di Croce, L.,et al. (2011). DNA methylation of the gonadal aromatase (cyp19a) promoter

is involved in temperature-dependent sex ratio shifts in the European sea bass. *PLoS Genet.* 7:e1002447. doi: 10.1371/journal.pgen.1002447

82. Nef, S., and Vassalli, J. D. (2009). Complementary pathways in mammalian female sex determination. *J. Biol.* 8:74. doi: 10.1186/jbiol173

83. Nelson, J. S. (2006). *Fishes of the World*, 4th edn. New York: John Wiley and Sons, Inc., 601.

84. Noleto, R. B., Vicari, M. R., Cestari, M. M., and Artoni, R. F. (2012). Variable B chromosome frequencies between males and females of two species of pufferfishes (*Tetraodontiformes*). *Rev. Fish Biol. Fisher.* 22, 343–349. doi: 10.1007/s11160-011-9231-9

85. Oliveira, C., Foresti, F., and Hilsdorf, A. W. (2009). Genetics of neotropical fish: from chromosomes to populations. *Fish Physiol. Biochem.* 35, 81–100. doi: 10.1007/s10695-008-9250-1

86. Orban, L., Sreenivasan, R., and Olsson, P. E. (2009). Long and winding roads: testis differentiation in zebrafish. *Mol. Cell. Endocrinol.* 312, 35–41. doi: 10.1016/j.mce.2009.04.014

87. Ospina-Álvarez, N., and Piferrer, F. (2008). Temperature-dependent sex determination in fish revisited: prevalence, a single sex ratio response pattern, and possible effects of climate change. *PLoS ONE* 3:e2837. doi: 10.1371/journal.pone.0002837

88. Palaiokostas, C., Bekaert, M., Khan, M. G. Q., Taggart, J. B., Gharbi, K., McAndrew, B. J.,et al. (2013a). Mapping and validation of the major sex-determining region in Nile Tilapia (*Oreochromis niloticus* L.) using RAD sequencing. *PLoS ONE* 8:e68389. doi: 10.1371/journal.pone.0068389

89. Palaiokostas, C., Bekaert, M., Khan, M. G. Q., Taggart, J. B., Gharbi, K., McAndrew, B. J.,et al. (2013b). Mapping the sex determination locus in the Atlantic halibut (*Hippoglossus hippoglossus*) using RAD sequencing. *BMC Genomics* 14:566. doi: 10.1186/1471-2164-14-566

90. Parise-Maltempi, P. P., da Silva, E. L., Rens, W., Dearden, F., O'Brien, P. C., Trifonov, V.,et al. (2013). Comparative analysis of sex chromosomes in *Leporinus* species (*Teleostei, Characiformes*) using chromosome painting. *BMC Genet.* 14:60. doi: 10.1186/1471-2156-14-60

91. Parker, G. A. (1992). The evolution of sexual size dimorphism in fish. *J. Fish Biol.* 41, 1–20. doi: 10.1111/j.1095-8649.1992.tb03864.x

92. Parnell, N. F., and Streelman, J. T. (2013). Genetic interactions controlling sex and color establish the potential for sexual conflict in Lake Malawi cichlid fishes. *Heredity (Edinb)* 110, 239–246. doi: 10.1038/hdy.2012.73

93. Pavlidis, M., Koumoundouros, G., Sterioti, A., Somarakis, S., Divanach, P., and Kentouri, M. (2000). Evidence of temperature-dependent sex determination in the European sea bass (*Dicentrarchus labrax* L.). *J. Exp. Zool.* 287, 225–232. doi: 10.1002/1097-010X(20000801)287:3<225::AID-JEZ4>3.0.CO;2-D

94. Penman, D. J., and Piferrer, F. (2008). Fish gonadogenesis. Part I: genetic and environmental mechanisms of sex determination. *Rev. Fish. Sci.*16(Suppl. 1), 16–34. doi: 10.1080/10641260802324610

95. Pereiro, P., Balseiro, P., Romero, A., Dios, S., Forn-Cuni, G., Fusté, B.,et al. (2012). High-throughput sequence analysis of turbot (*Scophthalmus maximus*) transcriptome using 454-pyrosequencing for the discovery of antiviral immune genes. *PLoS ONE* 7:e35369. doi: 10.1371/journal.pone.0035369

96. Person-Le Ruyet, J., Mahé, K., Le Bayon, N., and Le Delliou, H. (2004). Effects of temperature on growth and metabolism in a Mediterranean population of European sea bass, *Dicentrarchus labrax*. *Aquaculture* 237, 269–280. doi: 10.1016/j.aquaculture.2004.04.021

97. Phillips, R. B. (2013). Evolution of the sex chromosomes in salmonid fishes. *Cytogenet. Genome Res.* 141, 177–185. doi: 10.1159/000355149

98. Phillips, R. B., Konkol, N. R., Reed, K. M., and Stein, J. D. (2001). Chromosome painting supports lack of homology among sex chromosomes in*Oncorhynchus*, *Salmo*, and *Salvelinus* (*Salmonidae*). *Genetica* 111, 119–123. doi: 10.1023/A:1013743431738

99. Piferrer, F. (2001). Endocrine sex control strategies for the feminization of teleost fish. *Aquaculture* 197, 229–281. doi: 10.1016/S0044-8486(01)00589-0

100. Piferrer, F. (2013). Epigenetics of sex determination and gonadogenesis. *Dev. Dyn.* 242, 360–370. doi: 10.1002/dvdy.23924

101. Piferrer, F., Beaumont, A., Falguière, J. C., Flajshans, M., Haffray, P., and Colombo, L. (2009). The use of induced polyploidy in the aquaculture of bivalves and fish for performance improvement and genetic containment. *Aquaculture* 293, 125–156. doi: 10.1016/j.aquaculture.2009.04.036

102. Piferrer, F., Blázquez, M., Navarro, L., and González, A. (2005). Genetic, endocrine, and environmental components of sex determination and differentiation in the European sea bass, *Dicentrarchus labrax* L. *Gen. Comp. Endocrinol.* 142, 102–110. doi: 10.1016/j.ygcen.2005.02.011

103. Piferrer, F., Ribas, L., and Díaz, N. (2012). Genomic approaches to study genetic and environmental influences on fish sex determination and differentiation. *Mar. Biotechnol.* 14, 591–604. doi: 10.1007/s10126-012-9445-4

104. Ribas, L., Pardo, B. G., Fernandez, C., Alvarez-Dios, J. A., Gómez-Tato, A., Quiroga, M. I.,et al. (2013). A combined strategy involving Sanger and 454 pyrosequencing increases genomic resources to aid in the management of reproduction, disease control and genetic selection in the turbot (*Scophthalmus maximus*). *BMC Genomics* 14:180. doi: 10.1186/1471-2164-14-180

105. Rice, W. R. (1986). On the instability of polygenic sex determination: the effect of sex-specific selection. *Evolution* 40, 633–639. doi: 10.2307/2408584

106. Roberts, R. B., Ser, J. R., and Kocher, T. D. (2009). Sexual conflict resolved by invasion of a novel sex determiner in Lake Malawi cichlid fishes.*Science* 326,

998–1001. doi: 10.1126/science.1174705

107. Ross,J.A.,andPeichel,C.L.(2008).Molecularcytogenetic evidence of rearrangements on the Y chromosome of the threespine stickleback fish.*Genetics* 179, 2173–2182. doi: 10.1534/genetics.108.088559

108. Ross, J. A., Urton, J. R., Boland, J., Shapiro, M. D., and Peichel, C. L. (2009). Turnover of sex chromosomes in the stickleback fishes (*Gasterosteidae*). *PLOS Genet.* 5:e1000391. doi: 10.1371/journal.pgen.1000391

109. Saillant, E., Fostier, A., Haffray, P., Menu, B., Thimonier, J., and Chatain, B. (2002). Temperature effects and genotype-temperature interactions on sex determination in the European sea bass (*Dicentrarchus labrax* L.). *J. Exp. Zool.* 292, 494–505. doi: 10.1002/jez.10071

110. Saito, D., and Tanaka, M. (2009). Comparative aspects of gonadal sex differentiation in medaka: a conserved role of developing oocytes in sexual canalization. *Sex. Dev.* 3, 99–107. doi: 10.1159/000223075

111. Schlueter, P. J., Sang, X., Duan, C., and Wood, A. W. (2007). Insulin-like growth factor receptor 1b is required for zebrafish primordial germ cell migration and survival. *Dev. Biol.* 305, 377–387. doi: 10.1016/j.ydbio.2007.02.015

112. Schultheis, C., Böhne, A., Schartl, M., Volff, J. N., and Galiana-Arnoux, D. (2009). Sex determination diversity and sex chromosome evolution in poeciliid fish. *Sex. Dev.* 3, 68–77. doi: 10.1159/000223072

113. Ser, J. R., Roberts, R. B., and Kocher, T. D. (2010). Multiple interacting loci control sex determination in lake Malawi cichlid fish. *Evolution* 64, 486–501. doi: 10.1111/j.1558-5646.2009.00871.x

114. Shikano, T., Natri, H. M., Shimada, Y., and Merilä, Y. (2011). High degree of sex chromosome differentiation in stickleback fishes. *BMC Genomics*12:474. doi: 10.1186/1471-2164-12-474

115. Shinomiya, A., Otake, H., Hamaguchi, S., and Sakaizumi, M. (2010). Inherited XX sex reversal originating from wild medaka populations.*Heredity (Edinb)* 105, 443–448. doi: 10.1038/hdy.2010.51

116. Siegfried, K. R. (2010). In search of determinants: gene expression during gonadal sex differentiation. *J. Fish Biol.* 76, 1879–1902. doi: 10.1111/j.1095-8649.2010.02594.x

117. Siegfried, K. R., and Nusslein-Volhard, C. (2008). Germ line control of female sex determination in zebrafish. *Dev. Biol.* 324, 277–287. doi: 10.1016/j.ydbio.2008.09.025

118. Smith, C. A., and Sinclair, A. H. (2004). Sex determination: insights from the chicken. *Bioessays* 26, 120–132. doi: 10.1002/bies.10400

119. Sreenivasan, R., Jiang, J., Wang, X., Bártfai, R., Kwan, H. Y., Christoffels, A.,et al. (2014). Gonad differentiation in zebrafish is regulated by the canonical Wnt signaling pathway. *Biol. Reprod.* 90:45. doi: 10.1095/biolreprod.113.110874

120. Sun, F., Liu, S., Gao, X., Jiang, Y., Perera, D., Wang, X.,et al. (2013). Male-biased genes in catfish as revealed by RNA-Seq analysis of the testis transcriptome. *PLoS ONE* 8:e68452. doi: 10.1371/journal.pone.0068452

121. Taboada, X., Hermida, M., Pardo, B. G., Vera, M., Piferrer, F., Viñas, A.,et al. (2014). Fine mapping and evolution of the major sex determining region in turbot (*Scophthalmus maximus*). *G3 (Bethesda)* doi: 10.1534/g3.114.012328 [Epub ahead of print].

122. Tanaka, K., Takehana, Y., Naruse, K., Hamaguchi, S., and Sakaizumi, M. (2007). Evidence for different origins of sex chromosomes in closely related *Oryzias* fishes: substitution of the master sex-determining gene. *Genetics* 177, 2075–2081. doi: 10.1534/genetics.107.075598

123. Tao, W., Yuan, J., Zhou, L. Y., Sun, L. N., Sun, Y. L., Yang, S. J.,et al. (2013). Characterization of gonadal transcriptomes from Nile Tilapia (*Oreochromis niloticus*) reveals differentially expressed genes. *PLoS ONE* 8:e63604. doi: 10.1371/journal.pone.0063604

124. Tripathi, N., Hoffmann, M., Weigel, D., and Dreyer, C. (2009). Linkage analysis reveals the independent origin of Poecilid sex chromosomes and a case of atypical sex inheritance in the guppy, *Poecilia* reticulata. *Genetics* 182, 365–374. doi: 10.1534/genetics.108.098541

125. Uller, T., and Helanterä, H. (2011). From the origin of sex-determining factors to the evolution of sex-determining systems. *Q. Rev. Biol.* 86, 163–180. doi: 10.1086/661118

126. Valenzuela, N., and Lance, V. A. (2004). *Temperature DependentSexDeterminationinVertebrates.* Washington: Smithsonian Books.

127. Vandeputte, M., Dupont-Nivet, M., Chavanne, H., and Chatain, B. (2007). A polygenic hypothesis for sex determination in the European sea bass*Dicentrarchus labrax*. *Genetics* 176, 1049–1057. doi: 10.1534/genetics.107.072140

128. Vicari, M. R., Artoni, R. F., Moreira-Filho, O., and Bertollo, L. A. C. (2008). Diversification of a ZZ/ZW sex chromosome system in *Characidium*fish (*Crenuchidae, Characiformes*). *Genetica* 134, 311–317. doi: 10.1007/s10709-007-9238-2

129. Vicente, V. E., Moreira-Filho, O., and Camacho, J. P. (1996). Sex-ratio distortion associated with the presence of a B chromosome in Astyanax scabripinnis (*Teleostei, Characidae*). *Cytogenet. Cell Genet.* 74, 70–75. doi: 10.1159/000134385

130. Wu, G. C., Tomy, S., Lee, M. F., Lee, Y. H., Yueh, W. S., Lin, C. J.,et al. (2010). Sex differentiation and sex change in the protandrous black porgy, Acanthopagrus schlegeli. *Gen. Comp. Endocrinol.* 167, 417–421. doi: 10.1016/j.ygcen.2009.11.003

131. Yano, A., Guyomard, R., Nicol, B., Jouanno, E., Quillet, E., Klopp, C.,et al. (2012). An immune-related gene evolved into the master sex-determining gene in rainbow trout, *Oncorhynchus mykiss*. *Curr. Biol.* 22, 1423–1428. doi: 10.1016/j.cub.2012.05.045

132. Yano, A., Nicol, B., Jouanno, E., Quillet, E., Fostier, A., Guyomard, R.,et al. (2013). The sexually dimorphic on the Y-chromosome gene (sdY) is a conserved male-specific Y-chromosome sequence in many salmonids. *Evol. Appl.* 6, 486–496. doi: 10.1111/eva.12032

133. Yoshimoto, S., Okada, E., Umemoto, H., Tamura, K., Uno, Y., Nishida-Umehara, C.,et al. (2008). A W-linked DM-domain gene, DM-W, participates in primary ovary development in *Xenopus laevis*. *Proc. Natl. Acad. Sci. U.S.A.* 105, 2469–2474. doi: 10.1073/pnas.0712244105

134. Zanuy, S., Carrillo, M., Felip, A., Rodríguez, L., Blázquez, M., Ramos, J.,et al. (2001). Genetic, hormonal and environmental approaches for the control of reproduction in the European sea bass (*Dicentrarchus labrax* L.). *Aquaculture* 202, 187–203. doi: 10.1016/S0044-8486(01)00771-2

135. Zhou, L., and Gui, J. F. (2010). Molecular mechanisms underlying sex change in hermaphroditic groupers. *Fish Physiol. Biochem.* 36, 181–193. doi: 10.1007/s10695-008-9219-0

SECTION IV:
FISH HEALTH

CHAPTER 9 SHORT-TERM FEED DEPRIVATION ALTERS IMMUNE STATUS OF SURFACE MUCOSA IN CHANNEL CATFISH (ICTALURUS PUNCTATUS).

Lisa Liu, Chao Li, Baofeng Su, Benjamin H. Beck, Eric Peatman

INTRODUCTION

Alterations in feeding regimen are common practice in the modern aquaculture industry, predicated in part on research indicating that fasting is well tolerated by most fish species [1],[2]. Food deprivation strategies can be employed as part of a seasonal feeding pattern [3], in response to overproduction, or as a response to disease outbreak [4], [5]. When diseases prompt changes in feeding, producers must weigh potential savings on feed and the benefits of limiting fish-fish contact during feeding against reduced growth and heightened stress stemming from nutrient restriction. Traditionally missing in the equation has been any understanding on how short-term fasting may impact the immune status of farmed fish.

Most feed deprivation investigations to-date have focused on changes in intestinal cellular morphology and enzyme activity or on transcriptional changes in critical regulators of protein synthesis or glucose metabolism in tissues such as liver, muscle, and intestine [6]–[8]. Recent studies have documented changes in gene mediators of innate immunity following 4 week starvation studies in rainbow trout intestinal epithelia [9] and Atlantic salmon liver [10]. Comparatively little is known regarding the immune consequences of more common short-term fasting events. However, a small body of previous research indicates that short-term withdrawal of feed, while not visibly impacting growth parameters, can have striking effects, particularly at mucosal surfaces. Vieira et al. [11] reported broad impacts of a one week feed restriction regimen on skin healing and scale regeneration in sea bream utilizing an oligo microarray. Krogdahl and Bakke-McKellep [7] observed rapid 20–50% decreases in intestinal tissue mass and enzyme activities after two days of fasting in Atlantic salmon. Perhaps of greatest relevance here is the recent report that a seven day feed deprivation caused significant changes in microbial density and community composition in the cutaneous mucus of Atlantic salmon [12].

The impact of feed deprivation on mortality rates of channel catfish (*Ictalurus punctatus*) exposed to the bacterial pathogens *Edwardsiella ictaluri* and *Flavobacterium columnare* has previously been established [4], [5]. *F. columnare*, a widespread opportunistic pathogen of freshwater fish, and a leading cause of mortality in the U.S. catfish industry [13], causes higher losses in fish that have been withheld from feed for as little as seven days [4]. Attachment and entry of *F. columnare* into channel catfish is via surface mucosa, skin and gill. Recently, we identified a rhamnose-binding lectin (RBL) whose expression is strongly induced in channel catfish gill by *F. columnare* infection [14]. A more detailed subsequent study of RBL activity revealed

Citation: Liu L, Li C, Su B, Beck BH, Peatman E (2013) Short-Term Feed Deprivation Alters Immune Status of Surface Mucosa in Channel Catfish (Ictalurus punctatus). doi:10.1371/journal.pone.0074581

that expression levels in gill were correlated with columnaris susceptibility, and that saturation of the receptor with its putative ligands resulted in significantly decreased columnaris mortality. Furthermore, RBL expression increased greater than 100-fold in the gill tissue of channel catfish fingerlings fasted for seven days [15]. Given these results and the impact of feeding status on columnaris susceptibility, we wished to examine further the broader molecular effects of short-term fasting of channel catfish on surface mucosal health. Towards that end, here we utilized RNA-seq-based transcriptome profiling of skin and gill homogenates from fed and 7 d fasted channel catfish fingerlings to better understand immune-nutritional regulation in teleost fish.

RESULTS

Sequencing of Short Expressed Reads from Channel Catfish Gill and Skin

A total of 209 million 100 bp high quality reads were generated for the fasted and fed samples. Greater than 26 million reads were generated for each of the six libraries. Raw read data are archived at the NCBI Sequence Read Archive (SRA) under Accession **SRP017689**.

De novo Assembly of Channel Catfish Gill and Skin Transcriptome

Given the importance of assembly of long, accurate contigs to capture channel catfish genes and to correctly identify differential expression, we compared two prominent options for *de novo* transcriptome assembly: Trans-ABySS and Trinity. We had previously developed an in-house bioinformatics pipeline around Trans-ABySS [14], [16] and demonstrated its superior performance in comparison to use of CLC Genome Workbench or Velvet assemblers. However, we sought to determine whether use of Trinity [17] would improve assemblies further.

Trans-ABySS

Use of Trans-ABySS to merge ABySS multi-k-assembled contigs, resulted in approximately 301,500 contigs with average length of 1,036.1 bp and N50 size of 1,716 bp, with 24,345 contigs longer than 1,000 bp. After removing redundancy using CD-Hit and CAP-3, about 57.53% contigs were kept, resulting in a final assembly of 173,459 unique contigs with average length 887.3 bp (Table 1).

Table 1. Summary of *de novo* **assembly results of** Illumina RNA-seq data from channel catfish gill and skin using Trans-ABySS and Trinity assemblers

	Trans-ABySS	Trinity
Contigs	301,500	281,595
Large contigs (≥1000 bp)	24,345	30,202
N50 (bp)	1,716	2,270
Average contig length	1,036.1	1046.9
Contigs (After CD-HIT + CAP3)	173,459	272,229
Percentage contigs kept after redundancy removal	57.6%	96.7%
Average length (bp) (After CD-HIT + CAP3)	887.3	996.5
Reads mapped in pairs (%)	66.3%	79.9%
Reads mapped to final reference(%)	80.8%	85.7%

doi:10.1371/journal.pone.0074581.t001

Trinity

Trinity generated approximately 281,595 contigs in its initial contig assembly with average length of 1,046.9 bp and N50 size of 2,270 bp, with 30,202 contigs longer than 1,000 bp. After removing redundancy, about 96.67% contigs were kept, resulting in a final 272,229 contigs with average length 996.5 (Table 1).

Gene Identification and Annotation

BLAST-based gene identification was performed to annotate the channel catfish gill/skin transcriptome and inform downstream differential expression analysis. After gene annotation, 60,892 contigs from the Trans-ABySS assembly had significant BLAST hits against 16,610 unigene (unique gene) matches from zebrafish, the closest available reference genome to channel catfish (Table 2). Using a more stringent criteria of a BLAST score ≥100 and E-value ≤ 1e-20 (quality matches) identified 14,679 zebrafish unigene matches. The same BLAST criteria were used to annotate the Trans-ABySS assembly based on matches against the UniProt and NCBI nr (non-redundant) databases.

Table 2. Summary of gene identification and annotation of assembled catfish contigs based on BLAST homology searches against various protein databases (Zebrafish, UniProt, nr)

	Trans-ABySS			Trinity		
	Zebrafish	UniProt	nr	Zebrafish	UniProt	nr
Contigs with putative gene matches	60,892	50,766	64,788	82,365	68,797	83,972
Annotated contigs ≥500 bp	41,313	37,468	44,019	71,782	61,658	73,849
Annotated contigs ≥1000 bp	29,023	27,479	30,379	59,986	53,210	61,185
Unigene matches	16,610	19,341	25,416	17,892	19,970	26,549
Hypothetical gene matches	986	0	3,788	1,058	0	3,693
Quality Unigene matches	14,679	15,933	15,863	15,821	15,427	20,589

Putative gene matches were at E-value ≤ 1e-5. Hypothetical gene matches denote those BLAST hits with uninformative annotation. Quality unigene hits denote more stringent parameters, including score ≥100, E-value ≤ 1e-20.
doi:10.1371/journal.pone.0074581.t002

In contrast, 82,365 Trinity contigs had a significant BLAST hit against 17,892 unique zebrafish genes (Table 2). 15,821 unigenes were identified based on hits to the zebrafish database with the more stringent criteria of a BLAST score ≥100 and E-value ≤ 1e-20. As with the Trans-ABySS assembly, the same BLAST criteria were used in comparison of the Trinity reference contigs with the UniProt and nr databases. The largest number of matches was to the nr database with 83,972 contigs with putative gene matches to nr and 20,589 quality unigene matches (Table 2).

Best Assembly Selection

In a comparison of the assemblies generated by Trans-ABySS and Trinity (Table 1), it was clear that although Trans-ABySS consistently generated a larger initial number of contigs, redundancy was much higher than that observed with Trinity. CD-HIT/CAP3 removed over 40% of Trans-ABySS contigs due to this redundancy, while almost all Trinity contigs were carried forward after this process. The final Trinity assembly contained almost 100,000 more contigs than Trans-ABySS due to Trinity's superior ability to distinguish splicing isoforms and gene paralogs. Trinity contigs also had larger N50 and average length, 2,270 bp and 996.5 bp, respectively, than Trans-ABySS. These metrics reflect Trinity's superior ability to map paired end reads into the same contig. Trinity mapped 79.9% of reads in pairs versus 66.3% in Trans-ABySS (Table 1). The superior de novo assembly produced by Trinity was also reflected in the number of quality unigene matches against zebrafish (15,821 vs. 14,679) and nr (20,589 vs. 15,863) databases. Given these results, we utilized the Trinity assembly for subsequent analysis of differential expression.

Identification and Analysis of Differentially Expressed Genes

A total of 1,545 genes (unique annotated contigs with significant BLAST identities) were differentially expressed greater than 1.5-fold, with 412 up-regulated genes and 1,133 down-regulated genes (Table 3). Short read coverage within differentially expressed contigs is critical for accurate quantification of expression. We obtained good coverage of differentially expressed contigs, with an average of 784.6 reads/contig.

Table 3. Statistics of differently expressed genes between fasted and fed channel catfish with fed catfish serving as the control

Genes	Fasted Channel Catfish
Up-regulated	412
Down-regulated	1,133
Total	1,545
Reads per contig	784.6

Values indicate contigs/genes passing cutoff values of fold change ≥1.5 ($p<0.05$). Average contig size refers to reads/contig.
doi:10.1371/journal.pone.0074581.t003

Enrichment and Pathway Analysis

A total of 3,482 GO terms including 970 (27.86%) cellular component terms, 1,012 (29.06%) molecular functions terms and 1,491 (42.82%) biological process terms were assigned to 1,545 unique gene matches. The percentages of annotated channel catfish sequences assigned to GO terms are shown in Fig. S1. The differently expressed unique genes were then used as inputs to perform enrichment analysis using Ontologizer. A total of 125 terms with p-value (FDR-corrected) <0.05 were considered significantly overrepresented. Ten higher level GO terms were retained as informative for further pathway analysis (Table S2). The GO terms included response to stress, regulation of cell death, and cell cycle regulation.

Based on enrichment analysis and manual annotation and literature searches, representative key genes were arranged into three broad categories, including immune response, energy metabolism, and cell cycling (Table 4). Imputed putative functional roles of these genes are covered in the Discussion.

Table 4. Key differentially expressed genes in the gill and skin between fasted and fed channel catfish in different functional classifications

Gene Name	Contig_ID	Fold Change
Immune (Reads Number >0)		
Autoimmune regulator-like	comp99702_c0_seq1	−2.08
Beta-2-glycoprotein 1	comp100784_c1_seq4	9.82
cAMP-responsive element modulator	comp90058_c0_seq2	−6.47
CC chemokine SCYA113	comp82613_c0_seq1	−4.30
C-C motif chemokine 19-like precursor	comp81887_c0_seq1	−2.63
CD83	comp88858_c0_seq1	1.82
Chemokine CCL-C5a precursor	comp33700_c0_seq1	−3.30
Clusterin precursor	comp33405_c0_seq1	3.57
Complement C1q tumor necrosis factor-related protein 6	comp99810_c0_seq1	−3.04
Complement C4-B	comp99246_c0_seq3	4.40
C-X-C motif chemokine 10 precursor	comp90838_c0_seq1	−3.61
C-X-C motif chemokine 11-like	comp100129_c0_seq1	−3.46
Eosinophil peroxidase precursor	comp98419_c0_seq2	−2.14
Galectin-4-like isoform 1	comp93775_c0_seq1	−1.80
IgM chain C region, secreted form - channel catfish	comp94145_c1_seq2	−1.69
IgGFc-binding protein-like	comp100810_c0_seq1	−2.62
Immunoresponsive gene 1, like	comp55693_c0_seq1	−53.66
Interferon-induced protein 44	Contig6350	3.46
Interferon-induced very large GTPase 1-like	comp102878_c0_seq1	1.73
Interleukin 17a/f1 precursor	comp102117_c4_seq2	−3.27
Interleukin-22 receptor subunit alpha-2 precursor	comp77551_c0_seq1	−2.13
Intestinal-type alkaline phosphatase precursor	comp60600_c0_seq1	90.69
Lymphokine-activated killer T-cell-originated protein kinase	comp89030_c0_seq1	−2.99
Lysozyme G	comp91870_c0_seq1	−2.78
Lysozyme G-like 1	comp91870_c0_seq2	−2.07
Lysozyme C	comp75157_c0_seq1	−6.35
Matrix metalloproteinase-19-like	comp100787_c0_seq1	−1.74
MHC class II antigen	Contig2104	2.13
MHC class II beta chain	comp101401_c0_seq1	1.91
Microfibril-associated glycoprotein 4-like	comp87348_c0_seq1	−5.74
Myxovirus (influenza virus) resistance G	Contig5990	−2.95
NACHT, LRR and PYD domains-containing protein 14-like	comp95945_c1_seq1	1.64
NADPH oxidase organizer 1a	comp100364_c0_seq1	−1.81
Nitric oxide synthase 2b, inducible (iNOS2b)	comp93125_c0_seq1	−17.22
Olfactomedin-like	comp100788_c0_seq4	−2.08
Peptidoglycan recognition protein 6 (PGRP6)	comp88793_c0_seq1	−9.13
Polymeric immunoglobulin receptor (pIgR)	comp93688_c0_seq2	−1.70
Protein canopy homolog 2 precursor	comp93948_c0_seq1	−3.35
SAM domain and HD domain-containing protein 1-like	comp98561_c0_seq8	6.64
Serum amyloid P-component precursor	comp91375_c0_seq2	1.69
Toxin-1 precursor	comp112912_c0_seq1	−6.89
Vitelline membrane outer layer protein 1 homolog	comp101161_c0_seq1	−3.55
Complement C1q 3-like	comp85071_c0_seq1	−3.41
Immune (Reads Number contain 0)		
Chitinase, acidic.1 precursor	comp84271_c0_seq1	0/84
Chitinase, acidic.3 precursor	comp72677_c0_seq1	0/51
Energy Metabolism		

Gene Name	Contig_ID	Fold Change
6-phosphofructo-2-kinase/fructose-2,6-biphosphatase 4	comp97436_c0_seq2	2.77
Adipocyte enhancer-binding protein 1	comp97921_c0_seq1	3.26
Angiopoietin-related protein 4 precursor	comp95719_c0_seq1	3.16
Apolipoprotein Bb precursor	comp99989_c0_seq2	54.62
Apolipoprotein E precursor	comp53599_c0_seq1	3.80
Apolipoprotein L, 1	comp96519_c0_seq1	2.81
Carboxypeptidase A6-like	comp510161_c0_seq1	10.95
Carnitine O-acetyltransferase-like	comp97193_c0_seq1	2.35
Carnitine O-palmitoyltransferase 1, liver isoform	comp98534_c0_seq8	6.12
Corticosteroid 11-beta-dehydrogenase isozyme 2	comp84850_c0_seq1	5.72
Cytoplasmic phosphatidylinositol transfer protein 1	comp98091_c0_seq1	4.28
Diacylglycerol O-acyltransferase 2	Contig576	−5.78
Endothelial lipase precursor	comp118134_c0_seq1	−3.40
Fat storage-inducing transmembrane protein 2	comp85823_c0_seq1	−2.31
Fatty acid binding protein 1-B.1	comp143596_c0_seq1	11.26
Fatty acid-binding protein, brain	comp72920_c0_seq1	−3.57
Fatty acid-binding protein, intestinal	comp91705_c1_seq1	29.35
F-box only protein 32	comp83869_c0_seq1	9.07
Fructose-1,6-bisphosphatase 1b	comp99650_c0_seq4	3.96
Glutamine synthetase	comp90753_c0_seq1	−1.76
Glyceraldehyde 3-phosphate dehydrogenase 2	comp86158_c0_seq4	−2.03
Glycogen synthase 1	comp91600_c0_seq1	2.51
Growth hormone receptor b precursor	comp84382_c0_seq2	2.88
Hexokinase-2	comp93533_c0_seq1	−2.98
Krueppel-like factor 15	comp98940_c1_seq5	2.82
Lipoprotein lipase precursor	comp97781_c0_seq1	−2.15
Malate dehydrogenase	comp86729_c0_seq2	2.57
Mannose-1-phosphate guanyltransferase alpha-B	comp95181_c0_seq1	−5.29
Muscle RING finger 1	comp59139_c0_seq1	2.72
Probable fructose-2,6-bisphosphatase TIGAR A	comp91599_c0_seq3	−5.35
Relaxin-3 receptor 1-like	comp82094_c0_seq1	3.70
Solute carrier family 15 member 1	comp98443_c5_seq1	5.36
Solute carrier family 2, facilitated glucose transporter	comp91908_c0_seq2	−3.54
Squalene synthase	comp97820_c0_seq5	2.69
Star-related lipid transfer protein 4	comp102843_c2_seq1	−9.75
Stearoyl-coa desaturase 5	comp96892_c0_seq3	−8.22
Pyruvate dehydrogenase kinase, isozyme 4	comp101992_c0_seq2	5.99
Cell Cycling/Proliferation		
Anterior gradient protein 2 homolog precursor	comp74849_c0_seq1	−3.33
Antigen KI-67-like	comp93863_c1_seq1	−3.10
Borealin	comp98879_c0_seq1	−3.18
Cell division control protein 2 homolog	comp111617_c0_seq1	−3.88
Cell division cycle-associated protein 2	comp83275_c0_seq1	−3.08
Cell division cycle-associated protein 3	comp93142_c0_seq1	−3.18
Cell division cycle-associated protein 7	comp100916_c4_seq2	−2.89
Cyclin-A2	comp96163_c2_seq2	−3.71
Cyclin-dependent kinase inhibitor 1C-like	comp92450_c0_seq1	3.18
Cyclin-dependent kinase inhibitor 3	comp85776_c0_seq1	−2.89
DNA replication licensing factor MCM3	comp92554_c0_seq1	−3.36

Gene Name	Contig_ID	Fold Change
DNA replication licensing factor MCM4	Contig155	−2.92
DNA replication licensing factor MCM5	comp85130_c0_seq1	−3.14
DNA replication licensing factor MCM6	comp88864_c0_seq1	−3.42
Eukaryotic translation initiation factor 2-alpha	Contig6286	−1.94
G1/S-specific cyclin-E2	comp82150_c0_seq2	−3.78
G2/mitotic-specific cyclin-B1	comp80016_c0_seq1	−3.63
G2/mitotic-specific cyclin-B3	comp102297_c0_seq2	−2.39
Mitotic checkpoint serine/threonine-protein kinase BUB1	comp94936_c0_seq1	−3.49
PCNA-associated factor-like	comp115121_c0_seq1	−2.52
Protein regulator of cytokinesis 1	comp92273_c0_seq4	−7.87
Structural maintenance of chromosomes 2	comp101484_c0_seq1	−3.01

Positive/negative values indicate upregulation and downregulation, respectively, in fasted fish relative to the fed control. When reads number equaled to 0 in fed or fasted group, the fold change is presented by average normalized read number in fed/average normalized read number in fasted. All fold changes were significant at p-value <0.05.

doi:10.1371/journal.pone.0074581.t004

Validation of RNA-Seq Profiles by QPCR

We selected 10 genes for QPCR confirmation, choosing from those with differing expression patterns and from genes of interest based on functional enrichment and pathway results. Samples from fasted and fed channel catfish (with three replicate sample pools per timepoint) were used for QPCR. Melting-curve analysis revealed a single product for all tested genes. Fold changes from QPCR were compared with the RNA-seq expression analysis results. As shown in Fig. 1, QPCR results were significantly correlated with the RNA-seq results (average correlation coefficient 0.91, p-value <0.001; Figure 1). With the exception of anterior gradient protein 2 (AGR2), all examined genes had the same direction of differential expression by both methods indicating the reliability and accuracy of the Trinity reference assembly and RNA-seq-based transcriptome expression analysis.

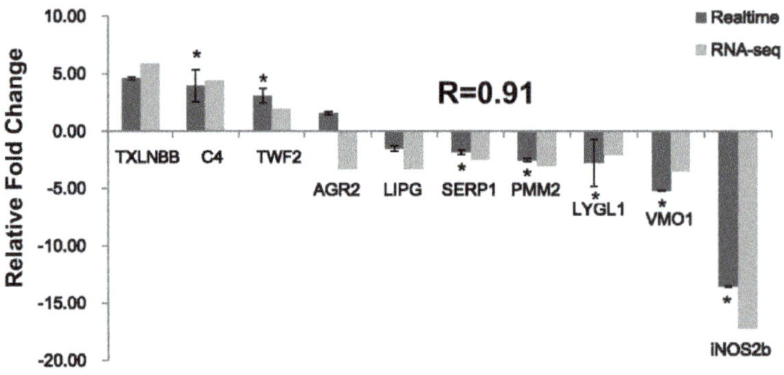

Figure 1. Comparison of relative fold changes between RNA-seq and QPCR results in channel catfish. Gene abbreviations are: Lysozyme g-like 1, LYGL1; Taxilin beta b, TXLNBB; Twinfilin-2, TWF2; Endothelial lipase precursor, LIPG; Stress-associated endoplasmic reticulum protein 2, SERP1; Anterior gradient protein 2 homolog precursor, AGR2; Phosphomannomutase 2, PMM2; Complement c4, C4; Nitric oxide synthase 2 b, inducible, iNOS2b; Vitelline membrane outer layer protein 1 homolog, VMO1.

DISCUSSION

While it is well understood that nutritional perturbations can modulate resistance to infectious diseases in agricultural animals [18], the molecular mechanisms by which dietary changes alter fish host immunity are largely unknown. The impact of these changes would be expected to differ depending on duration of feed withdrawal (fasting vs. starvation), species-specific evolutionary adaptation to natural levels of feed availability, season, and pathogen dynamics (prevalence, routes of infection, etc). Short (one week) periods of feed deprivation in channel catfish can decrease mortality rates to a Gram-negative bacterial pathogen, *E. ictaluri* [5], but, conversely, appear to lead to higher mortality to *F. columnare* [4]. Given that *F. columnare* gains entry through surface mucosa, we examined here transcriptomic changes in channel catfish gill and skin following a 7 d feed deprivation challenge. RNA-seq analysis revealed a total of 1,545 genes with expression perturbed by fasting. Fasting significantly altered expression of critical innate immune factors in a manner consistent with lower immune fitness as well as dysregulating key genes involved in energy metabolism and cell cycling/proliferation.

Studies of feed deprivation have traditionally focused on direct impacts on the digestive tract[19] or on systemic alterations in energy homeostasis [20]. Our understanding of the importance of mucosal surfaces as the front line in host immunity and the primary target for pathogen invasion continues to grow [21], [22]. However, the sensitivity of mucosal defensive barriers to nutritional changes is only now being explored in fish [12], [23]. Advances in next-generation sequencing now allow rapid, comprehensive analysis of the molecular underpinnings of these phenomena through RNA-seq approaches [24]. In order to gain a perspective on the molecular actors in the surface mucosa responding to fasting we pooled equal amounts of skin and gill from each fish contributing to the replicated tissue pools utilized for RNA-seq analysis. Clearly, by pooling heterogeneous cell populations stemming from the two tissues the potential exists for masking or confusing gene expression patterns from particular cellular subsets. However, we accepted this compromise in this initial study to more broadly capture novel genes and patterns in the surface mucosa. Future research will seek to ascertain which tissues and constituent cell types are contributing to key transcript profiles using laser capture micro-dissection and antibody-based cell sorting.

A total of 1,545 differentially expressed contigs could be annotated based on BLAST analysis. We attempted to classify key differentially expressed genes into broad functional categories based on GO annotation and manual imputation of putative function via literature search of studies in vertebrate model organisms (Table 4). Below we highlight several important pathways likely mediating the channel catfish mucosal response to short-term (7 d) fasting.

Immune Function

Recent work by our group examined how the gill transcriptome differed between *F. columnare* resistant and susceptible channel catfish both basally and at early timepoints following infection[25]. That dataset, combined with recent analysis of pathogen-mediated changes in the catfish skin and intestinal transcriptomes [16], [26], [27], provided a foundational reference of genes likely modulating immunity in mucosal tissues useful for comparison here. Initial analysis revealed enrichment of genes with immune functions among the mostly highly differentially expressed genes following fasting. Of the top 15 upregulated genes following fasting, 11 had immune functions according to a combination of GO annotation and manual literature searches. Similarly, 6 of the 15 most highly downregulated genes also had known immune functions. Among immune genes perturbed by fasting were acidic chitinase precursor genes which were not sequenced in fed fish (0 reads), but were present with between ~30–125 reads in the fasted replicates. Previously, we examined zero read data and concluded that although fold changes are difficult to predict, these data reflect true differential expression [25]. Chitinases have received little attention in fish to-date and have not been reported in the skin or gill of any species. However, in mammals, chitinases have the ability to cleave inhaled or ingested chitins from fungi, parasitic worms, crustaceans, and insects and are known to participate in the pathogenesis of allergic inflammation, particularly in the lung [28]. The acidic mammalian chitinase (AMCase) in particular has been found to be causative in lung inflammation and induced by IL-13 [29]. Also induced by fasting was an intestinal-type alkaline phosphatase (IAP; 90.69-fold). Known as a critical brush-border protein in mammals, its functions and expression patterns in fish have not been previously characterized. In mammals, IAP has the ability to detoxify lipopolysaccharide (LPS) and prevents infection across the gut mucosal barrier. Additionally, as here, IAP is reported to be sensitive to nutritional changes [30], [31]. However, IAP was silenced in the gut after 2 days of fasting in mice [30], contrasting with the induced expression seen here in skin and gill samples at 7 d. Clearly, additional study is required to delineate novel roles of IAP in modulating mucosal immune events in teleost fish.

Also induced following fasting were several genes whose expression had been perturbed in fish mucosa in previous studies in channel catfish. These included beta-2-glycoprotein 1, a novel mediator of innate immunity recently discovered to interact with LPS and activate macrophages[32]. We previously observed sharp upregulation of beta-2-glycoprotein 1 at 2 h after *Aeromonas hydrophila* infection of channel catfish skin [27]. Interestingly, beta-2-glycoprotein 1 was also reported to be upregulated following 7 d fasting in sea bream skin. Other examples such as SAM domain and HD domain-containing 1 (SAMHD1) and serum amyloid-P precursor (SAP) were induced here by fasting and upregulated in the gill following columnaris challenge[14], [25].

A suite of immune genes were also strongly downregulated following fasting (Table 4). Immunoresponsive gene 1 (IRG1), for example, was downregulated over 53-fold after the 7 d fasting period. IRG1 is a LPS-inducible gene linked to susceptibility to Marek›s Disease in chickens [33]. Additionally, we found basally lower levels of IRG1 in *F. columnare* susceptible channel catfish gill when compared with resistant channel catfish [25]. Also among downregulated immune genes following fasting was inducible nitric oxide synthase 2 b (iNOS2b), previously established as the most likely orthologue of mammalian iNOS [34]. iNOS generates nitric oxide (NO) from L-arginine. NO is a potent cytotoxic agent in immune defenses which can have beneficial antimicrobial activity at mucosal surfaces including respiratory epithelium [35]. Previously, we reported high constitutive expression of iNOS2b in healthy channel catfish gill and higher iNOS2b levels in *F. columnare* resistant catfish than susceptible channel catfish [25]. Here, following 7 d fasting, iNOS2b transcript levels plunged greater than 17-fold. Based on our previous studies, we would predict that these fasted levels of iNOS2b may open up channel catfish to heightened levels of pathogen colonization.

Another noteworthy alteration in immune status induced by food deprivation was the downregulation of peptidoglycan recognition protein 6 (PGRP6; -9.13-fold). Peptidoglycan recognition proteins are poorly-studied pathogen recognition molecules which can bind and kill commensal and pathogenic bacteria, often upstream of the better characterized Toll pathway[36]. Host species from insects to mammals rely on a diverse array of PGRP molecules, some mucus-secreted, to shape and coordinate responses to a wide range of microorganisms. They appear particularly adapted to local responses at mucosal epithelial interfaces rather than more systemic immune responses [37]. For example, in Drosophila PGRP proteins mediate tolerance of gut epithelia toward endogenous microbiota [38]. PGRP6 is one of four PGRP genes in zebrafish. Suppression of PGRP6 by RNAi in zebrafish downregulated the Toll pathway and resulted in markedly increased susceptibility to *F. columnare* [39]. Taken together, these studies point to the potential importance of PGRP6 as a nutrition-sensitive mediator of mucosal health. Its presence at the mucosal surface in fish may be important for the establishment of tolerance to commensal microbial populations through regulation of release of appropriate levels of antibacterial factors such as mucins and iNOS [37]. Future work should further characterize PGRP6 activity in catfish in the context of nutrition and disease.

In our previous study of patterns of gene expression linked to *F. columnare* resistance, levels of lysozyme C in channel catfish gill were markedly higher both pre-challenge and following infection in resistant fish [25]. Here, fasting for 7 d decreased lysozyme C levels greater than 6-fold. In mammals, lysozymes are critical mucosal enzymes secreted from the epithelium as well as a major component of granules of professional phagocytes. They help to kill bacterial pathogens through enzymatic and antimicrobial activity [40]. Previous research in zebrafish, in which a transgenic zebrafish strain expressing a chicken lysozyme gene under the control of a keratin promoter resulted in a 65% survival rate against *F. columnare* compared to 0% survival in wild-type fish, also points to the importance of lysozyme in fish mucosal immunity [41]. Additional genes exhibiting co-regulation in both the context of feed deprivation and *F. columnare* immunity included autoimmune-regulator like, CCL19, CD83, IgFc-binding protein-like, IL-17A, MFAP4, polymeric immunoglobulin receptor (pIgR), and toxin-1 precursor (Table 4; [25]). The fasting-induced modulation of genes previously indicated to be critical in resistance to *F. columnare* may explain in part the heightened sensitivity to *F. columnare* observed in fasted fish [4]. Fasting appears to decrease levels of innate immune mediators including iNOS2b, Lysozyme C, and PGRP6 at the surface barriers where *F. columnare* gains entry.

Energy Metabolism

While our primary focus here was to examine direct impacts of short-term feed deprivation on immune regulation in channel catfish surface mucosa, we also examined broader physiological impacts on cellular health and function. Changes in metabolic or stress parameters may be important indicators of reallocation of energy reserves and/or changes in barrier permeability which may, in turn, impact host pathogen susceptibility [18], [42]. Among the most highly induced genes following fasting were several apolipoproteins. Apoliproteins are proteins which classically bind and transport lipids through the blood but are also recognized for roles in immunity and acute phase responses [43]. Greatest changes were seen in the apolipoprotein ApoBb (Table 4). ApoB serum levels in humans have been reported to rise following fasting[44]. ApoE and ApoL1 levels also rose in channel catfish skin/gill following fasting, contrasting with starvation-induced decreases of apolipoproteins in the liver of rainbow trout and salmon[10], [45].

Expression of several fatty-acid binding proteins (FABP) were perturbed by fasting (Table 4), indicating changes in lipid metabolism. Largest changes were observed in the intestinal FABP, FABP2, which was induced greater than 29-fold. FABPs are well studied as intracellular fatty acid transporters in vertebrates, with limited studies to-date in fish [46]. Better studied in fish are genes linked to proteolytic gene expression during muscle atrophy [47], [48]. The ubiquitin ligases F-box only protein 32 (Fbx32) and muscle RING finger 1 (MuRF1) were upregulated following fasting, 9.07-fold and 2.72-fold respectively.

The absorption of oligopeptides following protein digestion in vertebrates is facilitated by a member of the solute carrier 15 family, SLC15A1 or PepT1. In our results, the transporter was upregulated 5.36-fold in channel catfish following fasting. Given that most luminal products of protein digestion in fish are di- and tripeptides rather than individual free amino acids, PepT1 is thought to be the major transporter of ingested protein across the gut mucosa [49]. Considerable research has been directed at understanding PepT1 function in a range of vertebrates including fish [50]. Studies in several teleost species have revealed that intestinal expression levels of PepT1 respond to fasting and refeeding, with expression levels rising in some species after short-term fasting, followed by a decline in long-term fasting/starvation. Re-feeding stimulates upregulation of PepT1 above pre-fasted levels, indicating sensitive responsiveness to food availability and a potential role for the transporter in the phenomenon known as compensatory growth [51]–[54]. Aquaculture researchers are keenly interested in use of PepT1 as an indicator of protein uptake and as a direct predictor of animal growth [51], [55]. PepT1 may also serve as another bridge between nutritional and immune regulation, given its roles in binding bacterial peptides and stimulation of gut innate immune activation and inflammation [56]. However, all the aforementioned studies focus on PepT1 function in the intestine. While Terova et al. [54] reported second highest tissue expression of PepT1 in sea bass gills (behind some but not all intestinal segments), no previous reports of a PepT1 response to fasting in teleost surface mucosa exist. Given the importance of skin and/or gill epithelia for nutrient absorption in invertebrates and at least one ancient vertebrate, the hagfish[57], [58], one is tempted to speculate that catfish PepT1 fasting-induced expression in the skin/gill may reflect conserved mechanisms of nutrient sensing and/or uptake previously unknown in fish. Further research is needed to pinpoint cellular sources of PepT1 expression in surface mucosa and examine responses to longer-term fasting and re-feeding.

Cell Cycling/Proliferation

A final notable category of genes perturbed in channel catfish skin/gill following short-term fasting were those involved in cell cycling and proliferation. Again here, while no previous study has examined how fasting may impact cell division in fish surface mucosa, our understanding of these processes in the mammalian gastrointestinal mucosa is fairly well refined [59]. There, fasting/starvation has been found to be one of the best models of a hypoproliferative response[60]. Subsequent studies have established a consistent pattern of feed deprivation reducing proliferation rates across multiple cell types, such that fasting and calorie restriction are ,now recommended as approaches to impede tumor growth [61]. We observed consistent downregulation of a large number of genes involved in cell cycling, proliferation, and differentiation (Table 4 and Table S2). These included Ki67 and PCNA-associated factors from channel catfish, cyclins, DNA replication licensing factors and checkpoint proteins (Table 4). The alteration of cell cycling indicated by these transcript changes may disrupt the integrity of catfish mucosal barriers with consequences for immune defenses at these surfaces [62].

CONCLUSION

RNA-seq-based transcriptome profiling in channel catfish revealed that short-term feed deprivation altered immune status in the surface mucosa. Changes in innate immune actors such as iNOS2b, Lysozyme C, and PGRP6 may impact the delicate recognition/tolerance balance for commensal and pathogenic bacteria on the skin and gill. Furthermore, our analysis identified critical regulators of metabolism, cell cycling, and transport previously unstudied and/or unreported in these tissues whose expression was perturbed by fasting. The highlighted expression profiles reveal potential mechanistic similarities between gut and surface mucosa and underscore the complex interrelationships between nutrition, mucosal integrity, and immunity in teleost fish.

MATERIALS AND METHODS

Experimental Animals and Tissue Collection

Animal care and experimental protocols were approved by the Stuttgart National Aquaculture Research Center Institutional Animal Care and Use Committee and conformed to Agricultural Research Service Policies and Procedures 130.4 and 635.1. Juvenile channel catfish (42.2±5.6 g) were stocked into four 600 L tanks with 30 fish per tank with forced air aeration and flow-through well water at 24.8±0.02°C, pH 7.7, and dissolved oxygen of 7.4±0.3 mg/L.

Channel catfish were subjected to two treatments with two replicate tanks per treatment. Fish in treatment group 1 were fed to satiation three times daily with a standard catfish ration (35% protein, 2.5% fat). In treatment 2, fish were withheld feed for 7 d. All fish were sacrificed on day 7, euthanized with tricaine methanesulfonate (MS 222) at 300 mg/L before tissues were collected.

Equivalent portions of the gill and skin were isolated from matching locations on each fish and stored in RNALater (Ambion, Life Technologies, Grand Island, New York) at −80°C until RNA extraction. Three pools (five fish each) of tissue were generated from each condition. Equal amounts of tissue were collected from each fish within a pool. Samples were immediately placed in RNAlater and stored at −80°C until extraction. Samples were homogenized with a mortar and pestle.

RNA Extraction, Library Construction and Sequencing

Extractions were performed according to the manufacturer's directions using an RNeasy Kit (Qiagen, Valencia, California). RNA concentration and integrity of each sample was measured on an Agilent 2100 Bioanalyzer. For each condition, equal amounts of RNA from skin and gill were pooled to generate three pooled replicates for RNA-seq library construction.

RNA-seq library preparation and sequencing was carried out by HudsonAlpha Genomic Services Lab (Huntsville, AL, USA) following the standard TruSeq protocols with 100 bp PE read chemistry on an Illumina HiSeq 2500 instrument [14].

De novo Assembly of Sequencing Read

The *de novo* assembly of short reads was performed on channel catfish using both ABySS and Trinity [17], [63], versions 1.3.2 and the 2012-10-05 editions, respectively. Before assembly, raw reads were trimmed by removing adaptor sequences and ambiguous nucleotides. Reads with quality scores less than 20 and length below 30 bp were all trimmed. The resulting high-quality sequences were used in the subsequent assembly.

In ABySS, briefly, the clean reads were first hashed according to a predefined k-mer length, the 'k-mers'. After capturing overlaps of length k-1 between these k-mers, the short reads were assembled into contigs. The k-mer size was set from 50 to 96, assemblies from all k-mers were merged into one assembly by Trans-ABySS (version 1.4.4).

In Trinity, briefly, the raw reads were assembled into the unique sequences of transcripts in Inchworm via greedy k-mer extension (k-mer 25). After mapping of reads to Inchworm contigs, Chrysalis incorporated reads into deBruijn graphs and the Butterfly module processed the individual graphs to generate full-length transcripts.

In order to reduce redundancy, the assembly results from different assemblers were passed to CD-Hit version 4.5.4 [64] and CAP3 [65] for multiple alignments and consensus building. The threshold was set as identity equal to 1 in CD-Hit, the minimal overlap length and identity equal to 100 bp and 99% in CAP3.

Gene Annotation and Ontology

The selected assembly contigs were used as queries against the NCBI zebrafish protein database, the UniProtKB/SwissProt database and the non-redundant (nr) protein database using the BLASTX program. The cutoff E-value was set at 1e-5 and only the top gene id and name were initially assigned to each contig. Gene ontology (GO) annotation analysis was performed using the zebrafish BLAST results in Blast2GO version 2.6.3 [66]. The zebrafish BLAST result or the nr BLAST result was imported to BLAST2GO. The final annotation file was categorized with respect to Biological Process, Molecular Function, and Cellular Component at level 2.

Identification of Differentially Expressed Contigs

The high quality reads from each sample were mapped onto the Trinity reference assembly using CLC Genomics Workbench software. During mapping, at least 95% of the bases were required to align to the reference and a maximum of two mismatches were allowed. The total mapped reads number for each transcript was determined and then normalized to detect RPKM (Reads Per Kilobase of exon model per Million mapped reads). The proportions-based test was used to identify the differently expressed genes between fed and fasted group with three replicates in each group with corrected p-value <0.05 [67]. After scaling normalization of the RPKM values, fold changes were calculated. Analysis was performed using the RNA-seq module and the expression analysis module in CLC Genomics Workbench [68]. Transcripts with absolute fold change values of larger than 1.5 were included in analysis as differently expressed genes.

Contigs with previously identified gene matches were carried forward for further analysis. Functional groups and pathways encompassing the differently expressed genes were identified based on GO analysis, pathway analysis based on the Kyoto Encyclopedia of Genes and Genomes (KEGG) database, and manual literature review.

Enrichment Analysis

Enrichment analysis of significantly expressed GO terms was performed using Ontologizer 2.0 using the Parent-Child-Intersection method with a Benjamini-Hochberg multiple testing correction [69], [70]. GO terms for each gene were obtained by utilizing zebrafish annotations for the unigene set. The threshold was set as FDR value <0.1.

Experimental Validation—QPCR

Ten significantly expressed genes with different expression patterns were selected for validation using real time QPCR with gene specific primers designed using Primer3 software. Primers were designed based on contig sequences (Table S1). Total RNA was extracted using the RNeasy Plus kit (Qiagen) following manufacturer's instructions. All the cDNA products were diluted to 250 ng/μl and utilized for the quantitative real-time PCR reaction using the SsoFast™ EvaGreen® Supermix on a CFX96 real-time PCR Detection System (Bio-Rad Laboratories, Hercules, CA). The thermal cycling profile consisted of an initial denaturation at 95°C (for 30 s), followed by 40 cycles of denaturation at 94°C (5 s), an appropriate annealing/extension temperature (58°C, 5 s). Results were expressed relative to the expression levels of 18S rRNA in each sample using the Relative Expression Software Tool (REST) version 2009 [71] as described in (14).

SUPPORTING INFORMATION

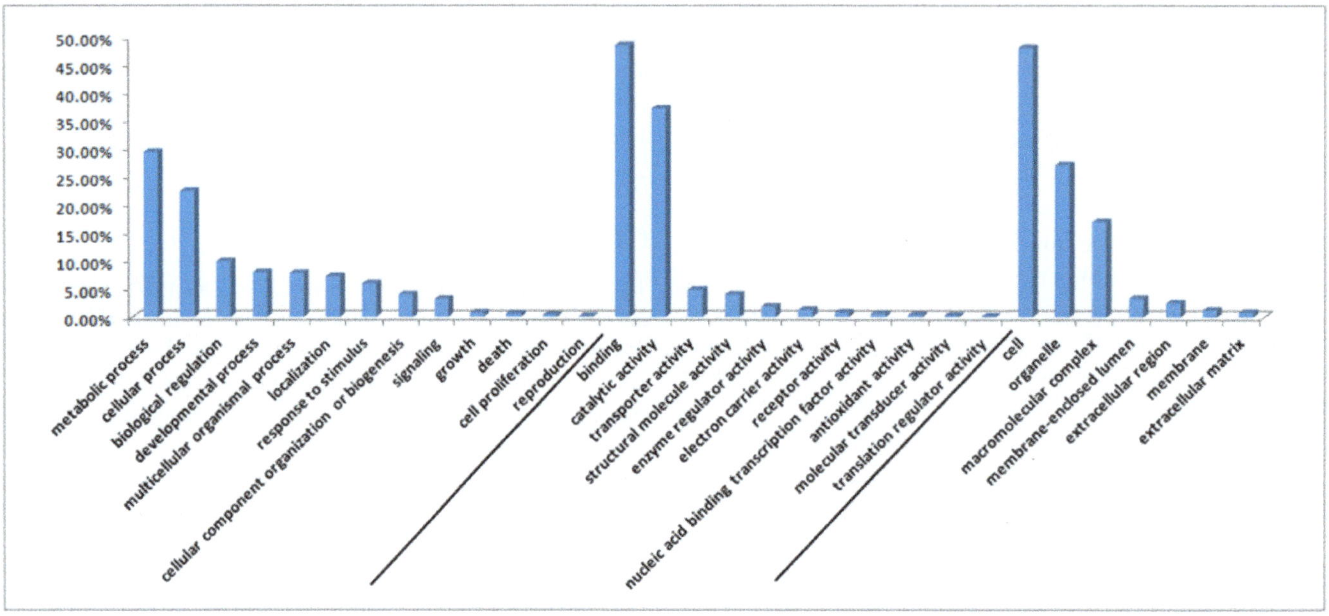

Figure S1. Gene ontology (GO) term categorization and distribution of differently expressed genes in channel catfish. GO-terms were processed by Blast2GO and categorized at level 2 under three main categories (cellular component, molecular function and biological process).

Table S1. Primers used for QPCR validation (5′ to 3′)

Gene	Forward	Reverse
Anterior Gradient Protein 2	TCCTTCTTGACCGCAGTCTTGT	AACCTGAGAGCAGCCTGTGAA
Complement C4-B	ACTCTGCTGCTTGTAGTTGAAC	GTCGCTCCTGACTCTATAACCA
Endothelial Lipase Precursor	CTCACAACGACACCAACGACAT	CCAGCTCACACGGATCTTCAAC
Lysozyme G-Like 1	GCTTCGGACTCATGCAGGTT	TTCTTACTCCAGCGTTGTAGGC
Nitric Oxide Synthase 2b,	TCAGCAGATGTCCGATGTCA	AGGAGTTCATTGGTGGAAGGT
Phosphomannomutase 2	GAGATACGCCTGGATGCTCTGT	CGTCGGAGGTTCTGATCTGGAT
Stress-Associated Endoplasmic Reticulum Protein 2	TGATGATGATGGTCCGGTTGTC	ACGAGAAGGCAGCAGTAGGA
Taxilin Beta B	ACGGCATCGCTGAAGTGTCT	ACCGCTCAGATCAGAAGCAGTT
Twinfilin-2	ACAGACACTCCACGCTATCACT	GCATCCTCTCCTTAACGCTACA
Vitelline Membrane Outer Layer Protein 1 Homolog	AGCCCACAAATCCCAGTTCCA	GATGACACCGCTGCTAACAACA
18S rRNA	GAGAAACGGCTACCACATCC	GATACGCTCATTCCGATTACAG

Table S2. Summary of GO term enrichment result of GO-based functional categories of significantly expressed genes in channel catfish after fasting. The 1,545 differentially expressed genes were analyzed as the study set in comparison to all the catfish unigenes. P-value ≤0.1 was considered significant. Population count is the number of genes associated with the term in the population set. Study count is the number of genes associated with the term in the study set. GO names were retained only from GO terms of levels >2

GO ID	GO Name	p-Value(FDR)	Population count	Study count
GO:0006950	Response To Stress	6.46E-11	557	61
GO:0006520	Cellular Amino Acid Metabolic Process	4.53E-04	189	33
GO:0005198	Structural Molecule Activity	2.78E-06	347	50
GO:0010941	Regulation Of Cell Death	0.002554	172	18
GO:0035556	Intracellular Signal Transduction	0.010998	749	29
GO:0051726	Regulation Of Cell Cycle	0.018269	149	15
GO:0016265	Death	0.016633	247	26
GO:0008219	Cell Death	0.022513	247	26
GO:0006259	DNA Metabolic Process	0.038383	340	39
GO:0022402	Cell Cycle Process	0.047922	196	21

ACKNOWLEDGMENTS

All authors read and approved the final manuscript. The authors wish to thank Matt Barnett for technical assistance. The USDA is an equal opportunity provider and employer. Mention of trade names or commercial products in this publication is solely for the purpose of providing specific information and does not imply recommendation or endorsement by the US Department of Agriculture.

AUTHOR CONTRIBUTIONS

Conceived and designed the experiments: EP BHB. Performed the experiments: LL CL BFS BHB. Analyzed the data: CL LL EP. Wrote the paper: CL EP.

REFERENCES

1. Larsson Å, Lewander K (1973) Metabolic effects of starvation in the eel, Anguilla anguilla L. Comp Biochem Physiol A Mol Integr Physiol 44: 367–374. doi: 10.1016/0300-9629(73)90489-1
2. Navarro I, Gutiérrez J (1995) Fasting and starvation. Biochemistry and molecular biology of fishes 4: 393–434. doi: 10.1016/s1873-0140(06)80020-2
3. Pottinger T, Rand-Weaver M, Sumpter J (2003) Overwinter fasting and re-feeding in rainbow trout: plasma growth hormone and cortisol levels in relation to energy mobilisation. Comp Biochem Physiol B Biochem Mol Biol 136: 403–417. doi: 10.1016/s1096-4959(03)00212-4
4. Shoemaker C, Klesius P, Lim C, Yildirim M (2003) Feed deprivation of channel catfish, Ictalurus punctatus (Rafinesque), influences organosomatic indices, chemical composition and susceptibility to Flavobacterium columnare. J Fish Dis 26: 553–561. doi: 10.1046/j.1365-2761.2003.00489.x
5. Wise DJ, Greenway T, Li MH, Camus AC, Robinson EH (2008) Effects of variable periods of food deprivation on the development of enteric septicemia in channel catfish. J Aquat Anim Health 20: 39–44. doi: 10.1577/h07-008.1
6. Krogdahl Å, Nordrum S, Sorensen M, Brudeseth L, Rosjo C (1999) Effects of diet composition on apparent nutrient absorption along the intestinal tract and of subsequent fasting on mucosal disaccharidase activities and plasma nutrient concentration in Atlantic salmon Salmo salar L. Aquacult Nutr 5: 121–134. doi: 10.1046/j.1365-2095.1999.00095.x
7. Krogdahl A, Bakke-McKellep AM (2005) Fasting and refeeding cause rapid changes in intestinal tissue mass and digestive enzyme capacities of Atlantic salmon (Salmo salar L.). Comp Biochem Physiol A Mol Integr Physiol 141: 450–460. doi: 10.1016/j.cbpb.2005.06.002
8. Cleveland BM, Weber GM, Blemings KP, Silverstein JT (2009) Insulin-like growth factor-I and genetic effects on indexes of protein degradation in response to feed deprivation in rainbow trout (Oncorhynchus mykiss). Am J Physiol Regul Integr Comp Physiol 297: R1332–1342. doi: 10.1152/ajpregu.00272.2009
9. Baumgarner BL, Bharadwaj AS, Inerowicz D, Goodman AS, Brown PB (2013) Proteomic analysis of rainbow trout (Oncorhynchus mykiss) intestinal epithelia: physiological acclimation to short-term starvation. Comp Biochem Physiol Part D Genomics Proteomics 8: 58–64. doi: 10.1016/j.cbd.2012.11.001
10. Martin SA, Douglas A, Houlihan DF, Secombes CJ (2010) Starvation alters the liver transcriptome of the innate immune response in Atlantic salmon (Salmo salar). BMC genomics 11: 418. doi: 10.1186/1471-2164-11-418
11. Vieira FA, Gregorio SF, Ferraresso S, Thorne MA, Costa R, et al. (2011) Skin healing and scale regeneration in fed and unfed sea bream, Sparus auratus. BMC genomics 12: 490. doi: 10.1186/1471-2164-12-490
12. Landeira-Dabarca A, Sieiro C, Alvarez M (2013) Change in food ingestion induces rapid shifts in the diversity of microbiota associated with cutaneous mucus of Atlantic salmon Salmo salar. J Fish Biol 82: 893–906. doi: 10.1111/jfb.12025
13. Arias CR, Cai W, Peatman E, Bullard SA (2012) Catfish hybrid Ictalurus punctatus x I. furcatus exhibits higher

resistance to columnaris disease than the parental species. Dis Aquat Organ 100: 77–81. doi: 10.3354/dao02492

14. Sun F, Peatman E, Li C, Liu S, Jiang Y, et al. (2012) Transcriptomic signatures of attachment, NF-kappaB suppression and IFN stimulation in the catfish gill following columnaris bacterial infection. Dev Comp Immunol 38: 169–180. doi: 10.1016/j.dci.2012.05.006
15. Beck BH, Farmer BD, Straus DL, Li C, Peatman E (2012) Putative roles for a rhamnose binding lectin in Flavobacterium columnare pathogenesis in channel catfish Ictalurus punctatus. Fish Shellfish Immunol 33: 1008–1015. doi: 10.1016/j.fsi.2012.08.018
16. Li C, Zhang Y, Wang R, Lu J, Nandi S, et al. (2012) RNA-seq analysis of mucosal immune responses reveals signatures of intestinal barrier disruption and pathogen entry following Edwardsiella ictaluri infection in channel catfish, Ictalurus punctatus. Fish Shellfish Immunol 32: 816–827. doi: 10.1016/j.fsi.2012.02.004
17. Grabherr MG, Haas BJ, Yassour M, Levin JZ, Thompson DA, et al. (2011) Full-length transcriptome assembly from RNA-Seq data without a reference genome. Nat Biotechnol 29: 644–652. doi: 10.1038/nbt.1883
18. Klasing K (1998) Nutritional modulation of resistance to infectious diseases. Poultry Science 77: 1119–1125. doi: 10.1093/ps/77.8.1119
19. Gisbert E, Fernández I, Alvarez-González C (2011) Prolonged feed deprivation does not permanently compromise digestive function in migrating European glass eels Anguilla anguilla. J Fish Biol 78: 580–592. doi: 10.1111/j.1095-8649.2010.02879.x
20. Costas B, Aragão C, Ruiz-Jarabo I, Vargas-Chacoff L, Arjona FJ, et al. (2011) Feed deprivation in Senegalese sole (Solea senegalensis Kaup, 1858) juveniles: effects on blood plasma metabolites and free amino acid levels. Fish Physiol Biochem 37: 495–504. doi: 10.1007/s10695-010-9451-2
21. Ángeles Esteban M (2012) An Overview of the Immunological Defenses in Fish Skin. ISRN Immunology 2012.
22. Rombout JH, Abelli L, Picchietti S, Scapigliati G, Kiron V (2011) Teleost intestinal immunology. Fish Shellfish Immunol 31: 616–626. doi: 10.1016/j.fsi.2010.09.001
23. Florbela V, Silvia G, Serena F, Michael T, Rita C, et al. (2011) Skin healing and scale regeneration in fed and unfed sea bream, Sparus auratus.
24. Liu G (2011) Recent applications of DNA sequencing technologies in food, nutrition and agriculture. Recent patents on food, nutrition & agriculture 3: 187. doi: 10.2174/2212798411103030187
25. Peatman E, Li C, Peterson B, Straus DL, Bradley Farmer D, et al. (2013) Basal polarization of the mucosal compartment in Flavobacterium columnare susceptible and resistant channel catfish (Ictalurus punctatus). Mol Immunol http://dx.doi.org/10.1016/j.molimm.2013.04.014.
26. Li C, Beck B, Su B, Terhune J, Peatman E (2013) Early mucosal responses in blue catfish (Ictalurus furcatus) skin to Aeromonas hydrophila infection. Fish Shellfish Immunol 34: 920–928. doi: 10.1016/j.fsi.2013.01.002
27. Li C, Wang R, Su B, Luo Y, Terhune J, et al. (2013) Evasion of mucosal defenses during Aeromonas hydrophila infection of channel catfish (Ictalurus punctatus) skin. Dev Comp Immunol 39: 447–455. doi: 10.1016/j.dci.2012.11.009
28. Goldman DL, Vicencio AG (2012) The chitin connection. MBio 3
29. Zhu Z, Zheng T, Homer RJ, Kim YK, Chen NY, et al. (2004) Acidic mammalian chitinase in asthmatic Th2 inflammation and IL-13 pathway activation. Science 304: 1678–1682. doi: 10.1126/science.1095336
30. Goldberg RF, Austen WG, Zhang X, Munene G, Mostafa G, et al. (2008) Intestinal alkaline phosphatase is a gut mucosal defense factor maintained by enteral nutrition. Proc Natl Acad Sci U S A 105: 3551–3556. doi: 10.1073/pnas.0712140105
31. Lallès JP (2010) Intestinal alkaline phosphatase: multiple biological roles in maintenance of intestinal homeostasis and modulation by diet. Nutr Rev 68: 323–332. doi: 10.1111/j.1753-4887.2010.00292.x
32. Ağar Ç, de Groot PG, Mörgelin M, Monk SD, van Os G, et al. (2011) β2-Glycoprotein I: a novel component of innate immunity. Blood 117: 6939–6947. doi: 10.1182/blood-2010-12-325951
33. Smith J, Sadeyen JR, Paton IR, Hocking PM, Salmon N, et al. (2011) Systems analysis of immune responses in Marek›s disease virus-infected chickens identifies a gene involved in susceptibility and highlights a possible novel pathogenicity mechanism. J Virol 85: 11146–11158. doi: 10.1128/jvi.05499-11
34. Lepiller S, Franche N, Solary E, Chluba J, Laurens V (2009) Comparative analysis of zebrafish nos2a and nos2b genes. Gene 445: 58–65. doi: 10.1016/j.gene.2009.05.016
35. Bogdan C (2001) Nitric oxide and the immune response. Nat Immunol 2: 907–916. doi: 10.1038/ni1001-907
36. Kashyap DR, Wang M, Liu LH, Boons GJ, Gupta D, et al. (2011) Peptidoglycan recognition proteins kill bacteria by activating protein-sensing two-component systems. Nat Med 17: 676–683. doi: 10.1038/nm.2357
37. Royet J, Gupta D, Dziarski R (2011) Peptidoglycan recognition proteins: modulators of the microbiome and inflammation. Nat Rev Immunol 11: 837–851. doi: 10.1038/nri3089
38. Gendrin M, Welchman DP, Poidevin M, Hervé M, Lemaitre B (2009) Long-range activation of systemic immunity through peptidoglycan diffusion in Drosophila. PLoS pathogens 5: e1000694. doi: 10.1371/journal.ppat.1000694
39. Chang M, Nie P (2008) RNAi suppression of zebrafish peptidoglycan recognition protein 6 (zfPGRP6) mediated differentially expressed genes involved in Toll-like receptor signaling pathway and caused increased susceptibility to Flavobacterium columnare. Vet Immunol Immunopathol 124: 295–301. doi: 10.1016/j.vetimm.2008.04.003

40. Davis KM, Weiser JN (2011) Modifications to the peptidoglycan backbone help bacteria to establish infection. Infect Immun 79: 562–570. doi: 10.1128/iai.00651-10

41. Yazawa R, Hirono I, Aoki T (2006) Transgenic zebrafish expressing chicken lysozyme show resistance against bacterial diseases. Transgenic Res 15: 385–391. doi: 10.1007/s11248-006-0009-0

42. Polakof S, Panserat S, Soengas JL, Moon TW (2012) Glucose metabolism in fish: a review. J Comp Physiol B 182: 1015–1045. doi: 10.1007/s00360-012-0658-7

43. Bolanos-Garcia VM, Miguel RN (2003) On the structure and function of apolipoproteins: more than a family of lipid-binding proteins. Prog Biophys Mol Biol 83: 47–68. doi: 10.1016/s0079-6107(03)00028-2

44. Savendahl L, Underwood LE (1999) Fasting increases serum total cholesterol, LDL cholesterol and apolipoprotein B in healthy, nonobese humans. J Nutr 129: 2005–2008.

45. Salem M, Silverstein J, Rexroad CE (2007) Effect of starvation on global gene expression and proteolysis in rainbow trout (Oncorhynchus mykiss). BMC genomics 8: 328. doi: 10.1186/1471-2164-8-328

46. Lai YY, Lubieniecki KP, Phillips RB, Chow W, Koop BF, et al. (2009) Genomic organization of Atlantic salmon (Salmo salar) fatty acid binding protein (fabp2) genes reveals independent loss of duplicate loci in teleosts. Mar Genomics 2: 193–200. doi: 10.1016/j.margen.2009.10.003

47. Wang J, Salem M, Qi N, Kenney PB, Rexroad CE, et al. (2011) Molecular characterization of the MuRF genes in rainbow trout: Potential role in muscle degradation. Comp Biochem Physiol B Biochem Mol Biol 158: 208–215. doi: 10.1016/j.cbpb.2010.11.010

48. Tacchi L, Bickerdike R, Secombes CJ, Martin SA (2012) Muscle-specific RING finger (MuRF) cDNAs in Atlantic salmon (Salmo salar) and their role as regulators of muscle protein degradation. Mar Biotechnol (NY) 14: 35–45. doi: 10.1007/s10126-011-9385-4

49. Terova G, Cora S, Verri T, Rimoldi S, Bernardini G, et al. (2009) Impact of feed availability on PepT1 mRNA expression levels in sea bass (Dicentrarchus labrax). Aquaculture 294: 288–299. doi: 10.1016/j.aquaculture.2009.06.014

50. Verri T, Kottra G, Romano A, Tiso N, Peric M, et al. (2003) Molecular and functional characterisation of the zebrafish (Danio rerio) PEPT1-type peptide transporter. FEBS Lett 549: 115–122. doi: 10.1016/s0014-5793(03)00759-2

51. Verri T, Terova G, Dabrowski K, Saroglia M (2011) Peptide transport and animal growth: the fish paradigm. Biol Lett 7: 597–600. doi: 10.1098/rsbl.2010.1164

52. Bucking C, Schulte PM (2012) Environmental and nutritional regulation of expression and function of two peptide transporter (PepT1) isoforms in a euryhaline teleost. Comp Biochem Physiol A Mol Integr Physiol 161: 379–387. doi: 10.1016/j.cbpa.2011.12.008

53. Hakim Y, Harpaz S, Uni Z (2009) Expression of brush border enzymes and transporters in the intestine of European sea bass (Dicentrarchus labrax) following food deprivation. Aquaculture 290: 110–115. doi: 10.1016/j.aquaculture.2009.02.008

54. Terova G, Forchino A, Rimoldi S, Brambilla F, Antonini M, et al. (2009) Bio-Mos: an effective inducer of dicentracin gene expression in European sea bass (Dicentrarchus labrax). Comp Biochem Physiol B Biochem Mol Biol 153: 372–377. doi: 10.1016/j.cbpb.2009.04.008

55. Terova G, Robaina L, Izquierdo M, Cattaneo A, Molinari S, et al. (2013) PepT1 mRNA expression levels in sea bream (Sparus aurata) fed different plant protein sources. SpringerPlus 2: 1–14. doi: 10.1186/2193-1801-2-17

56. Ingersoll SA, Ayyadurai S, Charania MA, Laroui H, Yan Y, et al. (2012) The role and pathophysiological relevance of membrane transporter PepT1 in intestinal inflammation and inflammatory bowel disease. A Am J Physiol Gastrointest Liver Physiol 302: G484–492. doi: 10.1152/ajpgi.00477.2011

57. Meissner B, Boll M, Daniel H, Baumeister R (2004) Deletion of the intestinal peptide transporter affects insulin and TOR signaling in Caenorhabditis elegans. J Biol Chem 279: 36739–36745. doi: 10.1074/jbc.m403415200

58. Glover CN, Bucking C, Wood CM (2011) Adaptations to in situ feeding: novel nutrient acquisition pathways in an ancient vertebrate. Proc Biol Sci 278: 3096–3101. doi: 10.1098/rspb.2010.2784

59. Wong WM, Wright NA (1999) Cell proliferation in gastrointestinal mucosa. J Clin Pathol 52: 321–333. doi: 10.1136/jcp.52.5.321

60. Brown HO, Levine ML, Lipkin M (1963) Inhibition of intestinal epithelial cell renewal and migration induced by starvation. Am J Physiol 205: 868–872.

61. Lee C, Raffaghello L, Brandhorst S, Safdie FM, Bianchi G, et al. (2012) Fasting cycles retard growth of tumors and sensitize a range of cancer cell types to chemotherapy. Sci Transl Med 4: 124ra127. doi: 10.1126/scitranslmed.3003293

62. Knight DA, Rossi FM, Hackett TL (2010) Mesenchymal stem cells for repair of the airway epithelium in asthma. Expert Rev Respir Med 4: 747–758. doi: 10.1586/ers.10.72

63. Simpson JT, Wong K, Jackman SD, Schein JE, Jones SJ, et al. (2009) ABySS: a parallel assembler for short read sequence data. Genome Res 19: 1117–1123. doi: 10.1101/gr.089532.108

64. Li W, Godzik A (2006) Cd-hit: a fast program for clustering and comparing large sets of protein or nucleotide sequences. Bioinformatics 22: 1658–1659. doi: 10.1093/bioinformatics/btl158

65. Huang X, Madan A (1999) CAP3: A DNA sequence assembly program. Genome Res 9: 868–877. doi: 10.1101/gr.9.9.868

66. Gotz S, Garcia-Gomez JM, Terol J, Williams TD, Nagaraj SH, et al. (2008) High-throughput functional annotation

and data mining with the Blast2GO suite. Nucleic Acids Res 36: 3420–3435. doi: 10.1093/nar/gkn176

67. Baggerly KA, Deng L, Morris JS, Aldaz CM (2003) Differential expression in SAGE: accounting for normal between-library variation. Bioinformatics 19: 1477–1483. doi: 10.1093/bioinformatics/btg173

68. Robinson MD, Oshlack A (2010) A scaling normalization method for differential expression analysis of RNA-seq data. Genome Biol 11: R25. doi: 10.1186/gb-2010-11-3-r25

69. Bauer S, Grossmann S, Vingron M, Robinson PN (2008) Ontologizer 2.0–a multifunctional tool for GO term enrichment analysis and data exploration. Bioinformatics 24: 1650–1651. doi: 10.1093/bioinformatics/btn250

70. Grossmann S, Bauer S, Robinson PN, Vingron M (2007) Improved detection of overrepresentation of Gene-Ontology annotations with parent child analysis. Bioinformatics 23: 3024–3031. doi: 10.1093/bioinformatics/btm440

71. Pfaffl MW, Horgan GW, Dempfle L (2002) Relative expression software tool (REST) for group-wise comparison and statistical analysis of relative expression results in real-time PCR. Nucleic Acids Res 30: e36. doi: 10.1093/nar/30.9.e36

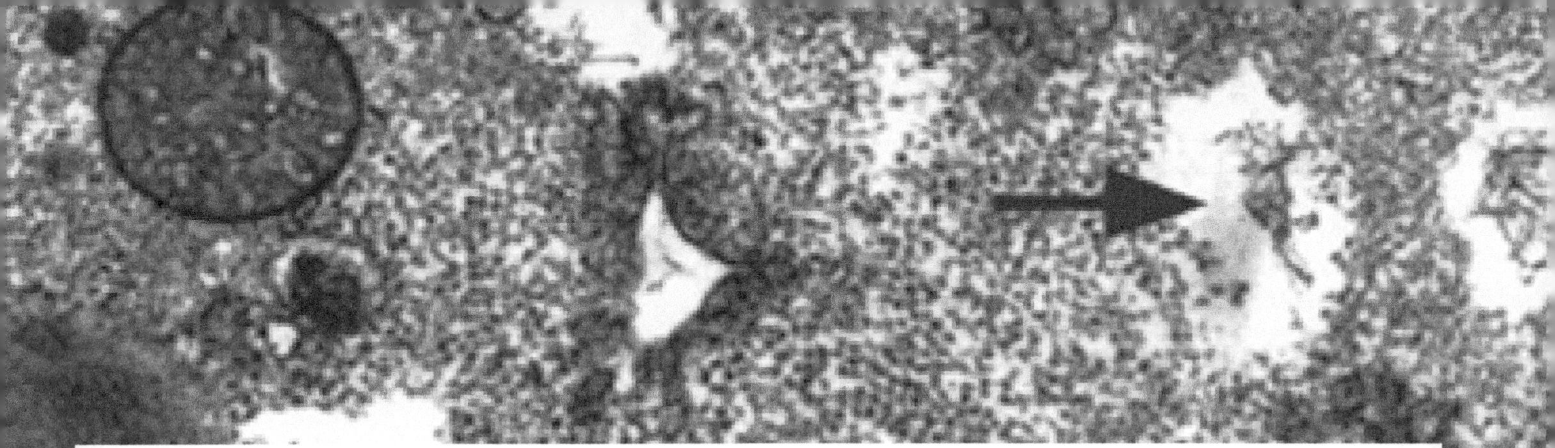

CHAPTER 10 VIRUS GENOMES AND VIRUS-HOST INTERACTIONS IN AQUACULTURE ANIMALS

Zhang Qiya and Gui Jian-Fang

Keywords: Aquaculture, viral genome, antiviral defense, iridoviruses, reoviruses, rhabdoviruses, herpesviruses, host-virus interactions

Aquaculture has become the fastest and most efficient agricultural production industry in the world over the last three decades, and China is believed to be a major contributor to it [1–3]. According to official figures, the production of aquatic products has reached 61.72 million tons, with the 45.6 million tons from aquaculture accounting for 73.9% of the total produced [4]. Over the last 20 years, China's aquaculture output has accounted for about 2/3 of the total global aquaculture production [5]. However, viral diseases, which have been frequently reported in aquaculture animals, have hampered aquaculture development [6–8]. Concurrently, a natural decline in populations of aquatic vertebrates, especially the global decline or extinction events seen with some frogs and amphibians, have been reported by ecologists; hence, the question "why are all the frogs 'croaking'?" has been asked [9–11]. To help resolve these problems, researchers have looked for and identified a large number of diverse pathogenic viruses in aquaculture and natural aquatic animals including iridoviruses, herpesviruses, reoviruses and rhabdoviruses; these pathogenic iridoviruses have been found to be the cause of epizootic diseases in aquaculture animals and the global decline of amphibian populations [12–15]. In the last 10 years especially, more than 100 viral genomes have been genetically characterized via the rapid advances in genome sequencing technologies [16–19]. These advances have enabled great progress to be made in understanding the mechanisms underlying interactions between viruses and their aquatic host animals [19]. Here, we review recent progress in the genomic and genetic characterization of some important pathogenic viruses, such as iridoviruses, herpesviruses, reoviruses, and rhabdoviruses, and virus–host interactions in aquaculture animals.

IRIDOVIRUSES AND THEIR GENOMES

Iridoviruses (family Iridoviridae) comprise the following five genera: Ranavirus, Lymphocystivirus, Megalocytivirus, Iridovirus and Chloriridovirus. The genomes of this family of viruses generally contain a single molecule of double-stranded DNA [19,20]. Ranavirus, Lymphocystivirus and Megalocytivirus infect more than 140 species of aquatic vertebrates including fish, amphibians and reptiles, and cause high mortality in aquaculture and problems with wildlife conservation [19–21]. For example, lymphocystis disease

Citation: Zhang QY, Gui JF. Virus genomes and virus host interactions in aquaculture animals.doi: 10.1007/s11427-015-4802-y

virus (LCDV), has been identified as the causative agent of lymphocystis disease in more than 100 different seawater and freshwater fish species [22,23]. In particular, diverse ranaviruses (genus Ranavirus) have been reported to infect about 70 amphibian species from at least 14 families, more than 100 fish species and dozens of reptiles; hence, some experts believe that ranaviruses infect not only frogs but also numerous different aquatic vertebrates, and are, therefore, promiscuous pathogens of cold-blooded vertebrates [24–33].

In total, 22 genomes from Ranavirus, Lymphocystivirus and Megalocytivirus have been completely sequenced; the smallest (105 kb) is that of the tiger frog virus (TFV), while the largest (186 kb) belongs to the Chinese strain of LCDV (LCDV-C) (Table 1).

Table 1: Known iridoviruses of aquatic animals and their genomes

No.	Genus and strain	Known host	Isolation region/time	Genome size (kb)	GC%	Potential ORFs	Accession No.	References
Ranavirus								
1	FV3 (Frog virus 3)	Frog	America, 1966	105.903	55	98	AY548484	[32–35]
2	ESV (European sheatfish virus)	Fish	Europe, 1985	127.732	54	136	JQ724856	[36]
3	EHNV (Epizootic haematopoietic necrosis virus)	Fish	Australia, 1986	127.011	54	100	FJ433873	[37]
4	RGV (*Rana grylio* virus)	Frog	China, 1995	105.791	55	106	JQ654586	[38–43]
5	ATV (*Ambystoma tigrinum* virus)	Salamander	America, 2003	106.332	54	96	AY150217	[44,45]
6	SGIV (Singaporegrouper iridovirus)	Fish	Singapore,1998	140.131	48	162	AY521625	[46,47]
7	STIV (Soft-shelled turtle iridovirus)	Turtle	China, 1999	105.890	55	105	EU627010	[48]
8	GIV (Grouper iridovirus)	Fish	Taiwan, 2000	139.793	49	120	AY666015	[49]
9	TFV (Tiger frog virus)	Frog	China, 2002	105.057	55	105	AF389451	[50]
10	CMTV (Common midwife toad ranavirus)	Toad	Europe, 2007	106.878	55	104	JQ231222	[51]
11	ADRV (*Andrias davidianus*ranavirus)	Giant salamander	China, 2013	106.734	55	101	KC865735	[52]
12	ADRV-2	Giant salamander	China, 2014	106.719	55	101	KF033124	[54]
13	CGSIV (Chinese giant salamander iridovirus)	Giant salamander	China, 2014	105.375	55	112	KF512820	[55]
Lymphocystivirus								
14	LCDV-1 (lymphocystis disease virus-1)	Fish	Red Sea, 1962	102.653	29	110	L63545	[22,56]
15	LCDV-C (lymphocystis disease virus-China)	Fish	China, 2004	186.247	27	240	AY380826	[23,43,57]
Megalocytivirus								
16	RSIV (Red seabream iridovirus)	Fish	Japan, 1992	112.415	53	116	BD143114	[61]
17	ISKNV (Infectious spleenand kidney necrosis virus)	Fish	China, 1998	111.362	55	124	AF371960	[60]
18	RBIV (Rock bream iridovirus)	Fish	Korea, 2000	112.080	53	118	AY532606	[62]
19	LYCIV (Large yellow croaker iridovirus)	Fish	China, 2001	111.767	54	–	AY779031	[63]
20	TRBIV (Turbot reddish body iridovirus)	Fish	China, 2004	110.104	55	115	GQ273492	[64]
21	OSGIV (Orange-spotted grouper iridovirus)	Fish	China, 2005	112.636	54	121	AY894343	[65]
22	RBIC-C1 (Rock bream iridovirus isolate from China)	Fish	China, 2012	112.333	55	119	KC244182	[66]

Ranaviruses and their Genomes

The following 13 ranavirus genomes have been completely sequenced:

- Frog virus 3 is a species of the genus Ranavirus. Frog virus 3 infection results in considerable morbidity and mortality in a wide range of wild and cultivated amphibian species [32–35].
- European sheatfish virus is a fish ranavirus isolated from moribund sheatfish (Silurus glanis) fry [36].
- Epizootic hematopoietic necrosis virus is a fish ranavirus that causes serious hematopoietic necrosis in redfin perch and rainbow trout, resulting in serious economic losses in aquaculture and severe decline in wild populations of these fish [37].
- Rana grylio virus (RGV). RGV, a ranavirus isolated from China, causes systemic hemorrhagic disease with a high mortality rate in frogs. It is also a model system for molecular characterization of ranaviruses [38–43].
- Ambystoma tigrinum virus is a lethal ranavirus originally isolated from Sonora tiger salamanders in southern Arizona, USA [44,45].
- Singapore grouper iridovirus is a fish ranavirus isolated from a diseased grouper in Singapore [46,47].
- Soft-shelled turtle iridovirus is a reptile ranavirus that

causes viral disease in cultured soft-shelled turtles [48].

- Grouper iridovirus is a fish ranavirus isolated from the spleen tissues of a diseased yellow grouper [49].
- TFV is a frog ranavirus isolated from diseased tiger frogs [50].
- Common midwife toad ranavirus is a toad ranavirus responsible for an outbreak of a systemic hemorrhagic disease that caused high mortality in toads from northern Spain [51].
- Andrias davidianus ranavirus (ADRV), the first sequenced ranavirus, is associated with high mortality in Chinese giant salamanders [52]. This ranavirus causes high mortality in wild and farmed Chinese giant salamanders [52,53].
- ADRV-2, another ranavirus strain isolated from Chinese giant salamanders after ADRV, shares a high level genome identity with ADRV [54].
- Finally, the Chinese giant salamander iridovirus, which is another ADRV, is the third sequenced ADRV strain [55].

Based on their genome sizes, gene contents and phylogenetic analyses, the sequenced ranaviruses (Table 1) have been divided into two subgroups: the amphibian subgroup and the fish subgroup. The amphibian subgroup includes ADRV, Common midwife toad ranavirus, RGV, Frog virus 3, TFV, and Ambystoma tigrinum virus, while the fish subgroup comprises Epizootic hematopoietic necrosis virus, European sheatfish virus, Grouper iridovirus, and Singapore grouper iridovirus [52]. From extensive analysis of the genome architectures and major genes of this diverse array of ranaviruses (especially ADRV and RGV genomes), we have proposed a hypothetical evolutionary model for ADRV [52]. In this model, ADRV is proposed to emerge (with its current genome) from a common ancestor of the amphibian ranavirus subgroup through changes in its genome architecture and variations in some of its major virulence-related genes (Figure 1). This hypothesis is based on the architectural changes observed in current ranavirus genomes; these include segment inversion, fragment insertion and deletion, and several variations in major genes, such as high diversification in two duplicate genes encoding the US22 family-like proteins, truncated domains in the virulence-related gene encoding vIF2α, and the appearance of novel genes with nuclear localization signal and nuclear export signal motifs [52]. Therefore, our current model provides possible routes leading to evolutionary genetic change and crossspecies transmission mechanisms in this diverse range of ranaviruses.

LCDV Genomes

LCDVs (genus Lymphocystivirus) cause lymphocystis disease in marine and freshwater fish. The genomes of two LCDVs, LCDV-1 and LCDV-C have been completely sequenced. LCDV-1, which infects plaice and flounder, was isolated from the propagated cell lines of bluegill and centrarchid fish in 1966 [32–35] and had its genome completely sequenced in 1997 [22]. LCDV-C was originally isolated from a flounder with lymphocystis disease by Chinese scientists in 2003, and its complete genome sequence was reported in 2004 [23,43,57]. The LCDV-C genome remains the largest among all known vertebrate iridoviruses sequenced thus far (Table 1).

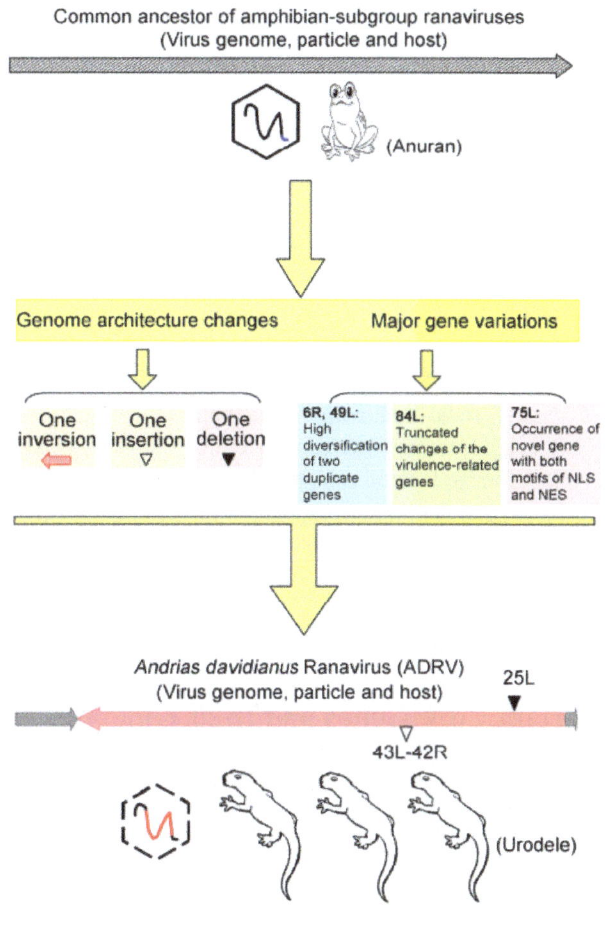

Figure 1: A hypothetical model of ADRV evolutionary emergence from a common ancestor of the amphibian subgroup of ranaviruses. During evolutionary processes, several changes leading to the current genome architecture and major virulence-related gene variations are proposed to have happened. Red hatched arrow: genome segment inversion; black triangles: fragment insertion; blank triangles: fragment deletion; 6R, 49L, 84L, 75L and 25L: different open reading frames (ORFs) in ADRV genome [52].

It is worth mentioning that LCDV-C is listed as a typical Iridoviridae strain in the "Virus Taxonomy Ninth Report" by the International Committee on Taxonomy of Viruses [20]. LCDV-C genomic data can be used as a reference resource for identifying other LCDVs and iridoviruses and for performing gene function analyses. This resource will enable virologists to explore the genetic characteristics of these large DNA viruses. In addition, several novel emerging LCDVs, such as LCDV-PF from the paradise fish Macropodus opercularis [58], and GLCDV, which was isolated from cultured grouper [59], have been identified.

Megalocytivirus Genomes

To date, seven megalocytiviruses have been subjected to complete genome sequencing:

- Infectious spleen and kidney necrosis virus—a megalocytivirus causing high mortality in mandarin fish—is characterized by cell hypertrophy in the spleen, kidney, cranial connective tissue and endocardium of this fish [60].
- Red sea bream iridovirus is a piscine iridovirus that

causes an acute and highly contagious disease in Red sea bream from Japan and Korea [61].

- Rock bream iridovirus is a megalocytivirus that occurs in cultured rock bream from Korea [62].
- Large yellow croaker iridovirus is a megalocytivirus causing gill paleness, liver congestion, spleen and kidney hypertrophy in cultured large croaker [63].
- Turbot reddish body iridovirus is a megalocytivirus that causes serious systemic diseases with high mortality in cultured turbot [64].
- Orange-spotted grouper iridovirus is the causative agent of serious systemic diseases with high mortality in the cultured orange-spotted grouper [65].
- Lastly, rock bream iridovirus isolated from China, is a megalocytivirus that caused a severe disease epidemic in Chinese farmed rock bream [66].

Iridoviruses of the Ranavirus and Lymphocystivirus genera have broadly similar genome sizes and potential gene contents as iridoviruses of the Megalocytivirus genus; their genome sizes range from 110 to 113 kb, while their potential number of genes range from 115 to 124 (Table 1).

Important Core Genes and Their Functions in Iridoviruses

Gene annotation and comparative genomic analysis have confirmed there are 26 core genes in iridoviruses [67]. Extensive comparisons of these important core genes has provided evidence for cross-species transmission in these iridoviruses, especially for the ranaviruses [68,69]. Moreover, some important genes encoding enzymes, structural proteins and immune-related proteins, such as the RGV 3β- hydroxysteroid dehydrogenase gene (RGV 3β-HSD) [70] and the RGV deoxyuridine triphosphatase gene (RGV dUTPase) [71], have been characterized and functionally analyzed. LCDV-C thymidylate synthase (LCDV-C TS) is able to promote cell cycle progression into S and G2/M phase. In comparison with control cells, TS-expressed cells have faster growth rates, and induce foci formation and anchorage-independent growth. These findings indicate that LCDV-C TS potentially exhibits the ability to transform cells (tumor formation) [72].

RGV 53R is a core gene in iridoviruses, and encodes a viral envelope protein that plays an important role in virus assembly and infection [73,74]. Recently, we have chosen RGV 53R as a target gene to construct a conditional lethal recombinant RGV (i53R-RGV-lacIO) containing the inducible lac repressor/operator system that can be regulated by IPTG, and have found that the 53R expression level, plaque formation ability and viral titers in i53R-RGV-lacIO are significantly reduced in the absence of IPTG. These results indicate that RGV 53R is not only essential for virus replication and assembly, but also contributes to virus infection and virion formation [75]. RGV 50L contains a nuclear localization signal and helix-extension-helix motif, and is an immediate-early gene. Immunofluorescence assays indicate that 50L expression occurs early during infection and persists in RGV-infected cells. RGV 50L exhibits a cytoplasm-nucleus-viromatrix distribution pattern and viromatrix distribution pattern, indicating that it encodes a structural protein, and plays an important role in viral assembly and life cycle [76]. RGV 2L is a core gene encoding an envelope protein. To investigate the role of 2L in viral infections, we constructed a conditional lethal mutant virus containing the lac repressor/operator system and dual fluorescent labeling. Significantly, when 2L expression is repressed, its plaque formation ability and virus titers were strongly reduced. Functional analysis indicates that the 2L protein is essential for iridovirus infection and its study has provided new insights into iridovirus envelope proteins [77].

Besides the controllable recombinant virus technique mentioned above [77], gene knockout methodology has been also used to investigate gene function in iridoviruses [78]. Expression inhibition of a structural protein gene and RNA polymerase gene by morpholino knockdown or gene-specific silencing has been observed to cause a significant reduction in the yield of virus progeny [79,80]. Additionally, analysis of gene expression timing and infection pathways have been undertaken in iridovirus using transcriptomics [81]. For example, transcriptional analysis of TFV infection (along with other approaches) has revealed that TFV entry into HepG2 cells occurs via a pH-dependent, atypical, caveola-mediated endocytosis pathway [82]. Also, miRNAs have been recently found to have a significant impact on interactions between iridoviruses and their host aquaculture animals [83]; indeed, some miRNAs may me- diate viral evasion [84].

AQUAREOVIRUS GENOMES

All members of the virus Reoviridae family are nonenveloped, and their genomes are composed of multiple (10, 11, or 12) segments of linear double-stranded RNAs housed within an icosahedral capsid. This family contains two subfamilies (Spinareovirinae and Sedoreovirinae) and 15 different genera (Aquareovirus, Orthoreovirus, Orbivirus, Rotavirus, Coltivirus, Seadornavirus, Cardoreovirus, Cypovirus, Idnoreovirus, Phytoreovirus, Fijivirus, Oryzavirus, Mycoreovirus, Mimoreovirus, and Dinovernavirus) [20].

Reoviruses that infect aquaculture fish belong to the Aquareovirus genus, and their genomes generally contain 11 segments of linear double-stranded RNA [85,86]. At least 15 reovirus genomes have been completely sequenced (Table 2). Of these, 11 reoviruses are different isolates from cultured grass carp obtained in various years and from different regions [85,87–92]; the other four reoviruses were isolated from cultured golden shiner, chum salmon, Atlantic salmon (Piscine reovirus), and turbot (Scophthalmus maximus reovirus), respectively [93–95].

Recently, our laboratory analyzed and compared the complete genome sequences and major core protein sequences of various grass carp reoviruses (GCRV), and this revealed significant genetic diversity among them [87,90]. GCRV can be divided into three groups. Most of them cluster into the first major group; these viruses are not cytopathic and contain a fiber-like protein. GCRV members of the second group are cytopathic and possess a fusionassociated small transmembrane (FAST) protein. GCRV 104, a lone member of the third group, is also cytopathic and has a fiber-like protein. However, the various genotypes are not associated with their regional distributions [90]. Therefore, more studies on the evolutionary and geographical relationships between genomic diversity and reovirus transmission should be performed on grass carp reoviruses.

Scophthalmus maximus reovirus, a novel reovirus isolated from marine fish, contains a FAST protein translated from a non-AUG start site that has been shown to partially contribute to the cytopathic effect caused by infection with this virus [94]. As a new reovirus equally related to members of the Orthoreovirus and Aquareovirus genera, Piscine reovirus, which is linked to heart and skeletal muscle inflammation in farmed Atlantic salmon (Salmo salar L.), has been suggested by whole genome comparisons to be more closely related to orthoreoviruses and, therefore, a new species of the Orthoreovirus genus [96]. Genome comparisons show that Piscine reovirus contains 10 genomic segments (not 11 as in all recognized aquareoviruses) and an outer-fiber protein that is present in most members of the Orthoreovirus genus [96,97]. Moreover, phylogenetic evidence of long distance dispersal and transmission has been revealed by comparing the protein coding sequences S1, S2 and S4 in Piscine reovirus genomic segments between farmed and wild Atlantic salmon [98].

Table 2: Known aquareoviruses and their genomes

No.	Virus strains	Host	Isolation region/time	Genome size (kb)	Segments/ ORFs	Accession No.	References
1	GCHV (Grass carp hemorrhage virus	Grass carp	China, 1980	21.366	11/11	AF260511–3 AF251262 F239175 AF239174 F259053 AF284504 F236688 F234321	[85]
2	GCRV-873 (Grass carp reovirus-873)	Grass carp	China, 1983	25.000	11/12	AF260511–3 AF403390–7	[85]
3	AGCRV (American grass carp reovirus)	Grass carp	America, 2001	23.576	11/12	EF589098–EF589108	[91]
4	GCRV-HuNan794 (Grass carp reovirus-HuNan794)	Grass carp	China, 2007	24.780	11/11	KC238676–KC238686	[92]
5	GCRV HZ08 (Grass carp reovirus HZ08)	Grass carp	China, 2008	24.707	11/11	GQ896334–7 GU350742–8	[87]
6	GCRV106 (Grass carp reovirus 106)	Grass carp	China, 2009	24.778	11/11	KC201166–KC201176	[97]
7	GCRV-HeNan988 (Grass carp reovirus-Henan988)	Grass carp	China, 2009	24.780	11/11	KC847320–KC847330	[97]
8	GCRV918 (Grass carp reovirus 918)	Grass carp	China, 2010	24.780	11/11	KC201177–KC201187	[97]
9	GCRV-109 (Grass carp reovirus-109)	Grass carp	China, 2014	24.625	11/11	KC201177–KC201187	[90]
10	GCRV-GD108 (Grass carp reovirus-GD108)	Grass carp	China, 2009	24.703	11/12	HQ231198–HQ231208	[88]
11	HGDRV (GCRV 104) (Hubei grass carp disease reovirus)	Grass carp	China, 2009	23.706	11/12	JN967629–JN967639	[89]
12	PRV(Piscine reovirus)	Atlantic salmon	Norway, 2012	23.320	10/11	GU994013–GU994022	[95]
13	GSRV (Golden shiner reovirus)	Golden shiner	America, 1979	23.695	11/12	AF403398–AF403408	[93]
14	CSRV (CHSRV) (chum salmon reovirus)	Salmon	Japan, 1981	23.015	11/12	AF418294–AF418304	[93]
15	SMReV (Turbot *Scophthalmus maximus* reovirus)	Turbot	China, 2012	24.042	11/12	HM989930–HM989940	[94]

As an ideal model system for studying the cell entry mechanism used by nonenveloped viruses, single-particle cryo-electron microscopy has been used to observe the 3.3 Å structure of the primed, infectious subvirion GCRV particle, thereby providing structural insight into the coupling of virion assembly [99,100]. Additionally, new insight into the mechanisms of viral factory formation and pathogenesis of aquareoviruses has been acquired from functional studies on aquareoviral genes where NS80, a nonstructural protein of fish reovirus, has been confirmed to be crucial for recruiting viral components to form aquareoviral factories [101].

RHABDOVIRUS GENOMES

Rhabdoviruses are a group of enveloped, single-stranded, negative-sense RNA viruses. The Rhabdoviridae family includes the following nine genera: Cytorhabdovirus, Ephemerovirus, Lyssavirus, Novirhabdovirus, Nucleorhabdovirus, Perhabdovirus, Sigmavirus, Tibrovirus and Vesiculovirus [20]. All known fish rhabdoviruses have been assigned to the following three genera: Vesiculovirus, Novirhabdovirus and Perhabdovirus [102,103].

Fish rhabdoviruses can cause severe hemorrhagic septicemia in freshwater and marine fish. In the last 10 years, virologists isolated and identified the following rhabdoviruses from aquaculture fish: Siniperca chuatsi rhabdovirus [104], Scophthalmus maximus rhabdovirus [105,106], Paralichthys olivaceus rhabdovirus [107], Monopterus albus rhabdovirus [108], snakehead rhabdovirus [109,110], Hirame rhabdovirus [111], and pike fry rhabdovirus [112]. Spring viremia of carp virus (SVCV), an earlier identified rhabdovirus, causes infectious hemorrhagic septicemia in common carp (Cyprinus carpio) [102,113]. Perch rhabdovirus causes lethal hemorrhagic disease in different farmed species of perch, bass, grayling and trout [114,115]. Viral hemorrhagic septicemia virus (VHSV) and infectious haematopoietic necrosis virus (IHNV) are two typical rhabdoviruses of the Novirhabdovirus genus. VHSV is a viral pathogen affecting both wild and cultured fish worldwide; infected species include salmon, trout, cod, herring, sole, catfish, pike, turbot, and flounder, among others [115]. IHNV causes severe losses to the salmon fish industry in the USA and Canada, and many other countries in Asia and Europe [18].

IHNV is the first fish rhabdovirus that has had its complete genome sequenced [116]. Currently, more than 100 fish rhabdovirus genomes have been completely sequenced, and over 80 of them are from different VHSV isolates or strains (Table 3). Fish rhabdovirus genomes are negative-sense, single-stranded RNA molecules, and their sizes range from 11 to 16 kb. In fish rhabdoviruses of the Vesiculovirus genus, such as SVCV, pike fry rhabdovirus, Siniperca chuatsi rhabdovirus, and Scophthalmus maximus rhabdovirus, their genomes encode the following five proteins: nucleoprotein (N), phosphoprotein (P), matrix protein (M), glycoprotein (G), and RNA-dependent RNA polymerase (L) in the order 3′-N-P-M-G-L-5′, whereas in fish rhabdoviruses of the Novirhabdovirus genus, such as VHSV, IHNV, Hirame rhabdovirus, snakehead rhabdovirus and Paralichthys olivaceus rhabdovirus, their genomes encode the following six proteins: N, P, M, G, non-virion protein (NV), and L, in the order 3′-N-P-M-G-NV-L-5′. In comparison with Vesiculovirus genus members, novirhabdoviruses possess an additional NV gene [107].

Because of their small genomes, short generation times and rapid mutation rates, fish rhabdoviruses, especially VHSV and IHNV, have been extensively used to analyze their evolutionary patterns, genetic diversity and biogeography of the numerous variants. Pierce and Stepien [117] evaluated the phylogenetic and biogeographic relationships of various VHSV isolates by comparing their corresponding genomic sequences, and depicted an evolutionary history of relatively rapid population diversifications in star-like patterns, following a quasispecies model. Furthermore, He et al. [118] applied the Bayesian coalescent method to the time-stamped entire coding sequences of each VHSV gene. Through age calculations on six genes, the first bifurcation event of the isolates they analyzed was estimated to have occurred within the last 300 years. Additionally, comprehensive phylogenetic analyses have been performed by comparing the corresponding gene sequences of worldwide VHSV or IHNV isolates [114,119,120].

In fish rhabdoviruses of the Vesiculovirus genus, SVCV genomes, which can be classified into different clades and genogroups, have been described as possessing high levels of diversity and plasticity [121,122]. Recently, Xiao et al. [123] performed recombination analysis of all known complete SVCV genomic sequences, and found evidence of homologous recombination in these genomes. This finding sheds light on recombination and the evolutionary process in various isolates of fish rhabdoviruses.

To allow functional studies to be conducted on fish rhabdovirus genes, several IHNV-VHSV chimeric recombinant viruses were constructed to allow the researchers to identify virulence genes through reverse genetics. Through comparative challenge experiments in rainbow trout fingerlings, recombinant IHNV gained higher virulence following substitution of the G gene with that of each individual VHSV strain [124]. Additionally, an in vivo superinfection assay has recently been developed to examine the role of virulence in IHNV of fish rhabdoviruses [125].

Table 3: Known fish rhabdoviruses and their genomes

No.	Virus strains	Date of isolation	Isolation region	Host	Genome size (knt)	Genotype (gene order)	Accession No.	References
Rhabdoviruses of different fish species								
1	IHNV (Infectious haemato-poietic necrosis virus)	1979	North America	Salmonids	11.137	3′N-M1-M2-G-NV-L-5′	X89213	[116, 124]
					11.131	3′N-M1-M (or M2)-G-NV-L-5′	NC_001652.1	
2	SCRV (*Siniperca chuatsi* rhabdovirus)	1999	China	Mandarin fish	11.545	3′-N-P-M/Ms-G-L-5′	DQ399789	[104]
3	SMRV (*Scophthalmus maximus*rhabdovirus)	2007	China	Turbot	11.492	3′N-Ps-P/C-M-G-L 5′	HQ003891	[105, 106]
4	PORV (*Paralichthys olivaceus* rhabdovirus)	2005	China	Flounder	11.182	3′-N-P-M-G-NV-L-5′	KC685626	[107]
5	SVCV (Spring viraemia of carp virus)	2007	China	Common carp	11.019	3′-N-P-M-G-L-5′	DQ097384	
					11.019	3′-N-P-M-G-L-5′	AJ318079	[102, 113,121–123]
					11.100	3′-N-P-M-G-L-5′	DQ097384	
					11.020	3′-N-P-M-G-L-5′	NC_002803.1	
6	PFRV (Pike fry rhabdovirus)	1973	Netherlands	Pike fry	11.097	3′ N-P-M-G-L-5′	FJ872827	[112]
7	EVEX (Eel virus European X)	2013	Europe	eel	11.778	3′ N-P-M-G-L-5′	JX827265	
8	TenRV (Tench rhabdovirus)	2013	Europe	tench	11.082	3′ N-P-M-G-L-5′	KC113517	
9	GrCRV (Grass carprhabdovirus)	2013	Europe	grass carp	11.096	3′ N-P-M-G-L-5′	KC113518	
10	SVCV_Fijan (Spring viraemia of carp virus-Fijan)	1971	Yugoslavia	Common carp	11.019	3′ N-P-M-G-L-5′	AJ318079	[102]
	HIRRV							

11	CA-9703 (Hirame rhabdovirus-CA-9703)	1986	Korea	Japanese flounder	11.034	3′ N-P-M-G-NV-L-5′	AF104985	[111]
12–30	PRV(Perch rhabdovirus)	1980 –2010	France, Denmark, Italy, Finland Baltic Sea, Sweden, Nether-lands	Pikeperch, Pike, Perch, Black bass, Grayling, Sea trout, Eel, Trout Brown	e.g. 11.487	e.g. 3′ N-P-M-G-L-5′	JX679246.1 JF502607, JF502603, JF502604 JF502609, JF502605, JF502596, JF502608, KC408701, KF146312 KC408697, KF146314, KC408700 KF146311, KC408699, KF146310 KC408698, KF146309, KF146308. KF146313, KF146315, AF434991 AF434992, FN557213,	[114]
Viral hemorrhagic septicemia virus (VHSV) from different fish species								
31–116	VHSV (Viral haemorrhagic septicaemia virus)	1962 –2007	Denmark, Norway France, Georgia English Baltic Sea, Kattegat Finland Archipel-ago Sea Scotland Ireland USA, Japan Italy, Sweden, Nether-lands	Rainbow trout Cod, Sprat Herring Lamprey Turbot Pout, Eel Coho salmon Japanese flounder	e.g. 11.184	e.g. 3′-N-P-M-G-NV-L-5′	GQ385941 AF345857, AY546621, AY546616, Z93412, Z93414, AY546619, U28800, AY356632, AF345859, AF345858, AF143862, AY546575, AY546612, AY546623, AM086355,AM086356, AM086357, AM086358, AM086359, M086360, AM086361, AM086362, AM086363, AM086364, AM086365, AM086366, AM086367, AM086368,AM086369, AM086370, AM086371, AM086372, AM086373, AM086374, AM086375, AM086376, AM086377, AM086378, AM086379, AM086380, AM086381, AB179621 AM086382, AM086383, FJ384761, AY546576, AY546577, AY546578, GQ504013, HQ112198-HQ112200, HQ112234, HQ112201- HQ112203, HQ112235, HQ112204- HQ112206, HQ112236, HQ112207-HQ112209, HQ112237, HQ112210-HQ112212, HQ112238, HQ112213-HQ112215, HQ112239,HQ112216, HQ112217, HQ112240, HQ112218, HQ112241, HQ112248, HQ112219-HQ112221, HQ112243, HQ112222-HQ112224 HQ112244, HQ112225-HQ112227, HQ112245, HQ112228-HQ112229, HQ112246 HQ112242, HQ112233, HQ112232, HQ112230, HQ112231, HQ112247, AY546628, AY546582, AY546620, AY546632, EU547740, Y546618, U28747	[117–119, 122]

HERPESVIRUS GENOMES

Herpesviruses are enveloped viruses containing large, double-stranded, linear DNA genomes. They are host-specific pathogens, and are wide spread among vertebrates such as mammals, birds, amphibians and fish [126]. After two new virus families, namely, Alloherpesviridae (incorporating fish and frog herpesviruses) and Malacoherpesviridae (containing mollusks) were recognized in 2009, the Herpesvirales order includes the following three families: Herpesviridae (containing mammals and birds), Alloherpesviridae, and Malacoherpesviridae. The Alloherpesviridae family is divided into the following four genera: Batrachovirus, Cyprinivirus, Ictalurivirus, and Salmonivirus [20]. The Cyprinivirus genus contains four species: three of them, cyprinid herpesvirus 1 (CyHV-1), cyprinid herpesvirus 2 (CyHV-2), and cyprinid herpesvirus 3 (CyHV-3), are associated with common carp or goldfish, while one of them, anguillid herpesvirus 1 (AngHV1), is associated with freshwater eels [127,128].

Cyprinid herpesviruses have been reported to cause serious mortality in common carp and crucian carp [129]. CyHV-1 is the cause of carp pox, CyHV-3 is fatal in carp and koi fish, and CyHV-2 is the etiological agent of herpesviral hematopoietic necrosis disease in common carp, goldfish, crucian carp, and gibel carp [130]. Through use of bioluminescence imaging, the skin covering a fish's fins and body has been shown to be the major port of entry for cyprinid herpesviruses [131]. Tissue culture and RT-PCR testing results indicate that herpesviruses may become latent in leukocytes and other tissues following a primary infection, and that they can be reactivated from latency by temperature stress [132]. Recently, B cells have been identified as a major site where CyHV-3 can become latent [133].

CyHV-3 has been observed to cause significant morbidity and mortality in koi and common carp. The pathological signs include epidermal abrasions, excess mucus production, necrosis of gills and internal organs, and lethargy. CyHV-3 propagates well in the intestines and kidneys, and high numbers of infectious viruses can be observed in the droppings of infected fish [134]. Several primary culture cell lines can be used to propagate cyprinid herpesviruses and for isolating these etiologic agents of disease [135].

So far, the complete genomes of 11 herpesviruses have been sequenced from aquatic animals, nine of which are members of the Alloherpesviridae family, while two are members of the Malacoherpesviridae family. The genomes of these herpesviruses range in size from about 134.2 kb for the smallest (Ictalurid herpesvirus 1) to about 295.2 kb for the largest (Koi herpesvirus-J); their potential numbers of genes range from 77 to 163 (Table 4).

The complete genome sequences of five cyprinid herpesviruses, the diseases of which are fatal in common carp, koi carp, goldfish, crucian carp, and gibel carp [130] have been reported. These genomes are characterized by a unique region flanked at each terminus by a sizeable direct repeat. About 120 orthologous genes are shared by these cyprinid herpesviruses, and 55 of them also share sequence conservation with AngHV1 of the Cyprinivirus genus. Significantly, only 12 genes were found to be conserved convincingly in all the sequenced alloherpesviruses [136–138].

Table 4: Known herpesviruses of aquatic animals and their genomes

No.	Genus/strain	Host/symptom	Isolation region/time	Genome size (kb)	Potential ORFs	Accession No.	References
Iloherpesviridae							
1	Ictalurivirus/Ictalurid herpesvirus 1 (IcHV-1)	Channel catfish	USA, 1971	134,226	77	M75136	[138,139]
2	Ictalurivirus/Anguillid herpesvirus 1 (AngHV-1)	Eels (Anguilla)	Netherlands, 1996	248.526	134	FJ940765	[140–142]
3	Cyprinivirus/Cyprinid herpesvirus1 (CyHV-1)	Common carp (*Cyprinus carpio*) and koi carp (a variety of *Cyprinus carpio*)/carp pox, papilloma	Japan, 1985	291,144	143	JQ815363	[91,128]
4	Cyprinivirus/Cyprinid herpesvirus 2 (CyHV-2)	Goldfish (*Carassius auratus*)/goldfish hematopoietic necrosis	Japan, 1992	290,304	154	JQ815364	
5	Cyprinivirus/Cyprinid herpesvirus 3 (CyHV-3) KHV-U	Koi and *Cyprinus carpio*/Bleeding gills, sunken eyes, pale patches or skin blistering	USA, 2003	295.146	163	DQ657948	[136]
6	KHV-J		Japan, 2004	295,271	156	AP008984	
7	KHV-I		Israel, 1998	295,138		DQ177346	
8	Batrachoviru/Ranid herpesvirus 1 (RaHV-1)	Leopard frog, *Rana pipiens*/Renal carcinoma or Lucke tumor	North American, 1964	220,859	132	DQ665917	[138,139]
9	Batrachoviru/Ranid herpesvirus 2 (RaHV-2) Frog virus 4	Leopard frog, *Rana pipiens*/Renal carcinoma or Lucke tumor	North American, 1964	231,801	147	DQ665652	
Malacoherpesviridae							
10	Ostreavirus/Oyster herpesvirus OsHV1	Oyster	New Zealand, 1995	207,439	124	AY509253	[128]
11	Ostreavirus/Acute viral necrosis virus (AVNV)	Chinese scallop *Chlamys farreri*	China, 1998	210,993	123	GQ153938	[144]

Ictalurid herpesvirus 1 is the type species of the Ictalurivirus genus, and is the first fish herpesvirus for which the complete genome is known [139]. AngHV1 also frequently causes fatal disease in freshwater eels. After complete genome sequencing [140] and deep-sequencing of the AngHV1 transcriptome [141] were finalized, a genomewide transcription analysis was performed using reverse transcription quantitative PCR, and a temporal regulation fashion similar to mammalian herpesviruses was observed in this fish herpesvirus [142].

Chelonid herpesvirus 5 is closely related to fibropapillomatosis, a neoplastic disease of marine turtles. Its genomic sequence has been shown to be largely collinear with the genomes of typical alphaherpesviruses [143]. In addition, the complete genome sequence of the acute viral necrosis virus, which belongs to the Malacoherpesviridae family, has also been reported recently [144].

In recent years, after cyprinid herpesvirus disease became widely reported in the world, (especially in China) [145–147], numerous studies on cyprinid herpesviruses and identification of immune-related genes have been conducted [148–150]. It is envisaged, therefore, that new insight and better understanding of these cyprinid herpesviruses will emerge and lead to efficient antiviral approaches being developed in the near future.

VIRUS-HOST INTERACTIONS IN AQUATIC ANIMALS

The diverse viruses discussed above are serious pathogens of aquatic animals, especially those used in aquaculture. To understand their pathogenetic mechanisms and thereby provide protective strategies again them, some significant experimental methods and high-throughput technologies such as transcriptomics and proteomics have been recently used to gain better knowledge of these viruses and their hosts. Such studies have greatly expanded our knowledge about the innate and acquired immune systems of aquatic animals [151–155]. In China especially, comparative immunological studies of aquaculture animals have flourished over the last 10 years, financial support for research in this area has increased, and progress in this field has been reported in several reviews [156–162]. Figure 2 is a schematic diagram outlining the interactions occurring between the diverse range of viruses discussed herein and their aquatic hosts. The diagram also shows how innate and acquired immunity, as well as related factors such as physical barriers, operate in aquaculture animals under attack by pathogenic viruses.

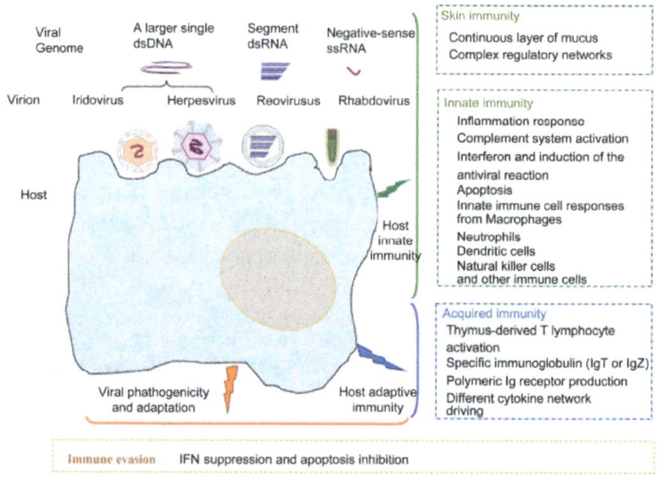

Figure 2: Schematic diagram illustrating the types of interactions that can occur between viruses and a host cell from an aquatic animal.

As shown in Figure 2, when viruses, such as iridoviruses, herpesviruses, reoviruses, or rhabdoviruses attempt to enter a host cell, they first meet physical barriers on the skin and interact with a continuous layer of mucus and the complex regulatory networks that control skin immunity [163–166]. After such viruses pass the first line of defense and enter the host cell, the infected cell immediately initiates a series of innate immune responses; these include an inflammatory response, complement system activation, interferon production, induction of antiviral reactions, cell apoptosis, and innate immune cell responses from macrophages, neutrophils, dendritic cells, natural killer and other immune cells that prevent viral replication and inhibit virus propagation [153,156,167]. Subsequently or concurrently, acquired (adaptive) immune responses, such as thymus-derived T lymphocyte activation, specific immunoglobulins (IgT or IgZ), polymeric Ig receptor production, and differential cytokine network activation, amongst others, enables collaboration between the cellular and humoral immune systems leading to destruction of the invading viruses [168–171]. Additionally, some aquatic viruses have developed immune evasion mechanisms and strategies to combat host immune systems through IFN suppression and apoptosis inhibition; such viruses have increased pathogenicity and have acquired the ability to cross species barriers in their transmission [153,167,172].

CONCLUSION AND OUTLOOK

Over the last decade, a large number of diverse pathogenic viruses, such as iridoviruses, herpesviruses, reoviruses, and rhabdoviruses, have been identified from aquaculture and natural aquatic animals, and many of their genomes have been completely sequenced. Comparative genomic and phylogenetic analyses have provided new insight into the origins of these viruses, as well as the different routes that have led to genetic change and evolutionary processes, and cross-species transmission mechanisms. Additionally, functional studies on genes have yielded crucial information about viral factory formation and pathogenesis in these viruses. Along with their genome architectures and genetic characterization, the interactions between these viruses and their aquatic animal hosts have become an important focus in aquaculture. Finally, significant progress has been made in understanding the following:

- the molecular mechanisms underlying virus-host interactions [173–176],
- innate antiviral immune responses in fish, and
- gene identification in the fish interferon system [177–179]. Armed with this knowledge, it is hoped that new drugs and strategies to protect aquaculture animals against pathogenic viruses will be developed in the near future.

ACKNOWLEDGEMENT

This work was supported by grants from the National Natural Science Foundation of China (31430091), the National Key Basic Research Program (2010CB126303), and the Project of State Key Laboratory of Freshwater Ecology and Biotechnology (2011FBZ12).

REFERENCES

1. Cressey D. Aquaculture: future fish. Nature, 2009, 458: 398–400
2. Bostock J, McAndrew B, Richards R. Aquaculture: global status and trends. Phil Trans R Soc B, 2010, 365: 2897–2912Central
3. Gui JF, Zhu ZY. Molecular basis and genetic improvement of economically important traits in aquaculture animals. Chin Sci Bull, 2012, 57: 1751–1760
4. Fisheries and Administration Agency, Ministry of Agriculture. China Fishery Statistical Yearbook. Beijing: China Agriculture Press, 2014
5. Tacon A, Metian M. Global overview on the use of fish meal and fish oil in industrially compounded aquafeeds: trends and future prospects. Aquaculture, 2008, 285: 146–158
6. Daszak P, Cunningham AA, Hyatt AD. Emerging infectious diseases of wildlife—threats to biodiversity and human health. Science, 2000, 287: 443–449
7. Walker PJ, Winton Jr. Emerging viral diseases of fish and shrimp. Vet Res, 2010, 41: 51Central
8. Collins JP, Crump ML. Extinction in Our Times: Global Amphibian Decline. New York: Oxford University Press, 2009
9. Fisher MC. Silent springs: why are all the frogs "croaking"? PLoS Biol, 2009, 7: e1000198Central
10. Echaubard P, Leduc J, Pauli B, Chinchar VG, Robert J, Lesbarrères D. Environmental dependency of amphibian-ranavirus genotypic interactions: evolutionary perspectives on infectious diseases. Evol Appl, 2014, doi:10.1111/eva.12169
11. Lips KR, Brem F, Brenes R, Reeve JD, Alford RA, Voyles J, Carey C, Pessier AP, Livo L, Collins JP. Emerging infectious disease and the loss of biodiversity in a Neotropical amphibian community. Proc Natl Acad Sci USA, 2006, 103: 3165–3170Central
12. Chinchar VG, Hyatt A, Miyazaki T, Williams T. Family Iridoviridae: poor viral relations no longer. Curr Top Microb Imm, 2009, 328: 123–170
13. Une Y, Sakuma A, Matsueda H, Nakai K, Murakami M. Ranavirus outbreak in North American bullfrogs (*Rana catesbeiana*), Japan, 2008. Emerg Infect Dis, 2009, 15:1146–1147Central
14. Hyatt AD, Gould AR, Zupanovic Z, Cunningham AA, Hengstberger S, Whittington RJ, Kattenbelt J, Coupar BE. Comparative studies of piscine and amphibian iridoviruses. Arch Virol, 2000, 145: 301–331
15. Kik M, Martel A, Spitzen-van der Sluijs A, Pasmans F, Wohlseind P, Grönea A, Rijksa JM. Ranavirus-associated mass mortality in wild amphibians, The Netherlands, 2010: A first report. Vet J, 2012, 190: 284–286
16. Zhang QY, Gui JF. Aquatic Virology. Beijing: Higher Education Press, 2008
17. Burnell G, Allan G. New technologies in aquaculture: Improving production efficiency, quality and environmental management. Cambridge: Woodhead Publishing Limited, 2009
18. Crane M, Hyatt A. Viruses of fish: an overview of significant pathogens. Viruses, 2011, 3: 2025–2046Central
19. Zhang QY, Gui JF. Atlas of Aquatic Viruses and Viral Diseases. Beijing: Science Press, 2012
20. King AMQ, Adams MJ, Carstens EB, Lefkowitz EJ, eds. Virus Taxonomy: Classification and Nomenclature of Viruses: Ninth Report of the International Committee on Taxonomy of Viruses. San Diego: Elsevier Academic Press, 2011

21. Williams T, Barbosa-Solomieu V, Chinchar VG. A decade of advances in iridovirus research. Adv Virus Res, 2005, 65: 173–248

22. Tidona CA, Darai G. The complete DNA sequence of lymphocystis disease virus. Virology, 1997, 230: 207–216

23. Zhang QY, Xiao F, Xie J, Li ZQ, Gui JF. Complete genome sequence of lymphocystis disease virus isolated from China. J Virol, 2004, 78: 6982–6994Central

24. Bandín I, Dopazo CP. Host range, host specificity and hypothesized host shift events among viruses of lower vertebrates. Vet Res, 2011, 42: 67Central

25. Whittington RJ, Becker JA, Dennis MM. Iridovirus infections in finfish—Critical review with emphasis on ranaviruses. J Fish Dis, 2010, 33: 95–122

26. Robert J, George E, De Jesus Andino F, Chen G. Waterborne infectivity of the *Ranavirus*frog virus 3 in *Xenopus laevis*. Virology, 2011, 417: 410–417Central

27. Geng Y, Wang KY, Zhou ZY, Li CW, Wang J, He M, Yin ZQ, Lai WM. First report of a ranavirus associated with morbidity and mor tality in farmed Chinese giant salamanders (*Andrias davidianus*). J Comp Pathol, 2011, 145: 95–102

28. Dong W, Zhang X, Yang C, An J, Qin J, Song F, Zeng W. Iridovirus infection in Chinese giant salamanders, China, 2010. Emerg Infect Dis, 2011, 17: 2388–2389Central

29. Brenes R, Gray MJ, Waltzek TB, Wilkes RP, Miller DL. Transmission of ranavirus between ectothermic vertebrate hosts. PLoS One, 2014, 9: e92476Central

30. Schloegel LM, Daszak P, Cunningham AA, Speare R, Hill B. Two amphibian diseases, chytridiomycosis and ranaviral disease, are now globally notifiable to the World Organization for Animal Health (OIE): an assessment. Dis Aquat Organ, 2010, 92: 101–108

31. Miller D, Gray M, Storfer A. Ecopathology of ranaviruses infecting amphibians. Viruses, 2011, 3: 2351–2373Central

32. Chinchar V. Ranavirus (family Iridoviridae): emerging cold-blooded killers. Arch Virol, 2002, 147: 447–470

33. Chinchar VG, Waltzek TB. Ranaviruses: not just for frogs. PLoS Pathog, 2014, 10: e1003850Central

34. Stuart SN, Chanson JS, Cox NA, Young BE, Rodrigues ASL, Fischman DL, Waller RW. Status and trends of amphibian declines and extinctions worldwide. Science, 2004, 306: 1783–1786

35. Morrison EA, Garner S, Echaubard P, Lesbarrères D, Kyle CJ, Brunetti CR. Complete genome analysis of a frog virus 3 (FV3) isolate and sequence comparison with isolates of differing levels of virulence. Virol J, 2014, 11: 46Central

36. Mavian C, López-Bueno A, Fernández Somalo MP, Alcamí A, Alejo A. Complete genome sequence of the European sheatfish virus. J Virol, 2012a, 86: 6365–6366Central

37. Jancovich JK, Bremont M, Touchman JW, Jacobs BL. Evidence for multiple recent host species shifts among the *Ranaviruses* (family Iridoviridae) J Virol, 2010, 84: 2636–2647Central

38. Lei XY, Ou T, Zhu RL, Zhang QY, Zhang QY. Sequencing and analysis of the complete genome of *Rana grylio* virus (RGV). Arch Virol, 2012, 157: 1559–1564

39. Tan WG, Barkman TJ, Chinchar VG, Essani, K. Comparative genomic analyses of frog virus 3, type species of the genus *Ranavirus* (family Iridoviridae). Virology, 2004, 323: 70–84

40. Zhang QY, Li ZQ, Gui JF. Studies on morphogenesis and cellular interactions of *Rana grylio* virus in an infected fish cell line. Aquaculture, 1999, 175: 185–197

41. Zhang QY, Xiao F, Li ZQ, Gui JF. Comparison of sequence homology of major capsid protein gene of *Rananvirus* frog virus 3 and *Rana grylio* virus isolates from China. Chin J Virol, 2001, 17: 372–374

42. Zhang QY, Xiao F, Li ZQ, Gui JF, Mai J, Chinchar VG. Characterization of an iridovirus from the cultured pig frog (*Rana grylio*) with lethal syndrome. Dis Aquatic Organ, 2001, 48: 27–36

43. Zhang QY, Zhao Z, Xiao F, Li ZQ, Gui JF. Molecular characterization of three *Rana grylio*virus (RGV) isolates and *Paralichthys olivaceus* lymphocystis disease virus (LCDV-C) in iridoviruses. Aquaculture, 2006, 251: 1–10

44. Jancovich JK, Davidson EW, Morado JF, Jacobs BL, Collins JP. Isolation of a lethal virus from the endangered tiger salamander *Ambystoma tigrinum stebbinsi*. Dis Aquat Org, 1997, 31: 161–167

45. Jancovich JK, Mao J, Chinchar VG, Wyatt C, Case ST, Kumar S, Valente G, Subramanian S, Davidson EW, Collins JP, Jacobs BL. Genomic sequence of a ranavirus (family Iridoviridae) associated with salamander mortalities in North America. Virology, 2003, 316: 90–103

46. Qin QW, Lam TJ, Sin YM, Shen H, Chang SF, Ngoh GH, Chen CL. Electron microscopic observations of a marine fish iridovirus isolated from brown-spotted grouper, *Epinephelus tauvina*. J Virol Method, 2001, 98: 17–24

47. Song WJ, Qin QW, Qiu J, Huang CH, Wang F, Hew CL. Functional genomics analysis of Singapore grouper iridovirus: complete sequence determination and proteomic analysis. J Virol, 2004, 78: 12576–12590Central

48. Huang Y, Huang X, Liu H, Gong J, Ouyang Z, Cui H, Cao J, Zhao Y, Wang X, Jiang Y, Qin Q. Complete sequence determination of a novel reptile iridovirus isolated from soft-shelled turtle and evolutionary analysis of Iridoviridae. BMC Genomics, 2009, 10: 224Central

49. Tsai CT, Ting JW, Wu MH, Wu MF, Guo IC, Chang CY. Complete genome sequence of the grouper iridovirus and comparison of genomic organization with those of other iridoviruses. J Virol, 2005, 79: 2010–2023Central

50. He JG, Lu L, Deng M, He HH, Weng SP, Wang XH, Zhou SY, Long QX, Wang XZ, Chan SM. Sequence analysis of the complete genome of an iridovirus isolated from the tiger frog. Virology, 2002, 292: 185–197

51. Mavian C, López-Bueno A, Balseiro A, Casais R, Alcamí A, Alejo A. The genome sequence of the emerging common midwife toad virus identifies an evolutionary

intermediate within ranaviruses. J Virol, 2012b, 86: 3617–3625Central

52. Chen ZY, Gui JF, Gao XC, Pei C, Hong YJ, Zhang QY. Genome architecture changes and major gene variations of *Andrias davidianus* ranavirus (ADRV). Vet Res, 2013, 44: 101Central
53. Zhou ZY, Geng Y, Liu XX, Ren SY, Zhou Y. Characterization of a ranavirus isolated from the Chinese giant salamander (*Andrias davidianus*, Blanchard, 1871) in China. Aquaculture, 2013, 384: 66–73
54. Wang N, Zhang M, Zhang L, Jing H, Jiang Y, Wu S, Lin X. Complete genome sequence of a ranavirus isolated from Chinese giant salamander (*Andrias davidianus*). Genome Announc, 2014, 2: e01032–13Central
55. Li W, Zhang X, Weng S, Zhao G, He J, Dong C. Virion-associated viral proteins of a Chinese giant salamander (*Andrias davidianus*) iridovirus (genus *Ranavirus*) and functional study of the major capsid protein (MCP). Vet Microbl, 2014, 172: 129–139
56. Darai G, Delius H, Clarke J, Apfel H, Schnitzler P, Flügel RM. Molecular cloning and physical mapping of the genome of fish lymphocystis disease virus. Virology, 1985, 146: 292–301
57. Zhang QY, Ruan HM, Li ZQ, Yuan XP, Gui JF. Infection and propagation of lymphocystis virus isolated from the cultured flounder *Paralichthys olivaceus* in grass carp cell lines. Dis Aquatic Org, 2003, 57: 27–34
58. Xu L, Feng J, Huang Y. Identification of lymphocystis disease virus from paradise fish*Macropodus opercularis* (LCDV-PF). Arch Virol, 2014, 159: 2445–2449
59. Huang X, Huang Y, Xu L, Wei S, Ouyang Z, Feng J, Qin Q. Identification and characterization of a novel lymphocystis disease virus isolate from cultured grouper in China. J Fish Dis, 2014, 10, doi:10.1111/jfd.12244
60. He JG, Deng M, Weng SP, Li Z, Zhou SY, Long QX, Wang XZ, Chan SM. Complete genome analysis of the mandarin fish infectious spleen and kidney necrosis iridovirus. Virology, 2001, 291: 126–139
61. Jeong JB, Jun LJ, Yoo MH, Kim MS, Komisar JL, Jeong HD. Characterization of the DNA nucleotide sequences in the genome of red sea bream iridoviruses isolated in Korea. Aquaculture, 2003, 220: 119–133
62. Do JW, Moon CH, Kim HJ, Ko MS, Kim SB, Son JH, Kim JS, An EJ, Kim MK, Lee SK, Han MS, Cha SJ, Park MS, Park MA, Kim YC, Kim JW, Park JW. Complete genomic DNA sequence of rock bream iridovirus. Virology, 2004, 325: 351–363
63. Ao JQ, Chen XH. Identification and characterization of a novel gene encoding an RGD-containing protein in large yellow croaker iridovirus. Virology, 2006, 355: 213–222
64. Shi CY, Jia KT, Yang B, Huang J. Complete genome sequence of a megalocytivirus (family Iridoviridae) associated with turbot mortality in China. Virol J, 2010, 7: 159Central
65. Lü L, Zhou SY, Chen C, Weng SP, Chan SM, He JG. Complete genome sequence analysis of an iridovirus isolated from the orangespotted grouper, *Epinephelus coiodes*. Virology, 2005, 339: 81–100
66. Zhang BC, Zhang M, Sun BG, Fang Y, Xiao ZZ, Sun L. Complete genome sequence and transcription profiles of the rock bream iridovirus RBIV-C1. Dis Aquat Org, 2013, 104: 203–214
67. Eaton HE, Metcalf J, Penny E, Tcherepanov V, Upton C, Brunetti CR. Comparative genomic analysis of the family Iridoviridae: re-annotating and defining the core set of iridovirus genes. Virol J, 2007, 4: 11Central
68. Abrams AJ, Cannatella DC, Hillis DM, Sawyer SL. Recent host-shifts in ranaviruses: signatures of positive selection in the viral genome. J Gen Virol, 2013, 94: 2082–2093Central
69. Currylow AF, Johnson AJ, Williams RN. Evidence of ranavirus infections among sympatric larval amphibians and box turtles. J Herpet, 2014, 48:117–121
70. Sun W, Huang YH, Zhao Z, Gui JF, Zhang QY. Characterization of the *Rana grylio* virus 3β-hydroxysteroid dehydrogenase and its novel role in suppressing virus-induced cytopathic effect. Biochem Biophys Res Commun, 2006, 351: 44–50
71. Zhao Z, Ke F, Gui JF, Zhang QY. Characterization of an early gene encoding for dUTPase in*Rana grylio* virus. Virus Res, 2007, 123: 128–137
72. Zhao Z, Shi Y, Ke F, Sun W, Gui JF, Zhang QY. Constitutive expression of thymidylate synthase from LCDV-C induces foci formation and anchorage-independent growth in fish cells. Virology, 2008, 372: 118–126
73. Zhao Z, Ke F, Huang YH, Zhao JG, Gui JF, Zhang QY. Identification and characterization of a novel envelope protein in *Rana grylio* virus. J Gen Virol, 2008, 89: 166–1872
74. Whitley DS, Yu K, Sample RC, Sinning A, Henegar J, Norcross E, Chinchar VG. Frog virus 3 ORF 53R, a putative myristoylated membrane protein, is essential for virus replication *in vitro*. Virology, 2010, 405: 448–456
75. He LB, Gao XC, Ke F, Zhang QY. A conditional lethal mutation in *Rana grylio* virus ORF 53R resulted in a marked reduction in virion formation. Virus Res, 2013, 177: 194–200
76. Lei XY, Ou T, Zhang QY. *Rana grylio* virus (RGV) 50L is associated with viral matrix and exhibited two distribution patterns. PLoS One, 2012, 7: e43033Central
77. He LB, Ke F, Wang J, Gao XC, Zhang QY. *Rana grylio* virus (RGV) envelope protein 2L subcellular localization and essential roles in virus infectivity revealed by conditional lethal mutant. J Gen Virol, 2014, 95: 679–690
78. Chen G, Ward BM, Yu EK, Chinchar VG, Robert J. Improved knockout methodology reveals that Frog virus 3 mutants lacking either the 18K immediate-early gene or the truncated vIF-2 alpha gene are defective for replication *in vivo*. J Virol, 2011, 85: 11131–11138Central
79. Sample R, Bryan L, Long S, Majji S, Hoskins G, Sinning A, Olivier J, Chinchar VG. Inhibition of iridovirus protein synthesis and virus replication by antisense morpholino oligonucleotides targeted to the major capsid protein, the

18 kDa immediate-early protein, and a viral homolog of RNA polymerase II. Virology, 2007, 358: 311–320

80. Kim YS, Ke F, Lei XY, Zhu R, Zhang QY. Viral envelope protein 53R gene highly specific silencing and iridovirus resistance in fish cells by a miRNA. PLoS One, 2010, 5: e10308Central
81. Majji S, Thodima V, Sample R, Whitley D, Deng Y, Mao J, Chinchar VG. Transcriptome analysis of Frog virus 3, the type species of the genus *Ranavirus*, family Iridoviridae. Virology, 2009, 391: 293–303Central
82. Guo CJ, Liu D, Wu YY, Yang XB, Yang LS, Mi S, Huang YX, Luo YW, Jia KT, Liu ZY, Chen WJ, Weng SP, Yu XQ, He JG. Entry of tiger frog virus (an Iridovirus) into HepG2 cells via a pH-dependent, atypical, caveola-mediated endocytosis pathway. J Virol, 2011, 85: 6416–6426Central
83. Yan Y, Cui H, Jiang S, Huang Y, Huang X, Wei S, Xu W, Qin Q. Identification of a novel marine fish virus, Singapore grouper iridovirusencoded microRNAs expressed in grouper cells by Solexa sequencing. PLoS One, 2011, 6: e19148Central
84. Cullen BR. microRNAs as mediators of viral evasion of the immune system. Nat Imm, 2013, 14: 205–210
85. Qiu T, Lu RH, Zhang J, Zhu ZY. Complete nucleotide sequence of the S10 genome segment of grass carp reovirus (GCRV). Dis Aquat Org, 2001, 44: 69–74
86. Zhang QY, Ruan HM, Li ZQ, Zhang J, Gui JF. Detection of grass carp hemorrhage virus (GCHV) from Vietnam and comparison with GCHV strain from China. High Technol Lett, 2003, 9: 7–13
87. Wang Q, Zeng W, Liu C, Zhang C, Wang Y, Shi C, Wu S. Complete genome sequence of a reovirus isolated from grass carp, indicating different genotypes of GCRV in China. J Virol, 2012, 86: 12466Central
88. Ye X, Tian YY, Deng GC, Chi YY, Jiang XY. Complete genomic sequence of a reovirus isolated from grass carp in China. Virus Res, 2012, 163: 275–283
89. Fan Y, Rao S, Zeng L, Ma J, Zhou Y, Xu J, Zhang H. Identification and genomic characterization of a novel fish reovirus, Hubei grass carp disease reovirus, isolated in 2009 in China. J Gen Virol, 2013, 94: 2266–2277
90. Pei C, Ke F, Chen ZY, Zhang QY. Complete genome sequence and comparative analysis of grass carp reovirus strain 109 (GCReV-109) with other grass carp reovirus strains reveals no significant correlate regional distribution. Arch Virol, 2014, 159: 2435–2440
91. Jaafar FM, Goodwin AE, Belhouchet M, Merry G, Fang Q, Cantaloube JF, Biagini P, de Micco P, Mertens PP, Attoui H. Complete characterisation of the American grass carp reovirus genome (genus *Aquareovirus*: family Reoviridae) reveals an evolutionary link between aquareoviruses and coltiviruses, Virology, 2008, 373: 310–321
92. Yan XY, Wang Y, Xiong LF, Jian JC, Wu ZH. Phylogenetic analysis of newly isolated grass carp reovirus. SpringerPlus, 2014, 3: 190Central
93. Attoui H, Fang Q, Mohd Jaafar F, Cantaloube JF, Biagini P, de Micco P, de Lamballerie X. Common evolutionary origin of aquareoviruses and orthoreoviruses revealed by genome characterization of Golden shiner reovirus, Grass carp reovirus, Striped bass reovirus and golden ide reovirus (genus *Aquareovirus*, family Reoviridae). J Gen Virol, 2002, 83: 1941–1951
94. Ke F, He LB, Pei C, Zhang QY. Turbot reovirus (SMReV) genome encoding a FAST protein with a non-AUG start site. BMC Genomics, 2011, 12: 323Central
95. Kibenge MJT, Iwamoto O, Wang Y, Morton A, Godoy MG, Kibenge FSB. Whole-genome analysis of piscine reovirus (PRV) shows PRV represents a new genus in family Reoviridae and its genome segment S1 sequences group it into two separate sub-genotypes. Virol J, 2013, 10: 230Central
96. Key T, Read J, Nibert ML, Duncan R. Piscine reovirus encodes a cytotoxic, non-fusogenic, integral membrane protein and previously unrecognized virion outer-capsid proteins. J Gen Virol, 2013, 94: 1039–1050
97. Nibert ML, Duncan R. Bioinformatics of recent aqua- and orthoreovirus isolates from fish: evolutionary gain or loss of FAST and fiber proteins and taxonomic implications. PLoS One, 2013, 8: e68607Central
98. Garseth ÅH, Ekrem T, Biering E. Phylogenetic evidence of long distance dispersal and transmission of piscine reovirus (PRV) between farmed and wild Atlantic salmon. PLoS One, 2013, 8: e82202Central
99. Zhang X, Jin L, Fang Q, Hui WH, Zhou ZH. 3.3 Å cryo-EM structure of a nonenveloped virus reveals a priming mechanism for cell entry. Cell, 2010, 141: 472–482Central
100. Trask SD, McDonald SM, Patton JT. Structural insights into the coupling of virion assembly and rotavirus replication. Nat Rev Microbiol, 2012, 10: 165–177Central
101. Ke F, He LB, Zhang QY. Nonstructural protein NS80 is crucial in recruiting viral components to form aquareoviral factories. PLoS One, 2013, 8: e63737Central
102. Hoffmann B, Beer M, Schutze H, Mettenleiter TC. Fish rhabdoviruses: molecular epidemiology and evolution. Curr Top Microb Imm, 2005, 292: 81–117
103. Dietzgen RG, Kuzmin IV, eds. Rhabdoviruses: molecular taxonomy, evolution, genomics, ecology, host-vector interactions, cytopathology and control. Norfolk: Caister Academic Press, 2012
104. Tao JJ, Zhou GZ, Gui JF, Zhang QY. Genomic sequence of mandarin fish rhabdovirus with an unusual small non-transcriptional ORF. Virus Res, 2008, 132: 86–96
105. Zhang QY, Tao JJ, Gui L, Zhou GZ, Ruan HM, Li ZQ, Gui JF. Isolation and characterization of *Scophthalmus maximus* (turbot) rhabdovirus. Dis Aquat Org, 2007, 74: 95–105
106. Zhu RL, Lei XY, Ke F, Yuan XP, Zhang QY. Genome of turbot rhabdovirus exhibits unusual non-coding regions and an additional ORF that could be expressed in fish cell. Virus Res, 2011, 155: 495–505
107. Zhu RL, Zhang QY. Determination and analysis of the complete genome sequence of*Paralichthys olivaceus* rhabdovirus (PORV). Arch Virol, 2014, 159:

817–820

108. Ou T, Zhu RL, Chen ZY, Zhang QY. Isolation and identification of a lethal rhabdovirus from farmed rice field eels *Monopterus albus*. Dis Aquat Org, 2013, 106: 197–206

109. Ahne W, Jorgensen PEV, Olesen NJ. Serological examination of a rhabdovirus isolated from snakehead fish (*Ophicephalus striatus*) in Thailand with ulcerative syndrome. J Appl Ichthyol, 1988. 4:194–196

110. Zeng W, Wang Q, Wang Y, Liu C, Liang H, Fang X, Wu S. Genomic characterization and taxonomic position of a rhabdovirus from a hybrid snakehead. Arch Virol, 2014, 159: 2469–2473

111. Kim DH, Oh HK, Eou JI, Seo HJ, Kim SK, Oh MJ, Nam SW, Choi TJ. Complete nucleotide sequence of the hirame rhabdovirus, a pathogen of marine fish. Virus Res, 2005, 107: 1–9

112. Chen HL, Liu H, Liu ZX, He JQ, Gao LY, Shi XJ, Jiang YL. Characterization of the complete genome sequence of pike fry rhabdovirus. Arch Virol, 2009, 154: 1489–1494

113. Teng Y, Liu H, Lv J, Fan WH, Zhang QY, Qin QW. Characterization of complete genome sequence of the spring viremia of carp virus isolated from common carp (*Cyprinus carpio*) in China. Arch Virol, 2007, 152: 1457–1465

114. Talbi C, Cabon J, Baud M, Bourjaily M, de Boisséson C, Castric J, Bigarré L. Genetic diversity of perch rhabdoviruses isolates based on the nucleoprotein and glycoprotein genes. Arch Virol, 2011, 156: 2133–2144

115. Gadd T. Fish rhabdoviruses: viral haemorrhagic septicaemia virus (VHSV) and perch rhabdovirus (PRV): study of viral strains and the disease epidemiology in Finland. Helsinki, 2013

116. Schütze H, Enzmann PJ, Kuchling R, Mundt E, Niemann H, Mettenleiter TC. Complete genomic sequence of the fish rhabdovirus infectious haematopoietic necrosis virus. J Gen Virol, 1995, 76: 2519–2527

117. Pierce LR, Stepien CA. Evolution and biogeography of an emerging quasispecies: diversity patterns of the fish Viral Hemorrhagic Septicemia virus (VHSv). Mol Phylogenet Evol, 2012, 63: 327–341

118. He M, Yan XC, Liang Y, Sun XW, Teng CB. Evolution of the viral hemorrhagic septicemia virus: divergence, selection and origin. Mol Phylogenet Evol, 2014, 77: 34–40

119. Kurath G. Fish novirhabdoviruses. In: Dietzgen RG et al., eds. Rhabdoviruses: Molecular Taxonomy, Evolution, Genomics, Ecology, Host-Vector Interactions, Cytopathology and Control. Norfolk: Caister Academic Press, 2012. 89–116

120. Kolodziejek J, Schachner O, D'Lirrwald R, Durrwald R, Latif M, Nowotny N. "Mid-G" region sequences of the glycoprotein gene of Austrian infectious hematopoietic necrosis virus isolates form two lineages within European isolates and are distinct from American and Asian lineages. J Clin Microbiol, 2008, 46: 22–30Central

121. Miller O, Fuller FJ, Gebreyes WA, Miller O, Fuller FJ, Gebreyes WA, Lewbart GA, Shchelkunov IS, Shivappa RB, Joiner C, Woolford G, Stone DM, Dixon PF, Raley ME, Levine JF. Phylogenetic analysis of spring virema of carp virus reveals distinct subgroups with common origins for recent isolates in North America and the UK. Dis Aquat Org, 2007, 76: 193–204

122. Stone DM, Kerr RC, Hughes M, Radford AD, Darby AC. Characterisation of the genomes of four putative vesiculoviruses: tench rhabdovirus, grass carp rhabdovirus, perch rhabdovirus and eel rhabdovirus European X. Arch Virol, 2013, 158: 2371–2377

123. Xiao Y, Shao L, Zhang C, An W. Genomic evidence of homologous recombination in spring viremia of carp virus: a negatively single stranded RNA virus. Virus Res, 2014, 189: 271–279

124. Einer-Jensen K, Harmache A, Biacchesi S, Bremont M, Stegmann A, Lorenzen N. High virulence differences among phylogenetically distinct isolates of the fish rhabdovirus viral hemorrhagic septicaemia virus are not explained by variability of the surface glycoprotein G or the non-virion protein Nv. J Gen Virol, 2014, 95: 307–316

125. Kell A, Wargo AR, Kurath G. The role of virulence in *in vivo* superinfection fitness of a vertebrate RNA virus, infections hematopoietic necrosis virus. J Virol, 2013, 87: 8145–8157Central

126. Hanson L, Dishon A, Kotler M. Herpesviruses that infect fish. Viruses, 2011, 3: 2160–2191Central

127. van Beurden SJ, Forlenza M, Westphal AH, Wiegertjes GF, Haenen OL, Engelsma MY. The alloherpesviral counterparts of interleukin 10 in European eel and common carp. Fish Shellfish Imm, 2011, 31: 1211–1217

128. Davison AJ, Kurobe T, Gatherer D, Cunningham C, Korf I, Fukuda H, Hedrick RP, Waltzek TB. Comparative genomics of carp herpesviruses. J Virol, 2013, 87: 2908–2922Central

129. Lovy J, Friend SE. Cyprinid herpesvirus-2 causing mass mortality in goldfish: applying electron microscopy to histological samples for diagnostic virology. Dis Aquat Org, 2014, 108: 1–9

130. Xu J, Zeng L, Zhang H, Zhou Y, Ma J, Fan YD. Cyprinid herpesvirus 2 infection emerged in cultured gibel carp, *Carassius auratus gibelio* in China. 2 infection emerged in cultured gibel carp, *Carassius auratus gibelio* in China. Vet Microb, 2013, 166: 138–144

131. Costes B, Raj VS, Michel B, Fournier G, Thirion M, Gillet L, Mast J, Lieffrig F, Bremont M, Vanderplasschen A. The major portal of entry of koi herpesvirus in *Cyprinus carpio* is the skin. J Virol, 2009, 83: 2819–2830Central

132. Eide KE, Miller-Morgan T, Heidel JR, Kent ML, Bildfella RJ, LaPatra S, Watson G, Jin L. Investigation of koi herpesvirus latency in koi. J Virol, 2011, 85: 4954–4962Central

133. Reed AN, Izume S, Dolan BP, LaPatra S, Kent M, Dong J, Jin L. Identification of B cells as a major site for cyprinid herpesvirus 3 latency. J Virol, 2014, 88: 9297–309Central

134. Gotesman M, Kattlun J, Bergmann SM, El-Matbouli M.

CyHV-3: the third cyprinid herpesvirus. Dis Aquat Org, 2013, 105: 163–174Central

135. Dong C, Weng S, Li W, Li X, Yi Y, Liang Q, He J. Characterization of a new cell line from caudal fin of koi, Cyprinus carpio koi, and first isolation of cyprinid herpesvirus 3 in China. Virus Res, 2011, 161: 140–149

136. Aoki T, Hirono I, Kurokawa K, Fukuda H, Nahary R, Eldar A, Davison AJ, Waltzek TB, Bercovier H, Hedrick RP. Genome sequencing of three koi herpesvirus isolates representing the expanding distribution of an emerging disease threatening koi and common carp worldwide. J Virol, 2007, 81: 5058–506Central

137. Kim HJ, Kwon SR. Evidence for two koi herpesvirus (KHV) genotypes in South Korea. Dis Aquat Org, 2013, 104: 197–202

138. Davison AJ, Kurobe T, Gatherer D, Cunningham C, Korf I, Fukuda H, Hedrick RP, Waltzek TB. Comparative genomics of carp herpesviruses. J Virol, 2013, 87:2908–2922Central

139. Davison AJ. Channel catfish virus: a new type of herpesvirus. Virology, 1992, 186: 9–14

140. van Beurden SJ, Bossers A, Voorbergen-Laarman MH, Haenen OL, Peters S, Abma-Henkens MH, Peeters BP, Rottier PJ, Engelsma MY. Complete genome sequence and taxonomic position of anguillid herpesvirus 1. J Gen Virol, 2010, 91 (Pt 4): 880–887

141. van Beurden SJ, Gatherer D, Kerr K, Galbraith J, Herzyk P, Peeters BP, Rottier PJ, Engelsma MY, Davison AJ. Anguillid herpesvirus 1 transcriptome. J Virol, 2012, 86: 10150–10161Central

142. van Beurden SJ, Peeters BP, Rottier PJ, Davison AJ, Engelsma MY. Genome-wide gene expression analysis of anguillid herpesvirus-1. BMC Genomics, 2013, 14: 83Central

143. Ackermann M, Koriabine M, Hartmann-Fritsch F, de Jong PJ, Lewis TD, Schetle N, Work TM, Dagenais J, Balazs GH, Leong JA. The genome of chelonid herpesvirus 5 harbors atypical genes. PLoS One, 2012, 7: e46623Central

144. Ren W, Chen H, Renault T, Cai Y, Bai C, Wang C, Huang J. Complete genome sequence of acute viral necrosis virus associated with massive mortality outbreaks in the Chinese scallop, *Chlamys farreri*. Virol J, 2013, 10:1–10

145. Ito T, Maeno Y. Susceptibility of Japanese Cyprininae fish species to cyprinid herpesvirus 2. Vet Microb, 2014, 169: 128–134

146. Marcos-Lopez M, Waltzek TB, Hedrick RP, Baxa DV, Garber AF, Liston R, Johnsen E, Backman S, Ferguson HW. Characterization of a novel alloherpesvirus from Atlantic cod (*Gadus morhua*). J Vet Diagn Invest, 2012, 24: 65–73

147. Zhang H, Zeng L, Fan Y, Zhou Y, Xu J, Ma J. A Loop-mediated isothermal amplification assay for rapid detection of cyprinid herpesvirus 2 in gibel carp (*Carassius auratus gibelio*). Sci World J, 2014, 2014: 716413

148. Xu L, Podok P, Xie J, Lu L. Comparative analysis of differential gene expression in kidney tissues of moribund and surviving crucian carp (*Carassius auratus gibelio*) in response to cyprinid herpesvirus 2 infection. Arch Virol, 2014, 159: 1961–1974

149. Sunarto A, McColl KA, Crane MS, Schat KA, Slobedman B, Barnes AC, Walker PJ. Characteristics of cyprinid herpesvirus 3 in different phases of infection: implications for disease transmission and control. Virus Res, 2014, 188: 45–53

150. Yi Y, Zhang H, Lee X, Weng S, He J, Dong C. Extracellular virion proteins of two Chinese CyHV-3/KHV isolates, and identification of two novel envelope proteins. Virus Res, 2014, 191C: 108–116

151. Sunyer JO. Fishing for mammalian paradigms in the teleost immune system. Nat Imm, 2013, 14: 320–326

152. Sun JC, Ugolini S, Vivier E. Immunological memory within the innate immune system. EMBO J, 2014, 33: 1295–1303

153. Adamek M, Steinhagen D, Irnazarow I, Hikima JI, Jung TS, Aoki T. Biology and host response to Cyprinid herpesvirus 3 infection in common carp. Dev Comp Immunol, 2014, 43: 151–159

154. Munang'andu HM, Mutoloki S, Evensen Ø. Acquired immunity and vaccination against infectious pancreatic necrosis virus of salmon. Dev Comp Immunol, 2014, 43: 184–196

155. Somamoto T, Koppang EO, Fischer U. Antiviral functions of CD8+ cytotoxic T cells in teleost fish. Dev Comp Immunol, 2014, 43: 197–204

156. Zhang YB, Gui JF. Molecular regulation of interferon antiviral response in fish. Dev Comp Immunol, 2012, 38: 193–202

157. Wang L, Qiu L, Zhou Z, Song L. Research progress on the mollusc immunity in China. Dev Comp Immunol, 2013, 39: 2–10

158. Li F, Xiang J. Recent advances in researches on the innate immunity of shrimp in China. Dev Comp Immunol, 2013, 39: 11–26

159. Wang XW, Wang JX. Diversity and multiple functions of lectins in shrimp immunity. Dev Comp Immunol, 2013, 39: 27–38

160. Zhu LY, Nie L, Zhu G, Xiang LX, Shao JZ. Advances in research of fish immune-relevant genes: a comparative overview of innate and adaptive immunity in teleosts. Dev Comp Immunol, 2013, 39: 39–62

161. Peng XX. Proteomics and its applications to aquaculture in China: infection, immunity, and interaction of aquaculture hosts with pathogens. Dev Comp Immunol, 2013, 39: 63–71

162. Zhang S, Wang Z, Wang H. Maternal immunity in fish. Dev Comp Immunol, 2013, 39: 72–78

163. Esteban MÁ. An overview of the immunological defenses in fish skin. ISRN Imm, 2012, Article ID 853470, 1-29

164. Xu Z, Parra D, Gómez D, Salinas I, Zhang YA, von Gersdorff Jørgensen L, Heinecke RD, Buchmann K, LaPatra S, Sunyer JO. Teleost skin, an ancient mucosal surface that elicits gut-like immune responses. Proc Natl Acad Sci USA, 2013, 110: 13097–13102Central

165. Bordon Y. Evolution: a gutsy defence of the skin. Nat Rev Immunol, 2013, 13: 616–617

166. Pasparakis M, Haase I, Nestle FO. Mechanisms regulating skin immunity and inflammation. Nat Rev Immunol, 2014, 14: 289–301

167. Collet B. Innate immune responses of salmonid fish to viral infections. Dev Comp Immunol, 2014, 43: 160–173

168. Castro R, Jouneau L, Pham HP, Bouchez O, Giudicelli V, Lefranc MP, Quillet E, Benmansour A, Cazals F, Six A, Fillatreau S, Sunyer O, Boudinot P. Teleost fish mount complex clonal IgM and IgT responses in spleen upon systemic viral infection. PLoS Pathog, 2013, 9: e1003098Central

169. Wang T, Secombes CJ. The cytokine networks of adaptive immunity in fish. Fish Shellfish Immunol, 2013, 35: 1703–1718

170. Rombout JH, Yang G, Kiron V. Adaptive immune responses at mucosal surfaces of teleost fish. Fish Shellfish Immunol, 2014, pii: S1050-4648 (14)00305-2

171. Ye J, Kaattari IM, Ma C, Kaattari S. The teleost humoral immune response. Fish Shellfish Immunol, 2013, 35: 1719–1728

172. Grayfer L, Andino Fde J, Chen G, Chinchar GV, Robert J. Immune evasion strategies of ranaviruses and innate immune responses to these emerging pathogens. Viruses, 2012, 4: 1075–1092Central

173. Chen ZY, Lei XY, Zhang QY. The antiviral defense mechanisms in mandarin fish induced by DNA vaccine against a rhabdovirus. Vet Microbiol, 2012, 157: 264–275

174. Shi Y, Zhao Z, Zhu XP, Chen KC, Zhang QY. Expression and functional characterization of a gene associated with retinoid-interferon-induced mortality 19 (GRIM-19) from orange-spotted grouper (*Epinephelus coioides*). Fish Shellfish Imm, 2013, 34: 273–279

175. Zhu R, Wang J, Lei XY, Gui JF, Zhang QY. Evidence for *Paralichthys olivaceus* IFITM1 antiviral effect by impeding viral entry into target cells. Fish Shellfish Immunol, 2013, 35: 918–926

176. Zhu R, Chen ZY, Wang J, Yuan JD, Liao XY, Gui JF, Zhang QY. Thymus cDNA library survey uncovers novel features of immune molecules in Chinese giant salamander (*Andrias davidianus*). Dev Comp Immunol, 2014, 46: 413–422

177. Sun F, Zhang YB, Liu TK, Gan L, Yu FF, Liu Y, Gui JF. Characterization of fish IRF3 as an IFN-inducible protein reveals evolving regulation of IFN response in vertebrates. J Immunol, 2010, 185: 7573–7582

178. Sun F, Zhang YB, Liu TK, Shi J, Wang B, Gui JF. Fish MITA activation serves as a mediator for distinct fish IFN gene activation dependent on IRF3 or IRF7. J Immunol, 2011, 187: 2531–2539

179. Liu TK, Zhang YB, Liu Y, Sun F, Gui JF. Cooperative roles of fish PKZ and PKR in IFN-mediated antiviral response. J Virol, 2011, 85: 12769–12780

CHAPTER 11

EFFICIENCY OF A HYBRID SOLID DIGESTION-DENITRIFICATION COLUMN IN SUSPENDED SOLID AND NITRATE REMOVAL FROM RECIRCULATING AQUACULTURE SYSTEM.

Wiboonluk Pungrasmi, Ratchadaporn Chaisri, Ekachai Malaphol, Sorawit Powtongsook

Keywords: Denitrification, Hybrid column, Nitrate, Recirculating aquaculture system, Sedimentation, Solid digestion

INTRODUCTION

Recently, aquaculture production has grown rapidly as a response to the demand of good quality protein food. Intensive aquaculture using the recirculating aquaculture system (RAS) is therefore the promising manner of the future for sustainable food production. With the RAS, fishes were grown at high density to produce large quantities in limited space. In addition, RAS reduces water consumption and waste discharge in land-based aquaculture where as the appropriate technologies are required to maintain good water quality in the pond [1]. However, the major problems of RAS are the gathering of suspended solids derived from fecal, uneaten feed and the accumulation of inorganic nitrogen compounds such as ammonia, nitrite and nitrate in the water [2]. Suspended solid can deposit at the bottom of the pond and decomposed by either aerobic or anaerobic microorganisms. As a result, hydrogen sulfide produced from the anaerobic sulfate reduction can harm the aquatic animals and even cause the mass mortality. Thus, it is necessary to control suspended solid within the optimum concentration [3]. The successful recommended ways to remove excess suspended solids from fish tank can be classified into three techniques which are gravity separation, filtration and flotation [4].

In the aquaculture pond, the accumulation of inorganic nitrogen compounds as ammonia, nitrite and nitrate is the results from the biological degradation of animal excretion and the uneaten feed. Nitrogen wastes have a deleterious impact on water quality and fish growth [5]. Typically, RAS must retain ammonia concentration less than 0.025 mg-N/L by nitrification process in which ammonia is oxidized to nitrite and nitrate, respectively. Likewise, accumulation of nitrate in the RAS to the concentration higher than 50 mg-N/L may result in stress of aquatic animal and decrease food intake [6]. Therefore, reducing nitrate to nitrogen gas by denitrification process is necessary for long-term water recycle [7]. The effectiveness of denitrification process depends on various factors such as temperature, pH, dissolved oxygen (DO), nitrite concentration [8], and also the amount of organic carbon substrate that available in the system [9, 10]. Among these factors, many previous reports were specified that organic carbon substrate is the major significant in aquaculture denitrification process [11-14]. In this research, we aim to develop the alternative approach to remove the suspended

Citation: Wiboonluk Pungrasmi, Ratchadaporn Chaisri, Ekachai Malaphol, Sorawit Powtongsook, Efficiency of a hybrid solid digestion-denitrification column in suspended solid and nitrate removal from recirculating aquaculture system., DOI : 10.4491/eer.2014.S1.011.

solids from indoor recirculating aquaculture pond by simple gravity sedimentation process. Since sedimentation generally requires low energy input and easy to operate. Moreover, instead of discarding the organic waste solids that remaining in the up-flow sedimentation column, we tried to degrade those particulate organic matters through anaerobic decomposition. The product yielded can be used as the sole carbon source and electron donors for biological denitrification without further need for the extra carbon supplementation as normally required in the aquaculture system [15, 16]. The treated water will have lower concentration of nitrate, in which can be safely recycled back to the aquaculture pond.

MATERIALS AND METHODS

The experimental system comprised of 3 equipment parts as (1) a fish tank, (2) an equalization tank and (3) a hybrid solid digestion-denitrification column as indicated in Fig. 1. The fish tank was a plastic container of 3.5 m in diameter and the total volume was 3,000 L. The Tilapia, Oreochromis niloticus, was cultivated at high density (17 kg/m^3). The fishes were fed with an artificial feed at 2% of the total body weight per day. The system was operated in a completely closed system without water discharge. The equalization tank was a 500 L plastic tank with the working volume of 430 L. Wastewater from Tilapia tank was pumped into the equalization tank then mixed well by air bubble and aquarium pump in order to prevent suspended solid sedimentation and sludge decomposition. The hybrid solid digestion-denitrification column was an up-flow suspended solid separation unit which made from a transparency acrylic tube with the working volume of 20 L. The unit consisted of 2 layers of acrylic tubes. The outer tube was 17 cm in diameter and 100 cm height. The inner tube (down-comer tube) diameter was 5.4 cm with shorter height of 97 cm. Water from the equalization tank was pumped into the inner tube. The water flowed down to the bottom of the column and rose up in the outer tube before returned back to the equalization tank. Thus the heavy solids were accumulated at the bottom of the inner separation unit.

The Effect of Flow Rate on Solid Removal

As shown in Fig. 1, the hybrid column was connected with the equalization tank. In this experiment, this column reactor will perform as a solid separation unit. The water from the fish pond was pumped into the equalization tank and the water from the equalization was moved to the separation unit by a submersible pump and then returned back into the equalization tank. The flow rate of the water pass through the solid separation unit was set up and maintained by 3 ball valves. With the settling velocity of suspended solids from fish pond is about 2.55 m/h5 , in this process, the flow rate was therefore varied with 5 levels including: 11.25, 25.71, 60, 105.88 and 360 L/h. These flow rates provided the different upflow velocity in the sedimentation column as 0.57, 1.29, 3.02, 5.33 and 18.13 m/h, respectively. The system was operated for 2 days for each flow rate. The water influent and effluent were collected 10 times/d as every hour interval for suspended solid analysis according to the standard method [17].

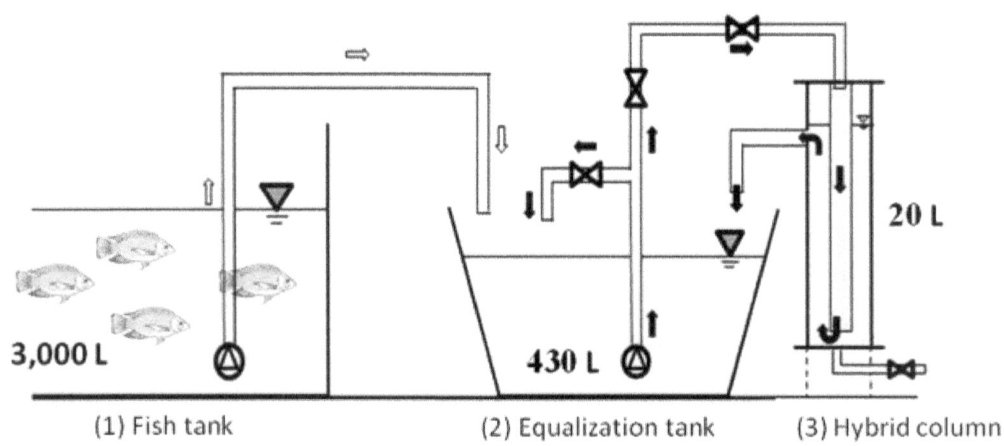

Figure 1. The experimental system for the suspended solid separation.

The Optimal Conditions for Denitrification Process

The similar procedures were performed as in the first experiment; wastewater with high suspended solids from the fish pond was pumped into the equalization tank. Sodium nitrate was added in this tank then adjusts to the final concentration of 100 mg-N/L in a total volume of 430 L. The well-mixed wastewater was transferred to the column reactor by a submersible pump with the optimal flow rate of the up-flow sedimentation column from the previous experiment. The experiment was operated with the optimum hydraulic retention time in a reactor by water from the equalization tank. Methanol was added as an external organic carbon source using a peristaltic pump. The oxidation-reduction potential (ORP) probe was used to measure and control the ORP in the reactor not less than -150 mV (Fig. 2). The 7 variable of COD:N ratios as 0.5:1, 1:1, 2:1, 3:1, 4:1, 5:1 and 6:1 were investigated and compare with a non-methanol addition system. The influent and effluent water were collected everyday for the water quality analysis (ammonia, nitrite, nitrate, alkalinity, pH, dissolved oxygen and temperature) according to the standard method [17]. Ammonia and nitrite concentrations were analyzed using a colorimetric method, while the nitrate concentration was measured with a spectrophotometric screening method according to APHA, 1992 [18]. Alkalinity was measured using a test kit (Aquatic Animal Medicine Division, Department of

Veterinary Medicine, Chulalongkorn University, Thailand) and residual methanol in the water from the denitrification tank on the last day was analyzed by gas chromatography (GC-2010; Shimadzu, Japan). Other parameters, such as pH, dissolved oxygen and temperature were measured using a pH meter (HI98240; Hanna Instruments, USA), DO meter (HI964400; Hanna Instruments, USA) and thermometer, respectively

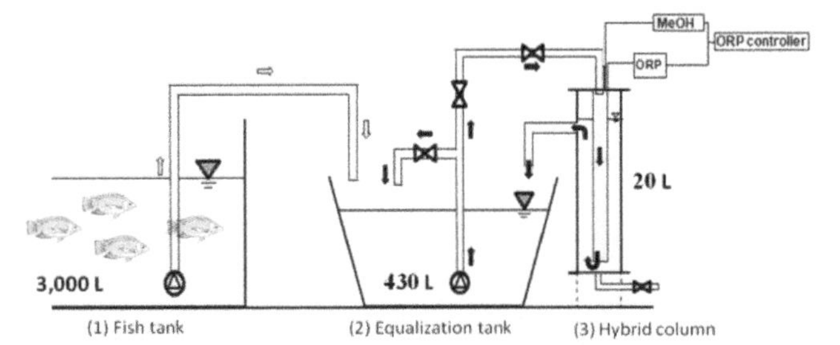

Figure 2. The experimental system for denitrification activity determination.

RESULTS AND DISCUSSION

The Effect of Flow Rate on Solid Removal

The solid removal efficiencies of the 2-days experiment for 5 upflow velocities at 0.57, 1.29, 3.02, 5.33 and 18.13 m/h are illustrated in Fig. 3. It was found that the slowest upflow velocity of 0.57 m/h (11.25 L/h) gave the highest solid removal efficiency of 72.72% ± 8.24% (Fig. 4). With low surface loading rate, small particles could be gravitated according to the Stokes' law. The settling velocity was controlled by diameter of the particle, dynamic viscosity, density of water and density of the particle. The result was comparable with the previous research by Nurit [19] that the higher efficiency of solid separation was occurred at the lower flow rate.

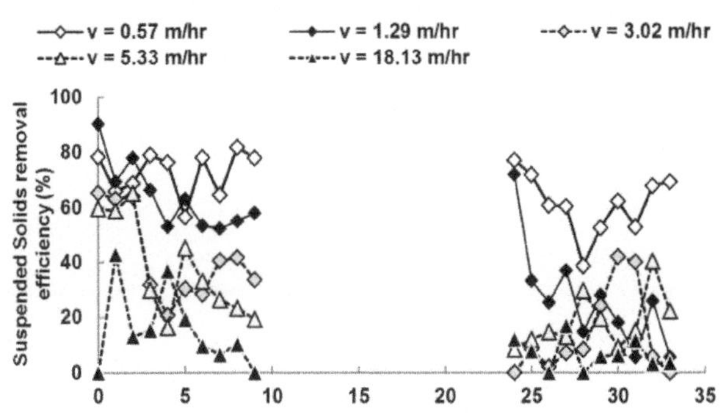

Figure 3. The solids removal efficiency of 5 variable upflow velocities for a 2 days operated system.

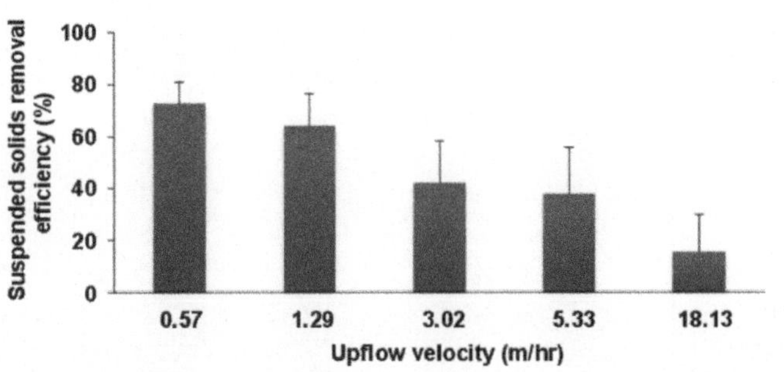

Figure 4. The average of solids removal efficiency during the first 10 hours at each upflow velocity.

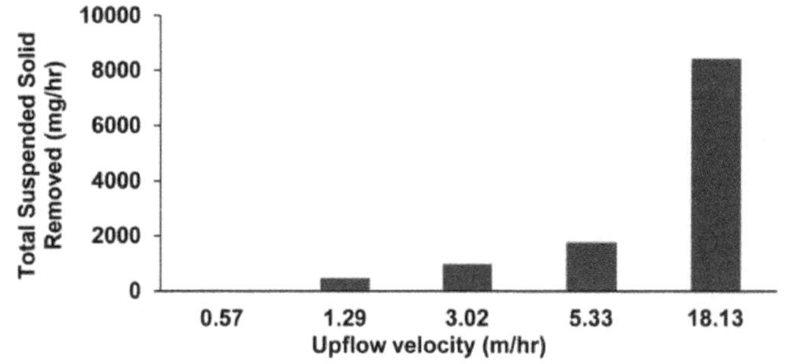

Figure 5. The average of total suspended solid (TSS) removal rate during the first 10 hours at each upflow velocity.

However, when considering the solid removal rate, it can be seen that the upflow velocity of 18.13 m/h (360 L/h) gave the highest total suspended solid (TSS) removal up to 8,439.84 mg/h (Fig. 5). The flow rate of 11.25 L/h, on the other hand, gave the highest efficiency but the total solids removed were only 26.93 mg/h. As a result, the upflow velocity of 18.13 m/h was therefore selected and applied for further studies in order to develop the solid separation device for aquaculture propose.

The Optimal Conditions for Denitrification Process

The effect of methanol addition and the COD:N ratio on nitrate removal efficiency by denitrification process in a hybrid solid digestion-denitrification column was evaluated. In this experiment, the up-flow velocity as 1.01 of m/h was controlled in order to prolong the hydraulic retention time of aquaculture wastewater in a reactor to about 1 hour. The HRT period of 3.33 minutes from the upflow velocity of 18.13 m/h was too short to sustain the biological digestion process. Moreover, this duration was not enough for denitrifiers to achieve the complete denitrification activity in the solid digestion column. The result in Fig. 6 indicated the alteration of nitrate concentration in the influent and effluent at each COD:N ratio. It was found that the nitrate concentration of both influent and effluent from a control reactor (a non-methanol addition system) was quite stable during the whole experiment. There was less different in nitrate concentration between the inlet and outlet water, which referred as no nitrate removal in this column reactor. The average nitrate concentration in the water inlet and outlet were constant as 92.56 ± 3.88 and 93.08 ± 3.80 mg-N/L, respectively. While in the test systems with the COD:N ratios of 0.5:1, 1:1, 2:1, 3:1, 4:1, 5:1 and 6:1, the largely diverse in nitrate concentration of the influent and effluent were found. At the lower COD:N ratios of 0.5:1, the system required a longest period of 19 days to diminish the nitrogen concentration in the effluent to the safety level for aquaculture (below 30 mg-N/L). Whereas, this duration became shorter in the higher COD:N ratios which were 9 days, 9 days, 6 days, 6 days and 4 days of the COD:N ratios as 1:1, 2:1, 3:1, 4:1 and 5:1, with the shortest of 3 days for the ratio as 6:1. The results clearly implied that the addition of methanol in a hybrid solid digestion-denitrification column significantly accelerated the denitrification process, although the rate of nitrate removal depends on the amount of methanol added.

The results in Table 1 referred the diverse denitrification rates, nitrate removal efficiencies and the COD depletion among the 8 experimental conditions performed. It suggested that the digestion column with the COD:N ratios of 6:1 gave the maximum denitrification rate (9,182.88 mg-N/day), while the lowest denitrification (70.27 mg-N/day) was found in the control system without methanol addition. In this reactor, the nitrate removal efficiency was only 0.55% ± 1.17%. The highest nitrate removal efficiency of 33.32% ± 21.18% was illustrated in the reactor with the COD:N ratios of 3:1. It was slightly higher than 31.44% ± 18.44% in the reactor with the COD:N ratios of 6:1. The results emphasized the significant of external carbon source addition for nitrate removal activity in a solid digester column. Furthermore, the result was comparable with the previous research by Sutthisee [20] which found that the higher efficiency of nitrate removal was occurred after the adequate addition of methanol. If the carbon source was not sufficient, the intermediate compounds of the reaction such as nitrite and nitrous oxide will take place due to the incomplete denitrification. In the opposite way, if methanol was added in the excess amount, not only the high cost of the system operation, the toxic effect will interfere the living of aquatic animals. In general, excess methanol must be avoided in an aquaculture system. Kaviraj et al. [21] reported that a methanol concentration greater than 47.49 mg/L (equivalent to 71.23 mg-COD/L) affects the growth and maturation of aquatic animals while a high concentration of methanol e.g., 1,527.60 mg/L (2,291 mg-COD/L) can induce acute toxic effects in fish. Moreover, the high carbon addition in the system may induce the ammonia reproduction from the dissimilatory nitrate reduction to ammonium or DNRA process [22]. Therefore, we can conclude from all the results that

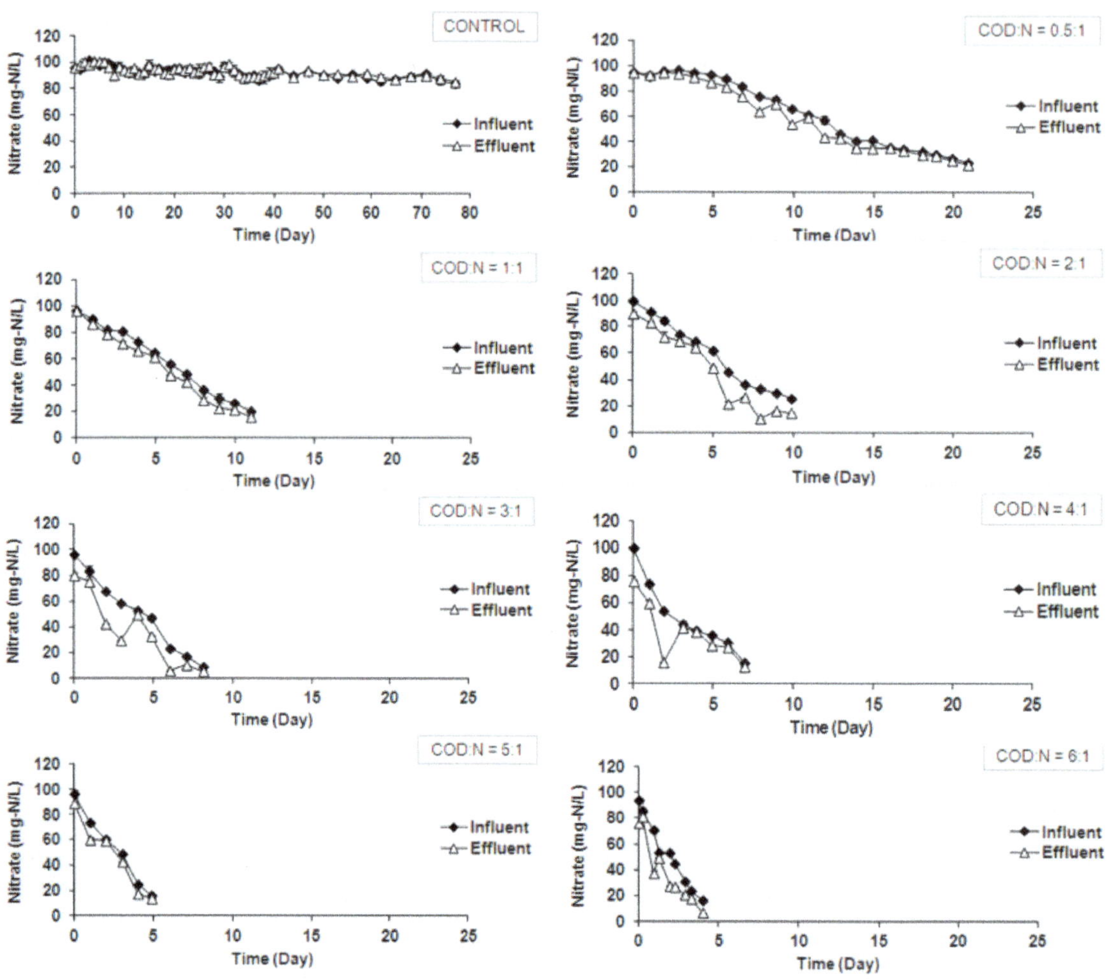

Figure 6. The alteration of nitrate concentration in the influent and effluent of each COD:N ratio.

Table 1. The nitrate Removal Rates and Efficiencies at Different COD:N Ratios

COD:N ratio	Denitrification rate (mg-N/day)	Nitrate removal efficiency (%)	The amount of methanol added (mg COD/day)	COD depletion (%)
No methanol addition	70.27	0.55 ± 1.17	-	-
0.5:1	1,933.10	7.47 ± 6.40	16.72	6.25
1:1	3,503.09	12.32 ± 8.17	29.2	26.82
2:1	3,806.93	27.27 ± 21.29	26.18	17.44
3:1	5,102.88	33.32 ± 21.18	39.46	-27.97
4:1	4,966.56	20.97 ± 21.29	41.84	-38.08
5:1	7,785.60	13.96 ± 9.42	59.78	-63.03
6:1	9,182.88	31.44 ± 18.44	81.63	-63.38

the proper COD:N ratio which promote the best denitrification activity in a solid digestion column was 3:1. The acceptable amount of excess methanol (as COD) was detected in the outlet water of the digestion column for this COD:N ratio. In comparison, this ratio was less than the number of 3.7:1 from stoichiometric reaction of nitrate removal using methanol as a carbon source [23]. It might be because a portion of the carbon source for the denitrification came from the digestion of the suspended solid retaining in the solid digestion column. Hence, the amount of external carbon source required could be reduced.

CONCLUSIONS

A single-reactor solid digestion-denitrification column can provide two successive performances of the suspended solid separation and nitrate removal. In the first step of solid removal from aquaculture pond using the simple gravity sedimentation column, the slowest upflow velocity of 0.57 m/h (11.25 L/h) gave the highest solid removal efficiency of 72.72% ± 8.24%. However, the total suspended solids removal was highest at 18.13 m/h (360 L/h). As a result, the upflow velocity of 18.13 m/h was therefore selected and applied to the solid separation purpose. For denitrification, the digestion of the suspended solid retaining in the solid digestion column was promoted under anaerobic condition. The optimum hydraulic retention time of 1 hour in a reactor by water from the equalization tank was operated, this duration was sufficient for the complete denitrification activity in a solid digestion column. The external carbon source addition (methanol) was essential for this process. The COD:N ratio, which promotes the best denitrification activity in the solid digestion column was 3:1. This was less than the stoichiometric number of 3.7:1 for the nitrate removal using methanol as a carbon source.

Acknowledgements

This research was financially supported by the Ratchadaphiseksomphot Endowment Fund of Chulalongkorn University (RES5605300-FW), with the additional subsidized budget from the National Research University Project, Office of the Higher Education Commission (WCU-014-FW-57). A special thank was refer to Center of Excellence for Marine Biotechnology, Department of Marine Science, Faculty of Science, Chulalongkorn University for equipment and research facilities providing. Some equipment (a microplate spectrophotometer) is provided by the Thai Government Stimulus Package2 (TKK2555) under the Project for Establishment of Comprehensive Center for Innovative Food, Health Products and Agriculture, Chulalongkorn University.

REFERENCES

1. Lawson TB. Fundamentals of aquaculture engineering. Massachusetts: Kluwer Academic Publishers; 2002.
2. Arbiv R, van Rijn J. Performance of a treatment system for inorganic nitrogen removal from intensive aquaculture systems. Aquac. Eng. 1995;14:189-203.
3. Cripps SJ, Bergheim A. Solids management and removal for intensive land-based aquaculture production systems. Aquac. Eng. 2000;22:33-56.
4. Timmons MB, Ebeling JM, Wheaton FW, Summerfelt ST, Vinci BJ. Recirculating aquaculture systems. 2nd ed. New York: Cayuca Aqua Ventures; 2002.
5. Crab R, Avnimelech Y, Defoirdt T, Bossier P, Verstraete W. Nitrogen removal techniques in aquaculture for a sustainable production. Aquaculture 2007;270:1-14.
6. Nootong K. Nitrogen treatment in closed-system aquaculture. J. King Mongkut's Institute of Technol. Ladkrabang. 2008;16: 11-12.
7. Gutierrez-Wing MT, Malone RF. Biological filters in aquaculture: Trends and research directions for freshwater and marine applications. Aquac. Eng. 2006;34:163-171.
8. Rivett MO, Buss SR, Morgan P, Smith JWN, Bemment CD. Nitrate attenuation in groundwater: A review of biogeochemical controlling processes. Water Res. 2008;42:4215-4232.
9. Mokhayeri Y, Riffat R, Takacs I, et al. Characterizing denitrification kinetics at cold Temperature using various carbon sources in lab-scale sequencing batch reactors. Water Sci. Technol. 2008;58:233-238.
10. Magram SF. Drinking water denitrification in a packed bed anoxic reactor: effect of carbon source and reactor depth. J. Appl. Sci. 2010;10:558-563.
11. Menasveta P, Panritdam T, Sihanonth P, Powtongsook S, Chuntapa B, Lee P. Design and function of a closed, recirculating seawater system with denitrification for the culture of black tiger shrimp broodstock. Aquac. Eng. 2001;25:35-49.
12. Lee PG, Lea RN, Dohmann E, et al. Denitrification in aquaculture systems: an example of a fuzzy logic control problem. Aquac. Eng. 2000;23:37-59.
13. Singer A, Parnes S, Gross A, Sagi A, Brenner A. A novel approach to denitrification processes in a zero-discharge recirculating system for small-scale urban aquaculture. Aquac. Eng. 2008;39: 72-77.
14. Pungrasmi W, Playchoom C, Powtongsook S. Optimization and evaluation of a bottom substrate denitrification tank for nitrate removal from a recirculating aquaculture system. J. Environ. Sci. 2013;25:1557-1564.
15. Aboutboul Y, Arviv R, van Rijn J. Anaerobic treatment of intensive fish culture effluents: volatile fatty acid mediated denitrification. Aquaculture 1995;133:21-32.
16. Tal Y, Schreier HJ, Sower KR, Stubblefield JD, Place AR, Zohar Y. Environmentally sustainable land –based marine aquaculture. Aquaculture 2009;286:28-35.
17. APHA, AWWA, WEF. Standard Methods for the Examination of Water and Wastewater. 21st ed. Washington DC: American Public Health Association; 2005.
18. APHA, AWWA, WEF. Standard Methods for the Examination of Water and Wastewater. 18th ed. Maryland, USA: American Public Health Association; 1992.
19. Nurit S. Development of solid separating unit and the effect of solid to water quality in closed aquaculture system. [Thesis]. Bangkok: Chulalongkorn Univ.; 2011.
20. Suttisri R. Efficiency of closed recirculating seawater

system for an indoor super-intensive shrimp culture. [Thesis]. Bangkok: Chulalongkorn Univ.; 2006.

21. Kaviraj A, Bhunia F, Saha NC. Toxicity of methanol to fish, crustacean, oligochaete worm and aquatic ecosystem. Int. J. Toxicol. 2004;23:55-63.

22. van Rjin J, Tal Y, Schreier HJ. Denitrification in recirculating systems: Theory and applications. Aquac. Eng. 2006;34: 364-376.

23. Surampalli RY, Tyagi KD. Advances in Water and Wastewater Treatment. Virginia: American Society of Civil Engineers; 2004.

SECTION V:
AQUACULTURE SYSTEMS AND FISH MANAGEMENT

CHAPTER 12 PRODUCTION OF FOOD SIZE PACIFIC WHITE SHRIMP LITOPENAEUS VANNAMEI

Joshua A. Haslun, Eudes Correia, Kevin Strychar, Timothy Morris, Tzachi Samocha

Keywords: Super-intensive; Foam fractionation; Settling tank; Flow cytometry; *Litopenaeus vannamei*; Biofloc

Over the last 30 years increases in epizootic diseases, eutrophication of freshwater resources, and environmental pollution have had a negative impact on the production of commercially available Pacific White Shrimp (*Litopenaeus vannamei*). Traditional production of food shrimp typically utilized outdoor earthen ponds with frequent water exchanges to maintain optimal dissolved oxygen and water quality. These practices increased the probability of release and introduction of pathogen and eutrophication of coastal waters. The use of zero exchange recirculating aquaculture systems (RAS) is an alternative method of shrimp production. It can minimize adverse environmental impacts and improve biosecurity, defined as "the sum of all procedures in place to protect living organisms from contracting, carrying, and spreading diseases and other non-desirable health conditions" (Moss et al., 1998). Significant reduction in released nutrient-rich water allows the system to be operated with native or exotic species with minimal risk of escapement into receiving waters while simultaneously decreasing the probability of disease introduction from water exchange (Moss, 2002). Although zero exchange systems represent an environmentally friendly form of shrimp production, the associated cost is difficult to offset unless super-intensive practices are used (Balasubramanian and Ravachandran, 2004; Stokes et al., 2009; Wasielesky Jr. et al., 2006). These practices decrease the probability of pathogen introduction, increase biosecurity while producing cost-effective margins of biomass production (Browdy and Moss, 2005; Wasielesky Jr. et al., 2006) demonstrated that zero exchange super-intensive methods enclosed in greenhouse structures decrease these risks while allowing for close monitoring of environmental parameters. Further, implementation of super-intensive zero exchange microbial dominated RAS in greenhouse settings resulted in 98% survival and no disease indicating improved biosecurity while increasing biomass compared to shrimp grown in clear water (Wasielesky Jr. et al., 2006). Similar results have been reported by Ballester et al (2010) with slightly lower survival (~89%), however the microbial-based super-intensive systems supported increased shrimp yield while decreasing operation costs compared to non-microbial based super-intensive systems. The results from Wasielesky Jr. et al (2006) and Ballester et al (2010) suggest that the presence of a complex community of microorganisms can remove potential toxicants from the water, promote shrimp growth through feed supplementation and improve shrimp health by preventing allochthonous pathogenic organisms from invading the system through competition.

The utilization of probiotic applications, as defined by Verschuere et al (2000), to RAS has become recognized as a beneficial practice to shrimp aquaculture see review by

Citation: Haslun et al., 2012, Production of Food Size Pacific White Shrimp Litopenaeus vannamei , doi: 10.5376/ija.2012. 02.0006).

Ninawe and Selvin, (2009). Generally, a single or combination of several known cultures isolated from aquacultured shrimp or water used for culture is used to inoculate the system. Organisms from the genus *Vibrio* (Austin et al., 1995; Balcázar et al., 2007), *Bacillus* (Gómez and Shen, 2008; Li et al., 2007; Rengipat et al., 2003; Vaseeharan and Ramasay, 2003; Solano and Soto, 2006), *Pseudomonas* (Vijayan et al., 2006), *Roseobacter* (Sandaa et al., 2003), and *Arthrobacter* (Li et al., 2006) have all been employed in shrimp aquaculture or display promise as potential probiotics (reviewed by Ninawe and Selvin, 2009). These bacteria once established within a RAS aid in the maintenance of water quality, most often with respect to nitrification (Kuhn et al., 2010). Further advantages to bacterial supplementation include microbial floc as natural food to improve growth (Burford et al., 2004; El-Haroun et al., 2006), prevention of pathogen proliferation through competitive exclusion (Verschuere et al., 2000), and immunostimulation of the immune system of target organisms (Ganguly et al., 2010). Although probiotic cultures have proved beneficial, most organisms rely on a complex community of microbial flora to grow most efficiently and aid in immune function. The use of pre-conditioned (established microbial community) RAS water has yet to become a common practice in aquaculture because of the difficulty to maintain and promote a microbial community which represents the target organism throughout its commercial growth period to food-size.

The current study assesses the efficacy of fully established heterotrophically dominated water from a 62-day nursery trial to act as a natural probiotic community to grow *L. vannamei* juveniles to food-size in greenhouse-enclosed super-intensive RAS under no water exchange. Further, we analyze the effectiveness of settling and foam fractionation to control microbial and algal communities within the system using flow cytometry.

RESULTS

Shrimp Grow-out Period

Biometric comparative analysis of settling tank (ST) versus foam fractionator (FF) salinity treatments (30.8 vs. 30.3, respectively) was significantly different throughout the course of this study. In addition, although the alkalinity in each of the raceways was adjusted at least two times per week to maintain a concentration of 160 mg/L as $CaCO_3$, mean alkalinity in the FF raceway was also significantly lower than ST raceways containing 124 vs. 129 mg/L $CaCO_3$. Similarly, we observed significant differences in daily nitrate concentrations in raceways using the FF method (1 027 vs. 855 NO_3 mg/L or 232 vs. 193 mg/L NO_3-N). On the last day of our study, we observed the FF raceway nitrate concentration at 2 028 NO_3 mg/L vs. the ST raceway concentration at 1 585 NO_3-N mg/L. We speculate that in the bottom of the ST, denitrification occurred under the anaerobic, sludge-water interface. The mean alkalinity in the total ammonia-nitrogen and nitrite levels in all four raceways remained very low throughout the study (>0.5 mg/L). Suspended solids were consistently maintained between 10 and 30 mL/L, although on Day 43 one raceway reached a concentration of 33 mL/L. We also maintained TSS concentrations at 400 to 500 mg/L and only observed a spike (790 mg/L) once in one raceway over the duration of the study. A large portion of the TSS was in the form of VSS. No signs of bacterial or viral pathogen infections were found in any of the shrimp samples sent to Texas Veterinary Diagnostic Laboratory for analysis. Our results from this study show that 7.5 kg/m of shrimp biomass can be sustained with periodic oxygen supplementation consisting of a mixture of 100% O_2 with ambient air at a rate of 1.0 L/min for 30 to 60 min after feeding. As the shrimp biomass exceeded 7.5 kg/m, we supplemented additional oxygen at a rate of 0.1~0.3 L/min during conditions in which power outage became cause for concern. In our zero water exchange study, management of super-intensive shrimp culture could not have been done without continual monitoring using YSI 5200 units, which provided real-time DO measurements and subsequently allowed us to use considerably less oxygen than compared to previous trials. During the final week of grow-out, we supplemented oxygen continually at a rate of 0.3~0.5 L/min allowing us to produce quality food-size shrimp with yields as high as 9.75 kg/m.

Microbial Community

The microorganisms associated with this system were characterized using FACS over a 1-month period throughout the middle of the study in order to determine the effectiveness of this technology as a general bacterial monitoring technique. Data identifying gram-positive vs. gram-negative bacteria indicated that the bacterial community present in all four raceways, foam fractionation or settling tank mediated, was dominated by gram-positive microorganisms (Figure 1). Differences in bacterial community structure were not identified between methods of particulate control or over the characterization period (two-way ANOVA p>0.05).

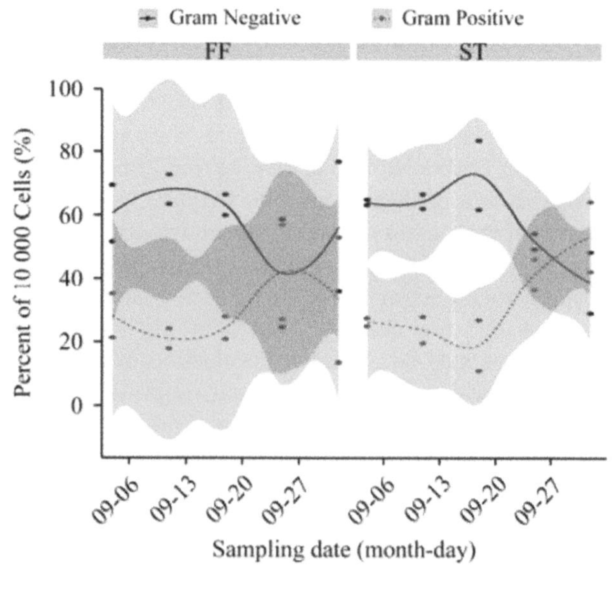

Figure 1: Gram-stained bacterial population change determined with flow cytometry.

General communities residing within the tanks including bacteria were characterized and found to consist of six independent populations based upon two auto-fluorescence parameters (red and green). Two-way ANOVA tables for each of the six community members with respect to treatment and week are provided. Changes in populations R1, R2, R3, R4, and R5 all indicated significant effect over time as well as displaying interactions between month and particulate control treatment with respect to populations R1 and R4 ($p<0.05$; (Figure 2).

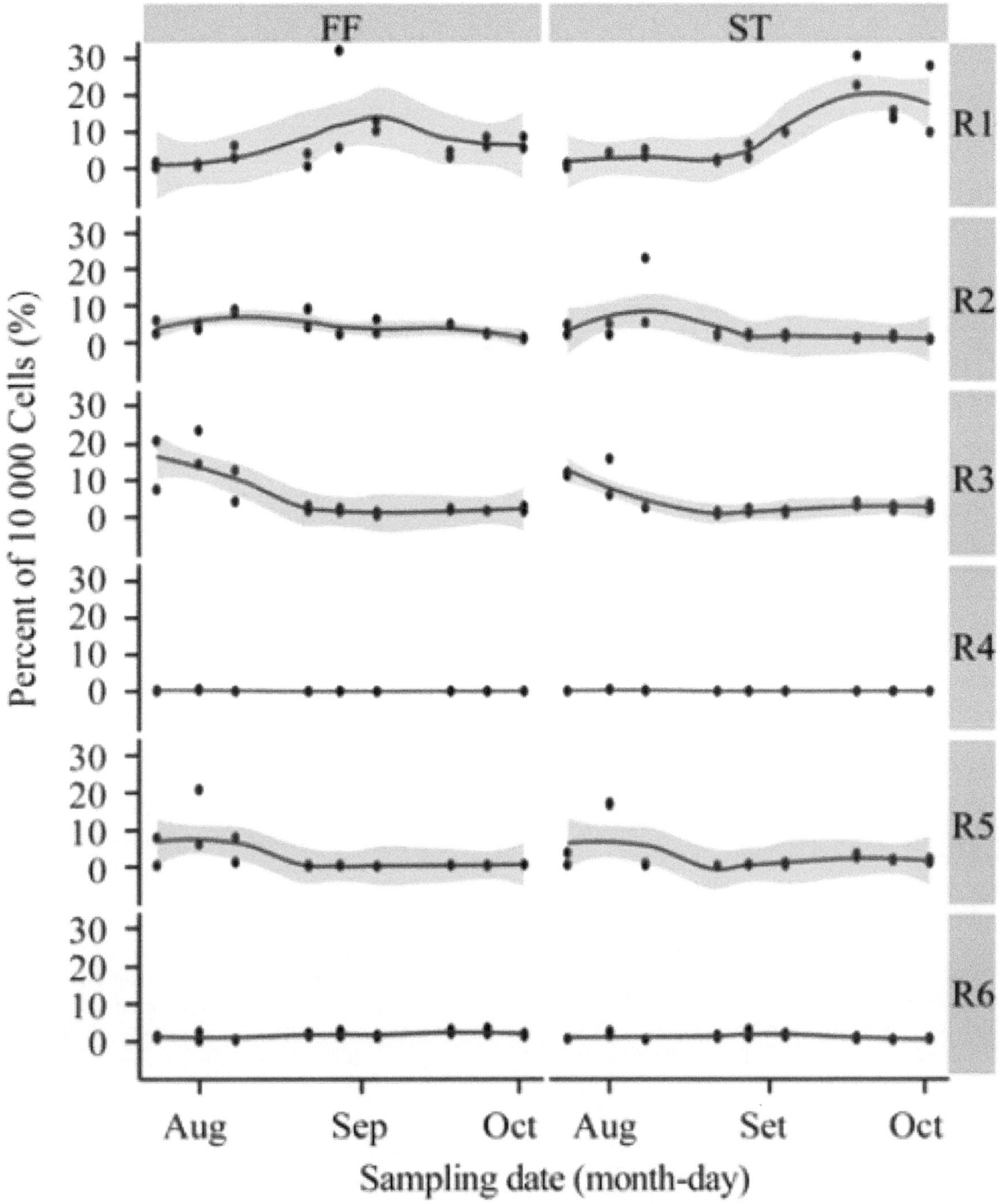

Figure 2: Characterization of fluorescent dependent water particulate.

Population R1 and R2 increased in prevalence within the community with regards to treatments (Figure 2). Populations R3, R4, and R5 displayed decreasing trends preceded by a peak occurring during the first 3 weeks (Figure 2). Population R6 was found to be least concentrated in the community as well as the possessing the most stable density ($p>0.05$) while population R1 wasmost prevalent (most likely bacteria due to low fluorescence and small forward scatter value). These six populations were further subdivided into three size categories based upon forward scatter paired with red fluorescence (this specific fluorescence parameter appeared to drive population differentiation). Picoplankton (0.2~2 μm) remained the greatest fraction of the population accounting for no less than 40% of the cumulative population at any sampling date and reached levels as high as 70% (Figure 3).

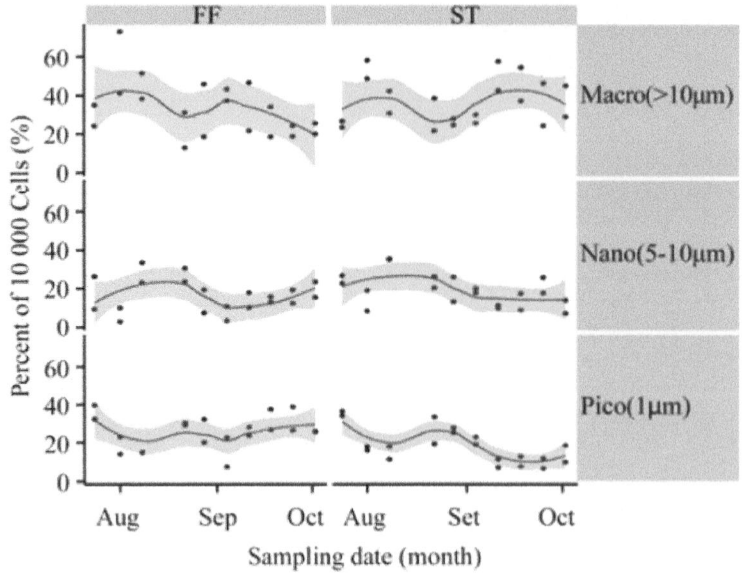

Figure 3: Fluctuation of size dependent biofloc determined with flow cytometry

Statistical analysis indicated no differences between FF or ST mediated particulate removal across sampling dates or between each individual population (p>0.05). In order to try to isolate water quality parameters that influenced the proliferation of microorganisms and particulate in super-intensive zero exchange systems, multiple linear regression analysis was performed with respect to the environmental parameters and flora data. The simplest linear model describing systems with foam fractionation as a solids control method included NO_2-N, $cBOD_5$, NO_3-N, and turbidity. This model significantly (p <0.05)

according to the adjusted R-squared value. The variability in gram-positive bacteria was negatively influenced by NO_2-N (p<0.01) and positively by NO_3-N (p<0.05). Turbidity and $cBOD_5$ displayed insignificant negative influence on gram-positive bacteria variability. The variability in macro, nano, and pico particles as well as the variability in autofluorescent regions could not be well described using multiple linear regression models of water quality parameters (p>0.05). The variability of gram-positive.parameters compared to that of foam fractionation. In particular VSS (p<0.05) and $cBOD_5$ (p<0.01) accounted for a majority of the influence. The full model (p<0.01) describing 95% of the variability also included PO_4+, TSS, and turbidity. Similarly to foam fractionation, size based particulates as well as autofluorescent regions could not be well described by multiple linear regression models (p<0.05).

Statistical analysis of shrimp harvest data indicate no differences exist in mean final weight between treatments. However, we have shown in support of previous research exist in mean final weight between treatments. However, we observed growth rates varying between 1.35 g/wk and 1.39 g/wk whilst final shrimp weights were 21.9 g and 22.4 g. It is also noteworthy to report that survival rates were high in all four raceways (94.5%~96.8%); FCR was between 1.53 and 1.60. Shrimp yields were 9.34 kg/m to 9.75 kg/m, although the latter weight represented one sample in the FF raceway. The volume of water per 1 kg of shrimp produced used in the raceways varied between 98 L to 126 L.

DISCUSSION

We have shown in support of previous researches (Menasveta, 2002; Wasielesky Jr. et al., 2006; Ballester et al., 2010), the efficacy of super-intensive zero exchange systems to produce market-sized *L. vannamei* at a very high culture density (450 shrimp/m^3) without the indication of disease. This technique is considered efficient, sustainable, biologically secure, environmentally sound, and therefore an alternative to traditional aquaculture methods, which have notable environmental impacts (Páez-Osuna, 2001a; 2001b). Unlike other super-intensive zero exchange shrimp rearing studies, we utilized water from a successful 62-day *L. vannamei* growth period from larvae to juvenile. We define successful as disease free with low incidence of mortality (<5%). We consider this water to be conditioned with a healthy community of microorganisms due to its success in the postlarvae to juvenile growth period. The bacterial communities associated with marine and brackish *L. vannamei* in culture have been shown to change with environment rather than diet (Luo et al., 2006; 2009) and this result has been reported in fish as well (Yan et al., 2012). We can infer from these results that the shrimp in our study would obtain a similar beneficial microbial community.

Shrimp culture has been shown to alter the pelagic bacterial community of intensive closed and open culture systems (Sakamiet al., 2008). Therefore, systems utilizing previously conditioned culture water may have a potential advantage over those starting from clean or non-conditioned water in that the initial establishment phase can be foregone resulting in a more stable and complex community of heterotrophic organisms. At the beginning of our sampling period the bacterial community associated with raceways managed by FF and ST solids control were dominated by gram-negative genera. This community transitioned into equal proportions of gram-positive and gram-negative organisms by the end of the grow-out period but transitioned in a polynomial fashion. Gram-negative bacteria generally represent r-selected species whilst gram-positive bacteria represent k-selected or slow-growing species. Fierer et al (2007) suggest that this delineation is referred to as copiotrophic/eutrophic or oligotrophic, respectively. Utilizing previously conditioned water would provide a possible avenue to maintain a complex culture of these organisms. Nutrient concentrations, particularly with respect to different forms of nitrogen (ammonium, nitrate, urea, and amino acids) tend to significantly impact the proliferation of bacteria in a eutrophic system such as an estuary (Middelburg and Nieuwenhuize, 2000). A complex community of organisms would therefore increase the ability of the system to handle greater nitrogen loads and result in a more efficient system. We did not explicitly test this idea in our study; however the exemplary growth we report is indicative of excellent culture conditions and these conditions are associated with conditioned water. The re-use of the same water used in the previous nursery trial along with specially formulated feed provided by Zeigler Brothers Inc. may have provided for the immediate establishment or continued maintenance of a healthy nitrifying microbial community in the grow-out study. This allowed for negligible amounts of TAN and NO2-N in the culture water throughout the grow-out study. Typically nitrification will require many weeks to attain a steady state in a closed aquaculture system (Sohier and Bianchi, 1985), however this system maintained steady state nitrification throughout the growth period. Very few studies have chose to use the aquaculture produced communities of bacteria in pre-conditioned water within enclosed zero exchange RAS. We believe that this avenue of shrimp culture deserves increased attention.

The use of an established community of bacteria will likely prevent the onset of disease through competitive exclusion or limitation of pathogenic strains. We did not observe a single case of disease related death throughout the study. Further, consumers and aquaculturists have become increasingly aware of the hazards of dosing intensively produced aquacultured stocks with antibiotics. Typically, shrimp farmers utilizing this disease prevention technique have incurred crop losses frequently in large production areas such as Thailand (Holmstrom et al., 2003). The acquisition of antibiotic resistance in potentially pathogenic bacteria such as Vibrio sp. is a typical side effect of antibiotic overuse (Le et al., 2005; Nakayama et al., 2006). Further, the use of the antibiotics oxyteracycline and oxolinic acid has been shown to decrease the expression of immune-related genes associated with antibacterial processes in shrimp (Fagutao et al., 2009). In each of the aforementioned studies, farming practices were carried-out in pond systems rather than RAS.

Probiotic strategies have been employed as an alternative to antibiotics in shrimp aquaculture for the past 20 years; however most of these studies have used a single organism to outcompete pathogenic organisms (Verschuere et al., 2000; Kuhn et al., 2010). A change in strategy from individual probiotic cultures to established probiotic communities may be essential to the future of super-intensive growth systems and the continued growth of an environmentally friendly and biologically secure aquaculture industry.

Isolating and culturing a majority of the associated flora of shrimp has proved difficult and therefore research has focused on growth independent identification techniques (Lau et al., 2002) or those techniques designed to monitor and identify particular microorganisms (Tendencia and de la Peña, 2001; Nakayama et al., 2006). We investigated the application of a rapid technique, flow cytometry, as a means to assess changes in bacterial as well as auto-fluorescent pelagic communities of an RAS system. We found that using the gram-stain like procedure developed by Holm and Jespersen (2003), and initially utilized for milk associated bacteria, the relative densities of gram-negative to gram-positive bacteria could be estimated. We were able to detect weekly fluctuations in this community of bacteria. Specifically, we were able to determine that the two solids control methods employed, foam fractionation and conic cylindrical settling tanks, and did not result in a significant change in the ratio of gram-positive to gram-negative bacteria when using pre-conditioned culture water. The fluctuations we observed within each treatment indicate a level of system homeostasis with respect to gram-negative and gram-positive bacteria. Fluctuations within a system were observed when the ratio of gram-positive:gram-negative bacteria reached 2:8. Gram-positive bacteria cultured from wild habitats (Rengpipat et al., 2003) as well as produced commercially (Wang et al., 2005), have received attention regarding their ability to kill or inhibit the growth of gram-negative organisms within marine aquaculture systems. It is therefore critical that a population of these organisms remain within the system to balance potential overgrowth of pathogenic gram-negative organisms. Gram-positive organisms also play an important role in minimizing the accumulation of phosphorous and nitrogen species. Decreased reactive phosphorous as well as total nitrogen and chemical oxygen demand have been reported in systems with gram positive probiotic use (Wang et al., 2005).

Additionally to bacterial staining, we were able to distinguish three major classes of particles in the biofloc using flow cytometry; pico particles (<2 m), nano particles (2~8 m), and macro particles (7~20 m). Pico particles displayed a range of red fluorescence spanning three log scale regions from 10^0~10^3 relative fluorescence units. Nano and macro particles were found within a more narrow range: 10^1~10^2 and 10^2~10^3, respectively. Although we found that macro particles were in the highest concentration, FACS does not differentiate between single and clustered cells or cells attached to debris. Flow cytometry methods will only detect these conglomerates as a single large "particle". Macro particles should therefore be considered a diverse group that may require greater resolution. Systems such as closed zero exchange super-intensive culture systems tend to produce suspended aggregates or biofloc. The larger the flocculated material the greater its porosity tends to be as well as it's potential to interact with the external environment producing a more efficient system. Uptake of nutrients by microorganisms has been found to be more efficient when flocculated than as a single cell (Logan and Hunt, 1987). It would not be surprising if these particles supported a diverse group of organisms.

The settling tank (ST) particulate control resulted in a slightly decreased but insignificant concentration of nitrate in the respective raceways. This was most likely a product of denitrification by bacteria occurring in the sludge within the STs. In a similar study raising *L. vannamei*, Holl et al (2011) reported that super-intensive systems relying on FF for particulate control resulted in greater accumulation of nitrogen in the system. Further, their study utilized a bead filter method of solids removal, which would produce a denitrifying community of bacteria similar to our STs. Particulate control methods such as settling tank clarifiers have also been demonstrated to promote increased shrimp growth (Lewis et al., 2009; Ray et al., 2010). These two groups also suggest that

STs and bead filters are more efficient than FF for maintaining adequate water quality with respect to solids and potentially harmful NO_2-N and NH_3. In our study there was a negligible difference between the capacity of the FF and the ST to change the microbial community within the raceway water column when general bacterial community characteristics were used as a proxy. This may have been an effect of using pre-conditioned water.

CONCLUSION

Heterotrophically dominated super-intensive zero exchange greenhouse covered RAS represent an environmentally friendly and biosecure practice for the production of food size *L. vannamei*. Pre-conditioned water with an established heterotrophically dominated community of microorganisms may serve to provide a more balanced system for the control of harmful nutrients such as NO_2-N and NH_3-N and help alleviate the risk of disease in these systems. This water may be an alternative to commercial probiotic supplementation and an alternative to individual culture treatments. The solid control methods used in this study behaved similarly unlike in other studies and this may be an effect of pre-conditioned culture water. Flow cytometry served as a cost-efficient high-throughput method for monitoring general community structure of microorganisms within a RAS. Due to the inherent limitations of flow cytometry for community biology we would recommend that future studies be performed in parallel with molecular techniques for a more comprehensive higher resolution investigation of the changes in bacterial communities.

MATERIALS AND METHODS

Experimental Design

A 108-d study was conducted in four 40 m^3 (25.4 × 2.7 m with 68.5 m^2 bottom area) raceways lined with EPDM (ethylene propylene diene monomer, Firestone Specialty Products, Indianapolis, IN, USA). All raceways were constructed with a longitudinal partition centered over a PVC pipe (5.1 cm long) containing spray nozzles. Six 0.91 m long air diffusers (1.9 cm OD, Aero-TubeTM, Tekni-plex Aeration, Austin, TX, USA) and six banks, each constructed with three 5.1 cm airlift pumps, were positioned at equal distances on both sides of the partition on each of the raceways used in this study. A 2 hp centrifugal pump and a 5.1 cm Venturi injector (model 2081-A, Mazzei Injector Corporation, Bakersfield, CA, USA) served to maintain dissolved oxygen concentrations consisting of atmospheric air or a mixture of 100% bottled oxygen and air. Water for this experiment was obtained from a preceding 62-d nursery study. Four raceways were used in which two were equipped with a commercial foam fractionator (VL65, Aquatic Eco System, Apopka, FL, USA) and the remaining two furnished with a 8.6 m^3 cylindroconical settling tank (4.5 m^3 working water volume). All four raceways were stocked (450 shrimp/m^3) with juveniles (0.99 g ± 0.17 g) of the Pacific White Shrimp, *Litopenaeus vannamei*, cultured at high density for 62-d in the same tanks. Shrimp were fed seven days a week specially formulated commercial 35% crude protein (CP) feed for intensive systems operated with limited discharge (Hyper-Intensive 35, Zeigler Bros., Gardners, PA, USA). Until Day 18, feed was offered at four equal portions during the day. From Day 19 onward, two-thirds ration partitioned into four equal portions were fed during the day with the remainder fed throughout the night using belt feeders (spring loaded, 12-hour clock 4.5 kg capacity, Zeigler Bros., Gardners, PA, USA). After stocking our tanks in the first week, shrimp feed was mixed with food using in the nursery trial (Fry #4, 30% CP, Rangen Inc., Buhl, ID). Each day rations were calibrated on a FCR of 1:1.4, approximating growth of 1.4 g/wk and assuming a mortality rate of ≤0.5%/wk. Settling tanks (ST) and foam fractionators (FF) were used 23 days after stocking. Both the ST and FF were used intermittently targeting concentrations between 400 to 500 mg/L of culture water total suspended solids (TSS) and settleable solids (SS) between 10mL/L.~14 mL/L. Water flow into the ST varied between 2 and 6 L/min to provide between 4.6 and 13.9 tank turnovers/day. Both the FF and the ST were operated via a side loop, which received the water from the pump installed in each RW. Raceways were maintained with zero water exchange throughout the study. Municipal chlorinated freshwater was added to compensate for water loss due to evaporation and operation of the FF.

Multiparameter Monitoring and Statistical Analyses

Water temperature, salinity, dissolved oxygen, and pH was simultaneously monitored twice daily using an YSI 650 Series multi-probe (YSI Inc., Yellow Springs, OH, USA). Alkalinity (Method # 2320 B, APHA, 1995), and SS (Imhoff Cone Method # 2540 F, APHA, 1995) were monitored every two to three days. Turbidity (Spectronic 21, Milton Roy Co, Ivyland, PA), TSS (Method # 2540 D, APHA, 1995), volatile suspended solids (VSS) (Method # 2540 E, APHA, 1995), and five-day carbonaceous biochemical oxygen demand ($cBOD_5$) (Method #5210 B, APHA, 1995) were monitored weekly. A Autoanalyzer (dual channel FIAlab-2600, FIAlab Instruments, Inc., Bellevue, WA, USA) was used to monitor weekly total ammonia-nitrogen (TAN) (salicylate method), nitrite-nitrogen (NO_2-N), nitrate-nitrogen (NO_3-N) (nitrate/nitrite method), and phosphate (water based samples, orthophosphate method). Sodium bicarbonate (160 mg/L as $CaCO_3$) was used to control alkalinity and pH. A YSI 5 200 multi-parameter sonde (YSI Inc., Yellow Springs, OH, USA) was placed in each raceway and data uploaded remotely. Daily and weekly water quality data from each treatment was analyzed using a repeated measures ANOVA. Mean survival, FCR, total yields, weekly growth and final shrimp weights were analyzed using one-way ANOVAs and an value of 0.05. All statistical analyses were conducted using SPSS statistical software (V. 15 for Windows, SPSS Inc., Chicago, IL, USA). Mixing of the ambient air with bottled oxygen did not begin until Day 68 of the study. For a period of 40 days (Days 68 through 108) air was enriched with oxygen only intermittently, specifically for 30~60 minutes following the day-time feeding and only if DO levels dropped below 3 mg/L, at a flow rate of 1.0 L/min. During the final week (Day 102 through termination at Day 108), air was constantly mixed with oxygen at a flow rate of 0.3~0.5 L/min.

Analysis of Microbial Community

The algal and bacterial communities associated with each raceway were monitored weekly in order to correlate particulate control methods (e.g., FF to ST) to the changes in the populations of algae and bacteria present within the system. Algal monitoring began on July 24 and ended October 2, 2009 while bacterial communities, specifically gram-negative and gram-positive, were monitored only during the final five weeks of the study. Water samples were collected from each raceway in sterile 50 mL conical FalconTM tubes in the morning at ~9:00 AM and transported to Texas A&M University-Corpus Christi to be analyzed within 2 h of collection. A fluorescence activated cell sorter (FACS) flow cytometer VantageSE (Becton, Dickinson and Company, San Jose, CA, USA) in combination with a COHERENT Innova Enterprise II Laser (488 nm) (COHERENT, Inc., Auburn, CA, USA) was used to analyze the water samples. Due to the interspersed use of the FACS, quality control occurred between sampling weeks with BioSure® chicken red blood cells (CRBC) in phosphate buffered saline (PBS) and standard instrument settings to insure that instrument parameters remained within acceptable limits between sampling dates.

Algal communities were distinguished using fluorescence characteristics inherent to the individual cells within (534±34) nm (FL1) and (630±22) nm (FL3) fluorescence ranges as well as the forward scatter (FSC) physical parameter, which is indicative of cell size. Prior to analysis, cells were passed through a 100 μm nylon mesh filter to decrease the prevalence of cell aggregates and remove debris. Ten thousand events (one event=one cell passing through the laser path) were acquired for each sample. The greatest community resolution was acquired with the FL1 and FL3 parameters when used on a 4 log-scale plot as well as the FL1 and FSC parameters (Trask et al., 1982; Olson et al., 1989).

A procedure analogous to gram-staining was used to determine the percentage of gram-positive vs. gram-negative cells following methods adapted from Holm and Jespersen (2003). Briefly, a 1.5 mL aliquot of filtered raceway water sample was transferred to a micro-centrifuge tube (~10^6 cells/mL) and centrifugated at 700×g for 10 min. The supernatant was decanted and re-suspended in 1.5 mL of a 3 M KCl solution (3 M KCl, 0.035 M EDTA [pH 7.0], Sigma Aldrich, St. Louis, MO, USA). This solution served to permeabalize the cell wall of gram-negative bacteria. Two hundred and fifty microliters of the suspension was stained with 50 μL of hexididum iodide (HI; 200 μg/mL in 0.1 M $NaHCO_3$, Invitrogen, Grand Island, NY, USA) and incubated at 50°C for 15 min. Since HI binds to all bacterial DNA (Holm and Jespersen, 2003) gram-positive bacteria resist HI staining due to their thicker cell wall. Twenty-five microliters of the secondary stain, Oregon Green 488 nm-conjugated wheat germ agglutinin (WGA; 200 μg/mL in 0.1 M $NaHCO_3$, Invitrogen, Grand Island, NY, USA), was then added to the suspension and incubated at room temperature for 4 min. WGA binds selectively to gram-positive cells (Holm and Jespersen, 2003). Ten thousand events were analyzed in the FL1 and FL3 parameters for each sample.

Statistical Analysis

FACS data was analyzed with FlowJo v. 8.8.6 (Tree Star, Inc., Ashland, OR, USA). In order to determine the effects of this system on the biofloc community experiments were carried out in fixed factor three-way ANOVA designs. Solid control method and sampling date were used with microorganism (gram-stain), particle (size) or region (autofluorescence and size) as explanatory variables. The percentage of total microorganisms, particles, or regions in each of these groups was used as the response variables. Data was checked for normality within the scope of ANOVA. In order to determine what water and solids parameters had the greatest impact on the ANOVA results, multiple linear regression models were developed. The percentage of gram-positive and gram-negative bacteria was highly negatively correlated (R2=0.99). Therefore models were created only for gram-positive bacteria. Pairs plots were used in combination with a variance inflation factor (Heiberger and Holland, 2004) and covariance matrix to remove variables with a high degree of collinearity. Further model simplification was performed using the drop1 function with F-test statistic (Chambers, 1992). Statistical analyses were performed with R v. 2.10.1 (The R Foundation for Statistical Computing, Vienna, Austria).

Author contributions

JH wrote the manuscript and analyzed the collected data using flow cytometry and statistics. TS designed and supervised the experiment, edited the manuscript, and coordinated the study activities. EC was in charge of the day-to-day operation and management of the experimental system. TM monitored and provided technical assistance. JW monitored water quality in the system. KS aided in flow cytometry, experimental design, and editing of the manuscript and proofs.

ACKNOWLEDGEMENT

The authors acknowledge the National Institute of Food & Agriculture (NIFA), USDA, AgriLife Research, National Academy of Sciences, and NOVUS International for their funding. Ziegler Bros. provided feed for the study. Aquatic Eco Systems, YSI, Colorite Plastics, and Firestone Specialty Products provided the foam fractionators, DO monitoring systems, air diffusers, and raceway EPDM liners, respectively used in our study. Post-larvae were supplied by Harlingen Shrimp Farms. Texas A&M University at Corpus Christi provided lab space to perform all flow cytometric analyses for this study. We would also like to acknowledge the staff of the AgriLife Research Mariculture Lab at Flour Bluff, Corpus Christi for their efforts throughout the study.

REFERENCE

1. Austin B., Stuckery L.F., Robertson P.A.W., Effendi I., and Griffith D.R.W., 1995, A probiotic strain of *Vibrio alginolyticus*effective in reducing diseases caused by *Aeromonas salmonicida, Vibrio anguillarum*, and *Vibrio ordalii*, Journal of Fish Disease, 18: 93-96 http://dx.doi.org/10.1111/j.1365-2761.1995.tb01271.x
2. Balasubramanian C.P., Pillai S.M., and Ravichandran P., 2004 Zero-water exchange shrimp farming systems (extensive) in the periphery of Chilka lagoon, Orissa, India, Aquaculture International, 12(6): 555-572
3. http://dx.doi.org/10.1007/s10499-004-0350-3
4. Balcázar J.L., Rojas-Luna T., and Cunningham D.P., 2007, Effect of the addition of four potential probiotic strains on the survival of pacific white shrimp *(Litopenaeus vannamei)* following immersion challenge

with *Vibrio parahaemolyticus*, Journal of Invertebrate Patholology, 96: 147-150 http://dx.doi.org/10.1016/j.jip.2007.04.008 PMid:17544437

5. Ballester E.L.C., Abreu P.C., Cavalli R.O., Emerenciano M., de Abreu L., and Wasielesky Jr. W., 2010, Effect of practical diets with different protein levels on the performance of *Farfantepenaeus paulensis* juveniles nursed in a zero exchange suspended microbial flocs intensive system, Aquaculture Nutrition, 16: 163-172 http://dx.doi.org/10.1111/j.1365-2095.2009.00648.x
6. Burford M.A., Thompson P.J., McIntosh R.P., Bauman R.H., and Pearson D.C., 2004, The contribution of flocculated material to shrimp (*Litopenaeus vannamei*) nutrition in a high-intensity, zero-exchange system, Aquaculture, 232: 525-537
7. http://dx.doi.org/10.1016/S0044-8486(03)00541-6
8. Browdy C.L., and Moss S.M., 2005, Shrimp culture in urban, super-intensive closed systems, In: Costa-Pierce B., Desbonnet A., Edwards P., and Baker D., eds., Urban Aquaculture, CABI Publishing, Oxfordshire, UK, pp. 173-186
9. http://dx.doi.org/10.1079/9780851998299.0173
10. Chambers J. M., 1992, Linear models, Chapter 4 of Statistical Models, In: Chambers J. M., and Hastie T. J., eds., Wadsworth & Brooks/Cole
11. El-Haroun E.R., Goda A., and Chowdhury M.A.K., 2006, Effect of dietary probiotic Biogen (R) supplementation as a growth promoter on growth performance and feed utilization of Nile tilapia *Oreochromis niloticus* (L.), Aquaculture Research, 37: 1473-1480
12. http://dx.doi.org/10.1111/j.1365-2109.2006.01584.x
13. Fagutao F.F., Yasuike M., Santos M.D., Ruangpan L., Sangrunggruang K., Tassanakajon A., Takahashi Y., Ueno R., Kondo H., Hirono I., and Aoki T., 2009, Differential gene expression in black tiger shrimp, *Penaeus monodon*, following administration of oxytetracycline and oxolinic acid, Developmental Comparative Immunolology, 33: 1088-1092
14. http://dx.doi.org/10.1016/j.dci.2009.05.010
15. Fierer N., Bradfored M.A., and Jackson R.B., 2007, Toward an ecological classification of soil bacteria, Ecology, 88: 1354-1364
16. http://dx.doi.org/10.1890/05-1839
17. Ganguly S., Paul I., and Mukhopadhayay S.K., 2010, Application and effectiveness of immunostimulants, probiotics, and prebiotics in aquaculture: A review, Israel Journal of Aquaculture-Bamidgeh, 62: 130-138
18. Gómez G.D., and Shen M.A., 2008, Influence of probiotics on the growth and digestive enzyme activity of white Pacific shrimp (*Litopenaeus vannamei)*, Journal of the Ocean University of China (English Edition), 7: 215-218
19. http://dx.doi.org/10.1007/s11802-008-0215-x
20. Heiberger R.M., and Holland B., 2004, Statistical Analysis and Data Display: An Intermediate Course with Examples in S-Plus, R, and SAS, Springer Texts in Statistics, Springer
21. Holl C.M., Glazer C.T., and Moss S.M., 2011, Nitrogen stable isotopes in recirculating aquaculture for super-intensive shrimp production: Tracing the effects of water filtration on microbial nitrogen cycling, Aquaculture, 311: 146-154
22. http://dx.doi.org/10.1016/j.aquaculture.2010.11.028
23. Holm C., and Jespersen L., 2003, A flow-cytometric gram-staining technique for milk-associated bacteria, Applied and Environmental Microbiology, 69(5): 2857-2863
24. http://dx.doi.org/10.1128/AEM.69.5.2857-2863.2003
25. Holmstrom K., Graslund S., Wahlstrom A., Poungshompoo S., Bengtsson B.E., and Kautsky N., 2003, Antibiotic use in shrimp farming and implications for environmental impacts and human health, International Journal of Food Science Technology, 38: 255-266
26. http://dx.doi.org/10.1046/j.1365-2621.2003.00671.x
27. Kuhn D.D., Drahos D.D., Marsh L., and Flick G.J., 2010, Evaluation of nitrifying bacteria product to improve nitrification efficacy in recirculating aquaculture systems, Aquaculture Engineering, 43: 78-82
28. http://dx.doi.org/10.1016/j.aquaeng.2010.07.001
29. Lau W.W.Y., Jumars P.A., and Armbrust E.V., 2002, Genetic diversity of attached bacteria in the hindgut of the deposit-feeding shrimp *Neotrypaea* (formerly *Callianassa*) *californiensis* (Decapoda: *Thalassinidae*), Microbial Ecology, 43(4): 455-466
30. http://dx.doi.org/10.1007/s00248-001-1043-3
31. Le T.X., Munekage Y., and Kato S., 2005, Antibiotic resistance in bacteria from shrimp farming in mangrove areas, Science of the Total Environment, 349: 95-105
32. http://dx.doi.org/10.1016/j.scitotenv.2005.01.006
33. Lewis B.L., Browdy C.L., Ray A.J., Lawson A., Shuler A., Venero J.A., Vinatea L., and Leffler J.W., 2009, Management of microbial biofloc communities using settling tank clarifiers in superintensive, zero-exchange shrimp production systems, Journal of Shellfish Research, 28: 710-710
34. Li P., Burr G.S., Gatlin D.M. 3rd, Hume M.E., Patnaik S., Castille G.L., and Lawrence A.L., 2007, Dietary supplementation of short-chain fructooligosaccharides influences gastrointestinal microbiota composition and immunity characteristics of Pacific white shrimp, *Litopenaeus vannamei,* cultured in a recirculating system, Journal of Nutrition, 137: 2763-2768PMid:18029496
35. Logan B.E., and Hunt J.R., 1987, Advantages to microbes of growth in permeable aggregates in marine systems, Limnology and Oceanography, 32(5): 1034-1048
36. http://dx.doi.org/10.4319/lo.1987.32.5.1034
37. Luo P., Hu C., Xie Z., Zhang L., Ren C, Xu Y., 2006, PCR-DGGE analysis of bacterial community composition in brackish water*Litopenaeus vannamei* culture system, Journal of Tropical Oceanography, 25(2): 50-53
38. Menasveta P., 2002, Improved shrimp growout systems for disease prevention and environmental sustainability in Asia, Reviews in Fisheries Science, 10: 391-402
39. http://dx.doi.org/10.1080/20026491051703

40. Middelburg J.J., Nieuwenhuize J., 2000, Uptake of dissolved inorganic nitrogen in turbid tidal estuaries, Marine Ecological Progress Series, 192: 79-88
41. http://dx.doi.org/10.3354/meps192079
42. Moss S.M., Reynolds W.J., and Mahler L.E., 1998, Design and economic analysis of a prototype biosecure shrimp growout facility, In: Moss S.M. ed., Proceedings of the US Marine Shrimp Farming Program Biosecurity Workshop, 14 February 1998, Hawaii, United States, The Oceanic Institute, Hawaii, United States, pp.5-14
43. Moss S.M., 2002, Marine shrimp farming in the Western Hemisphere: Past problems, present solutions, and future visions, In: Lee C.S., and O'Bryen P.J., (eds.), Proceedings of United States. Aquaculture Growout Systems-Challenges and Technological Solutions, Reviews in Fisheries Science, 101(3–4): 601–620
44. Nakayama T., Ito E., Nomura N., and Matsumura M., 2006, Comparison of *Vibrio harveyi* strains isolated from shrimp farms and from culture collection in terms of toxicity and antibiotic resistance, FEMS Microbiology Letters, 258: 194-199
45. http://dx.doi.org/10.1111/j.1574-6968.2006.00225.x
46. Ninawe A.S., and Selvin J., 2009, Probiotics in shrimp aquaculture: Avenues and challenges, Critical Reviews in Microbiology, 35: 43-66
47. http://dx.doi.org/10.1080/10408410802667202
48. Olson R.J., Zettler E.R., and Anderson O.K., 1989, Discrimination of eukaryotic phytoplankton cell types from light scatter and autofluorescence properties measured by flow cytometry, Cytometry, 10: 636-643
49. http://dx.doi.org/10.1002/cyto.990100520
50. Páez-Osuna F., 2001a, The environmental impact of shrimp aquaculture: a global perspective, Environmental Pollution, 112: 229-231
51. http://dx.doi.org/10.1016/S0269-7491(00)00111-1
52. Páez-Osuna F., 2001b, The environmental impact of shrimp aquaculture: causes, effects and mitigating alternatives, Environmental Mangagement, 28(1): 131-140
53. http://dx.doi.org/10.1007/s002670010212
54. Ray A.J., Lewis B.L., Browdy C.L., and Leffler J.W., 2010, Suspended solids removal to improve shrimp (*Litopenaeus vannamei*) production and an evaluation of a plant-based feed in minimal-exchange superintensive culture systems, Aquaculture, 299(1-4): 89-98
55. http://dx.doi.org/10.1016/j.aquaculture.2009.11.021
56. Rengpipat S., Tunyamum A., Fast A.W., Piyatiratitivoraku S., and Menasveta P., 2003, Enhanced growth and resistance to*vibrio* challenge in pond-reared black tiger shrimp *Penaeus monodon* fed a Bacillus probiotic, Diseases of Aquatic Organisms, 55: 169-173
57. http://dx.doi.org/10.3354/dao055169
58. Sakami T., Fujioka Y., and Shimoda T., 2008, Comparison of microbial community structures in intensive and extensive shrimp culture ponds and a mangrove area in Thailand, Fisheries Science, 74(4): 889-898
59. http://dx.doi.org/10.1111/j.1444-2906.2008.01604.x
60. Sandaa R.A., Magnesen T., Torkildsen L., and Bergh O., 2003, Characterization of the bacterial community associated with early stages of great scallop (*Pecten maximus*), using denaturing gradient gel electrophoresis (DGGE), Systematic Applied Microbiology, 26: 302-311
61. http://dx.doi.org/10.1078/072320203322346164
62. Solano J.L.O., and Soto J.O., 2006, The functional property of Bacillus for shrimp feeds, Food Microbiology, 23: 519-525
63. http://dx.doi.org/10.1016/j.fm.2005.10.004
64. Stokes A., Leffler J.W., Venero J.A., and Browdy C.L., 2009, Developing designing and operating a zero exchange bio-floc based shrimp production system, Journal of Shellfish Research, 28: 732-732
65. Tendencia E.A., and de la Peña L.D., 2001, Antibiotic resistance of bacteria from shrimp ponds, Aquaculture, 195(3-4): 193-204
66. http://dx.doi.org/10.1016/S0044-8486(00)00570-6
67. Trask B.J., van den Engh G.J., and Elgershuizen J.H.B.W., 1982, Analysis of phytoplankton by flow cytometry, Cytometry, 2(4): 258-264
68. http://dx.doi.org/10.1002/cyto.990020410
69. Vaseeharan B., and Ramasamy P., 2003, Control of pathogenic *Vibrio* spp. by *Bacillus subtilis* BT23, a possible probiotic treatment for black tiger shrimp *Penaeus monodon*, Letters in Applied Microbiology, 36(2): 83-87
70. http://dx.doi.org/10.1046/j.1472-765X.2003.01255.x
71. Verschuere L., Rombaut G., Sorgeloos P., and Verstraete W., 2000, Probiotic bacteria as biological control agents in aquaculture, Microbiology Molecular Biology Reviews, 64: 655-671
72. http://dx.doi.org/10.1128/MMBR.64.4.655-671.2000
73. Vijayan K.K., BrightSingh I.S., Jyapraksh N.S., Alavandi S.V., SomnathPai S., Preetha R., Rajan J.J.S., and Santiago T.C., 2006, A brackishwater isolate of *Pseudomonas* PS-102, a potential antagonistic bacterium against pathogenic vibrios in penaeid and non-penaeid rearing systems, Aquaculture, 251: 192-200
74. http://dx.doi.org/10.1016/j.aquaculture.2005.10.010
75. Wang Y., Xu Z., and Xia M., 2005, The effectiveness of commercial probiotics in northern white shrimp *Penaeus vannamei*ponds, Fisheries Science, 71(5): 1036-1041
76. http://dx.doi.org/10.1111/j.1444-2906.2005.01061.x
77. Wasielesky W., Atwood H., Stokes A., and Browdy C.L., 2006, Effect of natural production in a zero exchange suspended microbial floc based super-intensive culture system for white shrimp *Litopenaeus vannamei*, Aquaculture, 258: 396-403
78. http://dx.doi.org/10.1016/j.aquaculture.2006.04.030
79. Yan Q., van der Gast C.J., and Yu Y., 2012, Bacterial community assembly and turnover within the intestines of developing zebrafish, PLoS One, 7(1): e30603
80. http://dx.doi.org/10.1371/journal.pone.0030603
81. http://dx.doi.org/10.1371/journal.pone.0030603

CHAPTER 13

LIFE CYCLE ASSESSMENT OF AQUACULTURE SYSTEMS—A REVIEW OF METHODOLOGIES

Patrik J. G. Henriksson & Jeroen B. Guinée & René Kleijn & Geert R. de Snoo

Keywords: Aquaculture, FishFoodLCAL, Life cycle assessment, ReviewSeafood

INTRODUCTION

Historical increases of yields from capture fisheries have been achieved by increasing fishing efforts and exploring new fishing grounds. Around 1990, however, global fish landings levelled off, followed by increases in fuel consumption as fishing boats had to cover larger distances to reach productive fishing grounds and greater efforts were required to maintain catches (Tlusty and Lagueux 2009; FAO 2010a). As a result of this development, aquaculture has become increasingly important in meeting the rising global demand for aquatic food products, with annual per capita supply from aquaculture growing from 0.7 to 7.8 kg globally since the 1970s (FAO 2010a). Farming methods for aquatic organisms are highly diverse, and 91% of global production is based in Asia (FAO 2010b). Global aquaculture production is dominated by finfish (49% by weight), followed by aquatic plants (23%), bivalves (19%), and crustaceans (7%) (FAO2010b). Small-scale production of freshwater fish from ponds in Asia is the most common production system, with a general global trend towards intensification (Naylor et al. 2000; Muir2005; FAO 2010a).

The rapid expansion of the aquaculture sector has been associated with many sustainability concerns, such as emissions leading to climate change, eutrophication, toxic and ecotoxic impacts, use of antibiotics, use of land and water needed for feed production, loss of biodiversity, introduction of non-indigenous species, spread/amplification of parasites and disease, genetic pollution, dependence on capture fisheries, and socio-economic concerns (Naylor et al. 2000; Pelletier et al. 2007; Pelletier and Tyedmers 2008; Ayer and Tyedmers 2009; Naylor et al. 2009). In the process of better understanding the environmental impacts of aquaculture, life cycle assessment (LCA) has become more frequently used to identify best practices and to assess overall environmental performance (Pelletier and Tyedmers 2008). As part of the EU FP7 SEAT project (Sustaining Ethical Aquaculture Trade, www.seatglobal.eu), LCA studies of shrimp, freshwater prawn, tilapia, and catfish will be performed in Bangladesh, Thailand, Vietnam, and China during the upcoming years. To provide a starting point for these studies, we have reviewed ten articles found in ISI Web of Knowledge (accessed on 30-Nov-2010) and two PhD theses focusing on LCA of aquaculture systems. Although several other studies are available,[1] this

Citation: Patrik J. G. Henriksson & Jeroen B. Guinée & René Kleijn & Geert R. de Snoo, Life cycle assessment of aquaculture systems—a review of methodologies, DOI 10.1007/s11367-011-0369-4.

review only focuses on peer-reviewed literature. The present review aims to clarify the methodological choices made, identify possible data gaps, and provide recommendations for future developments in this field of research, as well as for the upcoming SEAT` LCA studies.

MATERIALS AND METHODS

Originally developed for industrial production and processes, LCA was later applied to food products with the first LCA studies of food production being published in the early 1990s (Roy et al. 2009). These allowed for the first aquaculture LCAs, where agricultural and livestock systems provide resources for fish-feed production. The first published aquaculture-related LCA was done by Papatryphon et al. (2004b) in order to evaluate salmon feed. This was followed by a number of publications on aquaculture production in the second half of that decade (Table 1). The most productive institutes in this field of research have been the French INRA-IFREMER and Dalhousie University, Canada. Ten of the LCAs focused on finfish production, while Mungkung (2005) studied shrimp farming and Iribarren et al. (2010) examined mussel production. Of the ten finfish studies, nine focused on intensive production, and one (Phong2010; Phong et al. 2011) described integrated semi-intensive farming systems. Six studies examined production systems based in Europe, three in Asia, and two in North America, while one study described global production.

Table 1: The ten articles and two PhD theses under review with general methodological choices highlighted

General				Functional unit	System boundary	Allocation method	Data and data quality		Institute
Reference	Species	System	Country				Software	Database	
Aubin et al. 2006	Turbot	Re-circulating	France	1 ton live weight	Farm gate	Economic value	SimaPro v.6.0	Buwal B250, Ecoinvent[a]	INRA/IFREMER
Aubin et al. 2009	Rainbow trout, sea-bass and turbot	Various	France	1 ton live weight	Farm gate	Economic value	SimaPro v.6.0	Buwal B250, Ecoinvent[a]	
d'Orbcaster et al. 2009	Trout	Various	(Model)	1 ton live weight	Farm gate	Economic value	SimaPro v.6.0	EDF 2004; etc.	
Pelletier and Tyedmers 2007	Atlantic salmon	Salmon feeds	Canada	1 ton live weight	Farm gate	Gross nutritional energy	SimaPro v.7.0	Ecoinvent v.2; Franklin[a]	
Ayer and Tyedmers 2009	Atlantic salmon and char	Various	Canada	1 ton live weight	Farm gate	Gross nutritional energy	SimaPro v.7.0	Ecoinvent 1.2; Franklin; IDEMAT 2001; LCA Food 2005	Dalhousie
Pelletier et al. 2009	Atlantic salmon	Cage	Global	1 ton live weight	Farm gate	Gross nutritional energy	Simapro v.7.1.8	Ecoinvent v.2	
Pelletier and Tyedmers 2010	Tilapia	Net cage and ponds	Indonesia	1 ton fillets	Market	Gross nutritional energy	SimaPro v.7.0	Ecoinvent v.2	
Mungkung 2005	Shrimps	Ponds	Thailand	1.8 kg shrimp tails	Waste	Economic value	SimaPro v.5.1	Buwal B250; etc[a]	
Grönroos et al. 2006	Rainbow trout	Net cage	Finland	1 ton dead weight	Farm gate	Mass	KCL-ECO 2003	Silvenius and Grönroos 2003	Other
Ellingsen and Aanondsen 2006	Atlantic salmon	Net cage	Norway	200 gram fillet	Market	Mass/Economic value	SimaPro v.6.0	ETH-ESU 96; Buwal 250	
Iribarren et al. 2010	Blue mussels	Rafts	Spain	1 kg of dry mussel flesh	Consumer	System expansion	SimaPro v.6.0	Ecoinvent v.2	
Phong 2010	Various[b]	Ponds, integrated	Vietnam	1 kg of live fish/1 kcal	Farm gate	Economic value	Excel[a]	Ecoinvent[a]	

[a]Data has been supplemented by personal communication

[b]Tilapia, kissing gourami, giant gourami, silver barb, common carp, silver carp and striped catfish

Below, we discuss the LCA methodology that exhibited the greatest difference amongst the LCA studies listed in Table 1 (Guinée et al. 2002):

- Functional unit
- System boundaries
- Data and data quality
- Allocation
- Impact assessment methods
- Interpretation methods

The selection of these six phases is based on an analysis of all methodological assumptions and choices made and data sources adopted for the different steps of Goal and Scope definition, Inventory Analysis, Impact Assessment, and Interpretation. An analysis of other methodological phases and a detailed discussion of the impact categories are published as online resource to this article. Each of the six phases listed above is analyzed below in terms of differences amongst the studies, followed by a summary of the current LCA literature and, where appropriate, recommendations for research or harmonization.

RESULTS

Functional Unit

The functional unit is the reference unit used to quantify the performance of a production system (ISO 14044 2006). The most commonly used functional unit in the twelve studies reviewed here is 1 ton of live fish at the farm gate (six studies; see Table 1). Two other studies also limit themselves to the farm gate, with Grönroos et al. (2006) adopting dead weight and Phong (2010) adopting 1 kcal alongside 1 kg as his two functional units. Four studies defined their functional unit in terms of edible yield, defined as the main part of the organism that was marketed (fillets, flesh, or tails).

The functional unit is the basis of comparison in comparative LCAs. The functional unit follows the goal of the study, since different goals may require different functional units. The goal of the study and the associated functional unit partially defines the system boundary of the inventory. For example, if frozen fillets in supermarkets are chosen as a functional unit, the system boundary needs to be defined so as to include processing, transportation, and distribution. The functional unit may, moreover, significantly influence comparative LCAs involving different species, since the edible portions and nutritional values of products can differ by an order of magnitude (Roy et al. 2009). Mussels and shrimp, for example, provide respectively, 13.6 and 140 kg of protein per ton of whole animals harvested (Mungkung 2005; Iribarren et al. 2010;www.nutraqua.com accessed 23-June-2010).

The choice of the functional unit is important for comparisons between species as well as across cultures, as the definition of edible will depend on cultural influenced consumer preferences. The choice of functional unit will also influence allocation decisions at the farm gate where more descriptive functional units, such as kilocalorie, may be more appropriate for comparisons between multi-output systems (Phong 2010). We therefore recommended to carefully choose a functional unit tailored to the goal and scope of the study.

System Boundaries

The system boundary determines which unit processes will be included within an LCA study and which ones are to be excluded. With respect to the diverse goals of the reviewed studies, few have considered supply chain impacts beyond the farm gate (Pelletier and Tyedmers 2010). Iribarren et al. (2010), however, did include the whole production chain and found that a significant part of the emissions from mussel production is related to processing and marketing, with dispatch centers contributing significantly to the overall emissions from live mussel production. Infrastructure is also often excluded, due to the large amount of time that has to be invested in calculating the total input in relation to the small impact that is considered (Ayer and Tyedmers 2009). Where included and distinguished (Aubin et al. 2006; Ayer and Tyedmers2009; Aubin et al. 2009; d'Orbcaster et al. 2009), however, infrastructure was found to contribute between 0% and 19.0% to the overall impacts in terms of global warming, eutrophication, and acidification indicators. Common cut-offs were based on the outcomes of previous studies, selection of impact categories, available data, and resource constraints (Mungkung 2005; Ellingsen and Aanondsen 2006; Grönroos et al. 2006; Pelletier et al. 2009; Pelletier and Tyedmers 2010).

The selection of the system boundary should be consistent with the goal of the study, and the criteria used to establish the system boundary should be identified and explained (ISO 140442006). In aquaculture systems, the length of the full production chain is largely dependent on the type of system (Fig. 1). For example, external inputs of feed and hatchery-reared juveniles may not be needed in extensive systems and if the product is sold fresh on the market it needs no processing (e.g., carp in China). Fish and seafood are also the most perishable of food products, and the level of processing will influence the longevity of the product, as well as the amounts wasted, hence the environmental impacts (Sonesson et al. 2005).

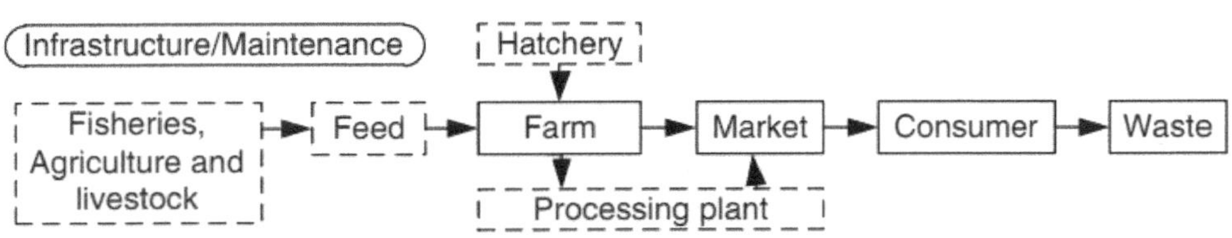

Figure 1: Simplified flow chart of aquaculture production. The inclusion of some processes (*dashed lines*) are dependent upon the system in focus.

The problem in defining cut-offs for the quantification of inventories is a lack of readily accessible data, implying disproportionate expenditure of funds and efforts on data collection. Limitations of time, funds, or data access will inevitably lead to the exclusion of processes and to less complete and accurate results. Nowadays, it is, however, possible to handle the cut-off problem better, by estimating the environmental interventions associated with flows for which no readily accessible data is available using environmentally extended input–output analysis (EIOA) (Suh et al. 2004). For the purpose of consumer guidance, we recommend a more extensive system boundary at or beyond the market, as impacts may otherwise be underestimated (Iribarren et al. 2010). Further efforts should also be directed towards expanding current knowledge about the contributions from infrastructure, as this has been reported to have a larger influence on agriculture than on most other industrial processes (Frischknecht et al. 2007).

Data and Data Quality

Although all of the studies reviewed here model relevant agriculture, fisheries and other related processes—to different extents—most of the inventory details of these modeling efforts remain unpublished. Articles that do extensively report the data mainly specify economic flows, with environmental flows often limited to nutrient balances. This may be due to the aquaculture-based background of most of the researchers for whom eutrophication has historically been a major concern. In various articles, it remains unclear whether background databases were used or whether real foreground data (site-samples) had been collected (ILCD 2010); neither is it always clear which processes that were included in the study. Consequently, reproducing their results is difficult or impossible. For example, Ellingsen and Aanondsen (2006) reported: "Data are generally collected from various sources by both literature surveys, a study of available data sources, telephone conversations, and meetings". This, unfortunately, provides no clue as to which processes, data, or data sources were included in the study.

Background data used in the studies were derived from a wide range of databases, including some which were quite out-dated (ETH 1996 and BUWAL 250) (see Table 1). Some authors used combinations of different databases or did not clearly specify the precise database(s) consulted. Several studies, for example, reported that they had used SimaPro software, with all of the databases included in it. As SimaPro includes many different databases (e.g., Ecoinvent, US LCI database, US IO dbase, Danish IO dbase, Dutch IO dbase, LCA food dbase, Industry data, Japanese IO dbase, IVAM dbase; seehttp://www.pre.nl/simapro/inventory_databases.htm), the actual data sources used in these studies remain unclear. All studies, moreover, rely on European databases (commonly different versions of Ecoinvent), even though various studies dealt with aquaculture in non-European countries. Although the authors of several studies did invest much effort in adapting inventories to regional conditions, there remains a real need for databases representing technologies of developing countries.

The focus of the studies ranged from single farms (Aubin et al. 2009; Ayer and Tyedmers 2009) to small samples of each farming system (Phong 2010), to aggregated industry averages representing significant parts of national outputs (Pelletier et al. 2009; Pelletier and Tyedmers2010). However, the quality of foreground data available for aquaculture systems often depends on the intensity of the system and the region of data collection. Highly intensive systems, such as land-based salmon systems, often keep more complete records of all inputs and outputs, while only general estimations are available for most extensive pond systems in rural areas. Accessibility to feed inventories may, moreover, be subject to the scale and nature of the feed mill, as exact mixtures of ingredients often are held confidential. Site-specific measurements are, moreover, dependent on the resources available. The articles offer limited reporting on other environmental flows beyond nutrient budgets (including methane, nitrous oxide emissions, copper-based anti-fouling agents, antibiotics, etc.).

The International Organization for Standardization (ISO 14044 2006) states that data quality requirements should be specified to enable the goal and scope of the LCA to be met, and also that the treatment of missing data should be documented, and that data sources as well as an assessment of the reproducibility of the study results by independent practitioners should be addressed as part of the data quality requirements. ISO does not, however, demand publication of all data used.

Transparency in the reporting of data and reproducibility of results are important for proper peer-reviewing and interpretation of background data, at least to the extent that this is possible with regards to sensitive industry inventory data. A good example of the way data can be published without compromising the focus of the article was given by Grönroos et al. (2006) and Pelletier et al. (2009), who both published supporting documents describing inventories (although with different coverage of environmental data; see above), core processes, assumptions, and calculations. Another solution to fitting large inventories to the often restricted format of scientific journals is to report which processes derived from a background database (e.g., ecoinvent) were included in the study without actually including the data of that process. Such processes could simply be reported using the process ID numbers, rather than the full process names. This kind of more open reporting of data is critical for developing specific LCA data sets for aquaculture-related processes, as much primary data currently are lost by aggregating results and by only presenting impacts, rather than inventories. It should, however, be pointed out that the data sourcing and reporting issues discussed here are not unique to aquaculture LCAs, but rather apply to the majority of LCA studies published, whether peer-reviewed or not.

Allocation

Some of the main differences amongst the studies reviewed here are related to allocation. While all of the most commonly applied procedures for allocation (including mass, economic value, gross energetic content, and system expansion) have been applied to aquaculture LCAs, economic value and gross nutritional energy content have more frequently been used in the more recent publications (see Table 1). This is also the main methodological difference between the two main publishing institutions, with INRA-IFREMER applying economic allocation, while researchers at Dalhousie University commonly prefer gross energy content as the basis for allocation (see Table 1).

Four publications applied system expansion to certain allocation situations (Ayer and Tyedmers2009; Pelletier et al. 2009; Pelletier and Tyedmers 2010; Iribarren et al. 2010). Iribarren et al. (2010), for example, used system expansion for Spanish mussel production (with mussel as the main product and shells as a co-product) with the assumption that mussel shells could be used to replace conventional calcium carbonate production. Grönroos et al. (2006) restricted their analysis to whole fish at the farm gate to avoid allocation in the processing phase, while mass allocation was used for feed inputs. Some authors did not report their allocation decisions in their articles.

Most industrial processes yield more than one product, and some recycle expanded products as raw materials. As a result, the materials and energy flows, as well as the associated environmental releases, have to be allocated to the different products according to clearly stated and justified procedures. In aquaculture, many of the feed inputs are co-produced in other production systems (e.g., rice bran, fisheries by-catch, and co-products from livestock processing), and co-products also occur in the processing phase.

It is our belief that the multi-functionality problem is an artefact of the desire to isolate one function out of many and as artefacts can only be resolved in an artificial way, there will always be more than one way of solving the multi-functionality problem. This is illustrated by the debate on methods to deal with the multi-functionality problem over the last two to three decades which still has not provided a generally accepted method. Depending on the application (e.g., policy or scientific publications), using alternative allocation methods could be seen as an opportunity to produce more realistic ranges of results and provide stronger conclusions. There are, however, certain requirements that need to be addressed when dealing with allocation issues, such as that the solution should be consistent, well justified and in-line with main methodological principles (Guinée et al. 2004; ILCD 2010). It is also important to always report on the allocation method(s) applied and perform a sensitivity analysis, as allocation plays a pivotal role in the performance of a production system (ISO 14044 2006).

Life Cycle Impact Assessment Methods

All reviewed studies applied one or more life-cycle impact assessment methods. The major impact assessment methodology used for characterization was the midpoint CML baseline method (Guinée et al. 2002) with only Ellingsen and Aanondsen (2006) applying an endpoint approach (eco-indicator 99 method; Goedkoop and Spriensma 1999). Grönroos et al. (2006) choose to use region-specific characterization factors for eutrophication and acidification, while making a distinction between aquatic and terrestrial emissions. Only climate change, acidification, and eutrophication were adopted as impact categories by all studies. In addition, a few novel methods were introduced for biotic resource use, water dependency, and land (surface) use (Table 2).

Table 2: Frequency of applying different impact categories in LCA studies on aquaculture and the impact assessment method used.

Impact category	Σ	**Impact assessment method**
Global warming potential	12	Houghton et al. 2001 (1, 2, 3, 4, 5, 6, 7, 8, 9, 10, 11, 12)
Acidification	12	Huijbregts 1999a(1, 2, 5, 6, 7, 8, 9, 10, 11); Goedkoop and Spriensma1999 (3); Seppälä et al. 2006 (4); Heijungs et al. 1992 (12)
Eutrophication	12	Heijungs et al. 1992 (1, 2, 5, 6, 7, 8, 9, 10); Goedkoop and Spriensma1999 (3); Weidema et al. 1996 (12), Seppälä et al. 2004 (4)[a]; Seppälä et al. 2006 (4)[a]
Energy use	8	VDI 1997 (2, 5, 6, 7, 8, 9, 11); Goedkoop and Spriensma 1999 (3); Article specific (12)
Biotic resource use	6	Papatryphon et al. 2004b (2, 6, 8); Pelletier and Tyedmers 2007 (5, 9, 11)
Marine aquatic ecotoxicity	4	Huijbregts 1999b (5, 7, 10); Meent and Klepper 1997 (3)[b]
Abiotic depletion potential	3	Guinée and Heijungs 1995 (1, 7, 10)
Ozone depletion potential	3	WMO 1999 (1, 10); Goedkoop and Spriensma 1999 (3)
Human toxicity	3	Huijbregts 1999b (1, 7, 10)
Water dependence	2	Own methodology (6, 8)
Photochemical oxidant formation	2	Derwent et al. 1998/Jenkin and Hayman 1999 (1, 10)
Freshwater aquatic ecotoxicity	2	Meent and Klepper 1997 (3)[b]; Huijbregts 1999b (10)
Terrestrial ecotoxicity	2	Meent and Klepper 1997 (3)[b]; Huijbregts 1999b (10)
Surface use	2	Own methodology (8, 12)
Respiratory impacts from inorganics	1	Goedkoop and Spriensma 1999 (3)
Carcinogenic effects on humans	1	Goedkoop and Spriensma 1999 (3)

Global warming, acidification and eutrophication were the only impact categories applied by all authors. References: (1) Mungkung 2005; (2) Aubin et al. 2006; (3) Ellingsen and Aanondsen 2006; (4) Grönroos et al. 2006; (5) Pelletier and Tyedmers 2007; (6) Aubin et al.2009; (7) Ayer and Tyedmers 2009; (8) d'Orbcaster et al. 2009; (9) Pelletier et al. 2009; (10) Iribarren et al. 2010; (11) Pelletier and Tyedmers 2010; (12) Phong 2010. For full references on the impact assessment methods, please refer to the Online Resource (ESM)

[a]Aquatic and terrestrial eutrophication was reported separately

[b]Ecotoxicity is summarized under one category

As regards climate change, the characterization factors suggested by the international panel on climate change (IPCC; Houghton et al. 2001) were the basis for all reviewed studies. This therefore enables for valid conclusions to be drawn amongst the studies, e.g., the great importance of feed inputs for aquaculture systems.

As regards acidification, all but three studies adopted the approach developed by Huijbregts (1999a). Apart from Ellingsen and Aanondsen (2006) and Grönroos et al. (2006), Phong (2010) chose alternative characterization factors, in this case, the older Heijungs et al. (1992) acidification method. As regards eutrophication, similar differences are found as for acidification, while Grönroos et al. (2006) chose to separate terrestrial and aquatic emissions due to their distinct association to feed production and feed application, respectively. Phong (2010), again, refers to an older alternative publication, Weidema et al. (1996).

Cumulative primary fossil energy demand was the fourth most commonly included impact category amongst the studies and showed a large overlap with abiotic resource depletion (Ayer and Tyedmers 2009). Strikingly, six studies adopted and quantified a novel impact category, biotic resource use. Its use aims to capture the ultimate carbon-based energy stemming from biological systems that support fed aquaculture production, although a standardized protocol for this impact category still remains to be developed (Pelletier et al. 2007). Marine exotoxicity, an impact category for which the existing characterization methods have been widely debated within the LCA community (Pettersen and Hertwich 2008; Gloria et al. 2006), was adopted and quantified in four studies. A range of other toxicity related impact categories were less frequently adopted, along with abiotic resource depletion and ozone depletion. Water dependency and land use were represented in only two studies each, using own methodology.

Little consideration was, however, given to the type of water use (e.g., marine or freshwater, degradative or consumptive; Bayart et al. 2010) on either the input or the output side, nor were emissions relating to land use and transformation considered (ILCD 2010). Other concerns not covered by the LCA methodologies reported in our review include impacts on the seafloor from capture fisheries, the introduction of invasive species, the spread of diseases, genetic pollution, and socio-economic concerns (Pelletier et al. 2007).

In summary, the current review of aquaculture LCAs shows that impact assessment methodologies have been applied to all studies reviewed. The range of impact categories covered is, however, limited, and the methods adopted for the various categories differ, hampering comparisons of study results. Some authors used old characterization factors, while others developed their own quantification methods. Future harmonization with the developments within the LCA community is therefore advised, focusing on the standardization efforts promoted by ILCD (the European Commission's Join Research Centre) and UNEP-SETAC (the United Nations Environment Programme and the Society of Environmental Toxicology and Chemistry) including the ILCD handbook (lct.jrc.ec.europa.eu), USEtox™ (www.usetox.org), land-use (lcinitiative.unep.fr, accessed: 17-Oct-2010) and freshwater use (Bayart et al. 2010) (for a complete overview of the life cycle impact assessment methods adopted by the different studies reviewed here, please see the Online Resource (ESM) to this paper).

Interpretation Methods

Although all studies performed a dominance or contribution analysis, many did not perform a complete set of sensitivity analyses, as is required by the current ISO standards. Ayer and Tyedmers (2009), however, conducted an extensive set of sensitivity analyses, one of which highlighted the importance of electricity sourcing. Another study by Pelletier and Tyedmers (2007) concluded that allocation factors strongly influence the impact of different feed inputs. Both d'Orbcaster et al. (2009) and Pelletier et al. (2009) drew a parallel between food conversion ratios (FCRs, defined as kilogram dry feed/kilogram live fish) and GHG (greenhouse gas) emissions, while Mungkung (2005) supported her conclusions by performing a sensitivity analysis on data assumptions for fishing practices as well as for different impact assessment methods. Ellingsen and Aanondsen (2006) also used two alternative impact assessment methods to strengthen their conclusions. Pelletier et al. (2009) evaluated the range of nitrous oxide emissions from nitrogen fertilizers, compared to the default value indicated by the IPCC. Only Phong (2010) applied statistical tools to different farming practices, in the form of one-way ANOVA (analysis of variance).

According to ISO (2006), the life cycle interpretation phase of an LCA comprises the identification of the significant issues based on the results of the LCI (life cycle inventory) and LCIA stages, an evaluation involving completeness, sensitivity, and consistency checks, and finally the formulation of conclusions, limitations, and recommendations. It is an important phase of any LCA study, where any weaknesses should be highlighted and results critically tested. Irregularities at temporal and spatial scales give rise to deviations in inventories of aquaculture production. Underlying models, moreover, rely on assumptions and methodological choices influence the results. Statistical tools and sensitivity analyses are therefore important to strengthen the arguments and conclusions in aquaculture LCAs. Treating farms individually, rather than as averages, would here allow for more extensive statistical comparisons to be made between farms. Known pivotal factors identified in the articles reviewed here include various inventory choices (feeds, raw materials, infrastructure, etc.), GHG emissions from agricultural fields and aquatic systems, nitrogen and phosphorus emissions, allocation factors, and characterization factors. Further efforts are therefore needed to account for the many degrees of freedom, using more extensive sensitivity analyses and implementing, e.g., Monte Carlo analysis.

DISCUSSION

Aquaculture is currently the fastest growing animal production sector, and ever larger amounts of farmed aquatic products are being traded on international markets. Increasing concerns about the sustainability of production have, however, been raised, and the standards and requirements imposed on the aquaculture sector are becoming ever stricter. In the search for best practice, LCA has proved to be a valuable tool to identify environmental hot spots and compare different production systems. To date, however, there has been limited LCA coverage of the various farming systems worldwide, especially in Asia from where the bulk of farmed aquatic products originate. The present review has identified a range of methodological and data sourcing approaches reported in existing publications, where methodological choices often govern the outcomes.

Nine of the twelve peer-reviewed publications included in this review focused on intensive finfish production, which represents a small share of the global aquaculture output. Eight of the studies were, moreover, limited to whole fish at the farm gate, which may give misleading results if consumer guidance is the objective. Distribution of fish and seafood to markets may, for example, contribute disproportionately to the overall impacts as these are highly perishable commodities with high value attached to their freshness (Tlusty and Lagueux 2009). More LCA-based research is therefore needed to guide this still expanding sector towards best practice. With the wide-spread of aquaculture in Asia, these studies should focus on Asian aquaculture and alternative farming practices, using a functional unit relevant to the aim of the study.

The greatest single methodological difference amongst the studies was in allocation, with monetary value and gross energy content being the most commonly applied allocation factors. However, the level of reasoning and consistency regarding choices made varied greatly amongst studies. As consensus, let alone scientific clarity, is not likely to be achieved soon, allocation choices should be clearly defined and justified. Inventory results with regard to the allocation method adopted should also be supported by thorough sensitivity analysis, as advocated by ISO. Databases and software should, moreover, simplify the application of alternative allocation decisions to enable more extensive sensitivity analyses.

All studies had adopted the IPCC recommendations for global warming, as it represents a highly resourceful centralized scientific body. Similar developments should be encouraged for other impact categories, following initiatives by ILCD and UNEP-SETAC. However, inventories of the characterized environmental flows need to be made available to allow alternative characterization factors to be implemented. Toxicological implications should also be given more attention as they have historical importance in the aquaculture sector (e.g., Malachite green). New characterization factors and standardized protocols need to be developed to address more aquaculture-specific concerns (e.g., seafloor disturbance and biotic resource use). A distinction between terrestrial and aquatic eutrophication may also have to be made, as these emissions usually have distinctly different origins.

CONCLUSIONS AND RECOMMENDATIONS

There is a need for more detailed LCA studies of non-finfish species, as well as of integrated, extensive, and semi-intensive production of finfish in developing countries (especially in Asia), in order to guide the industry towards best practice, highlight hot spots, and guide consumers. These studies should conform to up-to-date guidelines from, e.g., ISO, ILCD, and SETAC-UNEP in order to move towards a more harmonized methodology. The characterization factors and background databases selected should also be the latest available versions. There is also a need to develop impact categories more specifically related to aquaculture, such as seafloor disturbance, biotic resource depletion, and loss of biodiversity. Moreover, the reporting of methodological choices and data should be improved to allow for comprehensive critical analysis and the joint development of extensive inventories.

Sourcing of background data should be consistent and give consideration to the underlying methodological and geographical characteristics of the database used. More extensive reporting of inventory data as online resource and by defining process numbers is also recommended, as well as efforts to extend the coverage of environmental flows. This would assist the development of specific data (and databases) for aquaculture practices and feeds, which would further promote the quantity and quality of aquaculture LCAs. Finally, the contribution of infrastructure seems to be strongly influenced by the methodology and impact categories used, while applying EIOA to aquaculture systems would allow the importance of missing data in aquaculture LCAs to be estimated. Many of the improvement options mentioned here can be implemented by increasing knowledge exchange between the aquaculture community, from which most of the reviewed studies originate, and the LCA community. In the ongoing SEAT project, the ambition over the coming years is to describe four major aquaculture exports farmed in Asia. Detailed LCAs will be conducted of a representative sample of each major farming system, supported by a larger scoping survey collecting basic data for 1,600 grow-out farmers in the region. Foreground data will also be collected on other actors in the value chain, including feed producers, processing plants, hatcheries, nurseries, and fishmeal factories in each country. The results of this research are to be presented in inventory and impact assessment reports over the upcoming years, with the ambition to adopt the recommendations suggested above. These efforts together with several other LCA studies published after this review (e.g., Bosma et al. 2011; Cao et al. 2011) will hopefully improve our current knowledge of the impacts of the aquaculture sector and promote best practice.

Footnotes

1Including: Seppälä et al. 2002; Papatryphon et al. 2004a, etc.

ACKNOWLEDGEMENTS

This work is part of the Sustaining Ethical Aquaculture Trade (SEAT) project, which is co-funded by the European Commission within the Seventh Framework Programme—Sustainable Development Global Change and Ecosystem (project no. 222889). www.seatglobal.eu.

REFERENCES

1. Aubin J, Papatryphon E, van der Werf HMG, Petit J, Morvan YM (2006) Characterisation of the environmental impact of a turbot (*Scophtalmus maximus*) re-circulating production system using Life Cycle Assessment. Aquaculture 261:1259–1268
2. Aubin J, Papatryphon E, van der Werf HMG, Chatzifotis S (2009) Assessment of the environmental impact of carnivorous finfish production systems using life cycle assessment. J Clean Prod 17:354–361
3. Ayer NW, Tyedmers PH (2009) Assessing alternative aquaculture technologies: life cycle assessment of salmonid culture systems in Canada. J Clean Prod 17:362–373
4. Bayart J-B, Bulle C, Deschênes L, Margini M, Pfister S, Vince F, Koehler A (2010) A framework for assessing off-stream freshwater use in LCA. Int J Life Cycle Assess 15:439–453
5. Bosma R, Anh PT, Potting J (2011) Life cycle assessment of intensive striped catfish farming in the Mekong Delta for screening hotspots as input to environmental policy and research agenda. Int J Life Cycle Assess 9:903–915
6. Cao L, Diana JS, Keoleian GA, Lai Q (2011) Life cycle assessment of Chinese shrimp farming systems targeted for export and domestic sales. Environ Sci Technol 45:6531–6538
7. d'Orbcaster RE, Blancheton J-P, Aubin J (2009) Towards environmentally sustainable aquaculture: comparison between two trout farming systems using life cycle assessment. Aquacult Eng 40:113–119
8. Derwent RG, Jenkin ME, Saunders SM, Pilling MJ (1998) Photochemical ozone creation potentials for organic compounds in Northwest Europe calculated with a master chemical mechanism. Atmos Environ 32:2429–2441
9. Ellingsen H, Aanondsen SA (2006) Environmental impacts of wild caught cod and farmed salmon—a comparison with chicken. Int J Life Cycle Assess 1:60–65
10. FAO (2010a) State of world fisheries and aquaculture 2010. Rome, Italy
11. FAO Fisheries and Aquaculture Information and Statistics Service (2010b) Aquaculture production 1950–2008. Available at:http://www.fao.org/fishery/statistics/software/fishstat/en
12. Frischknecht R, Althaus H-J, Bauer C, Doka G, Heck T, Jungbluth N, Kellenberger D, Nemecek T (2007) The

environmental relevance of capital goods in life cycle assessments of products and services. Int J Life Cycle Assess 12:7–17

13. Gloria TP, Russell AJ, Atherton J, Baker SR, Cook M (2006) Ecological toxicity methods and metals—an examination of two case studies. Int J Life Cycle Assess 11:26–33
14. Goedkoop M, Spriensma R (1999) The Eco-indicator 99—A damage oriented method for Life Cycle Impact Assessment—Methodology Report, PRé Consultants. Amersfoort, Netherlands. Available at: http://www.pre.nl/eco-indicator99/ei99-reports.htm
15. Grönroos J, Seppälä J, Silvenius F, Mäkinen T (2006) Life cycle assessment of Finnish cultivated rainbow trout. Boreal Environ Res 11:401–414
16. Guinée JB, Heijungs R (1995) A proposal for the definition of resource equivalency factors for use in product life-cycle assessment. Environ Toxicol Chem 14:917–925
17. Guinée JB, Gorrée M, Heijungs R, Huppes G, Kleijn R, de Koning A, van Oers L, Sleeswijk AW, Sangwon S, Udo de Haes HA, de Bruijn JA, van Duin R, Huijbregts MAJ (2002) Handbook on life cycle assessment: operational guide to the ISO standards. Kluwer Academic Publishers, Dordrecht
18. Guinée JB, Heijungs R, Huppes G (2004) Economic allocation: examples and derived decision tree. Int J Life Cycle Assess 9:23–33
19. Heijungs R, Guinee JB, Huppes G, Lankreijer RM, Udo De Haes HA, Wegener Sleeswijk A, Ansems AMM, Eggels PG, van Duin R, De Goede HP (1992) Environmental lift-cycle assessment of products. CML, Leiden University, Leiden, The Netherlands, Guide and Backgrounds
20. Houghton JT, Ding Y, Griggs DJ, Noguer M, van der Linden PJ, Dai X, Maskell K, Johnson CA (2001) Climate Change 2001: The Scientific Basis, Intergovernmental Panel on Climate Change (IPCC). Cambridge Univ Press, Cambridge, UK
21. Huijbregts MAJ (1999a) Life cycle impact assessment of acidifying and eutrophying air pollutants. Interfaculty Department of Environmental Science, Faculty of Environmental Science, University of Amsterdam, Calculations of equivalency factors with RAINS-LCA
22. Huijbregts MAJ (1999b) Priority assessment of toxic substances in LCA—development and application of the multi-media fate, exposure and effect model USES-LCA. University of Amsterdam, IVAM environmental research
23. ILCD—International Reference Life Cycle Data System Handbook (2010) Analysis of existing environmental impact assessment methodologies for use in life cycle assessment. JRC, European Commission
24. Iribarren D, Moreira MT, Gumersindo F (2010) Revisiting the life cycle assessment of mussels from a sectorial perspective. J Clean Prod 18:101–111
25. ISO 14044 - International Organization for Standardization (2006) Environmental management—life cycle assessment—requirements and guidelines (ISO 14044:2006). International Organization for Standardization, Geneva, Switzerland
26. Jenkin ME, Hayman GD (1999) Photochemical ozone creation potentials for oxygenated volatile organic compounds: sensitivity to variations in kinetic and mechanistic parameters. Atmos Environ 33:1275–1293
27. Meent D van de, Klepper O (1997) Mapping the potential affected fraction (PAF) of species as an indicator of generic toxic stress, RIVM report 607504001, Bilthoven
28. Muir J (2005) Managing to harvest? Perspectives on the potential of aquaculture. Phil Trans Roy Soc B 360:191–218
29. Mungkung RT (2005) Shrimp aquaculture in Thailand: application of life cycle assessment to support sustainable development. Ph.D. thesis. Center for Environmental Strategy, School of Engineering, University of Surrey, Surrey, UK
30. Naylor RL, Goldburg RJ, Primavera JH, Kautsky N, Beveridge MCM, Clay J, Folke C, Lubchenco J, Mooney H, Troell M (2000) Effect of aquaculture on world fish supplies. Nature 405:1017–1024
31. Naylor RL, Hardy RW, Bureau DP, Chiu A, Elliott M, Farrell AP, Forster I, Gatlin DM, Goldburg RJ, Hua K, Nichols PD (2009) Feeding aquaculture in an era of finite resources. Proc Natl Acad Sci 106:15103–15110
32. Papatryphon E, Petit J, van der Werf HMG (2004a) The development of life cycle assessment for the evaluation of rainbow trout farming in France. In: Halberg N (ed) Life cycle assessment in the agri-food sector. Proceedings from the 4th international conference. Danish Institute of Agricultural Sciences, Horsens, pp 73–80
33. Papatryphon E, Petit J, Kaushik SJ, van der Werf HMG (2004b) Environmental impact assessment of salmonid feeds using life cycle assessment (LCA). Ambio 33:316–323
34. Pelletier NL, Tyedmers PH (2007) Feeding farmed salmon: is organic better? Aquaculture 272:399–416
35. Pelletier NL, Tyedmers PH (2008) Life cycle considerations for improving sustainability assessments in seafood awareness campaigns. Environ Manage 42:918–941
36. Pelletier N, Tyedmers PH (2010) A life cycle assessment of frozen Indonesian tilapia fillets from lake and pond-based production systems. J Ind Ecol 14:467–481
37. Pelletier NL, Ayer NW, Tyedmers PH, Kruse SA, Flysjo A, Robillard G, Ziegler F, Scholz AJ, Sonesson U (2007) Impact categories for life cycle assessment research of seafood production systems: review and prospectus. Int J Life Cycle Assess 12:414–421
38. Pelletier NL, Tyedmers PH, Sonesson U, Scholz A, Zeigler F, Flysjo A, Kruse SA, Cancino B, Silverman H (2009) Not all salmon are created equal: life cycle assessment (LCA) of global salmon farming systems. Environ Sci Technol 43:8730–8736
39. Pettersen J, Hertwich EG (2008) Critical review: life-cycle inventory procedures for long-term release of metals. Environ Sci Technol 42:4639–4647

40. Phong LT (2010) Dynamics of sustainability in Integrated Agriculture-Aquaculture Systems of the Mekong Delta. Ph.D thesis. Wageningen Institute of Animal Sciences, Wageningen University

41. Phong LT, de Boer IJM, Udo HMJ (2011) Life cycle assessment of food production in integrated agriculture–aquaculture systems of the Mekong Delta. Livest Sci 139:80–90

42. Roy P, Nei R, Orikasa T, Xu Q, Okadome H, Nakamura N, Shiina T (2009) A review of life cycle assessment (LCA) on some food products. J Food Eng 90:1–10

43. Seppälä J, Silvenius F, Grönroos J, Mäkkinen T, Silvo K, Storhammar E (2002) A life cycle assessment study of rainbow trout. Finnish Environmental Institute, Finish Game and Fisheries Research Institute. University of Jyväskylä, Finland

44. Seppälä J, Knuuttila S, Silvo K (2004) Eutrophication of aquatic ecosystems. A new method for calculating the potential contributions of nitrogen and phosphorus. Int J Life Cycle Assess 9:90–100

45. Seppälä J, Maximilian P, Johansson M, Hettelingh J-P (2006) Country-dependent characterisation factors for acidification and terrestrial eutrophication based on accumulated exceedance as an impact category indicator. Int J Life Cycle Assess 11:403–416

46. Silvenius F, Grönroos J (2003) Fish farming and the environment-Results of inventory analysis. Finnish Environment Institute 276. Helsinki, Finland

47. Sonesson U, Mattsson B, Nybrant T, Ohlsson T (2005) Industrial processing *versus* home cooking: an environmental comparison between three ways to prepare a meal. Ambio 34:414–421

48. Suh S, Lenzen M, Treloar GJ, Hondo H, Horvath A, Huppes G, Jolliet O, Klann U, Krewitt W, Moriguchi Y, Munksgaard J, Norris G (2004) System boundary selection in life-cycle inventories using hybrid approaches. Environ Sci Technol 38:657–664

49. Tlusty MF, Lagueux K (2009) Isolines as a new tool to assess the energy costs of the production and distribution of multiple sources of seafood. J Clean Prod 17:408–415

50. VDI (1997) Cumulative energy demand—terms, definitions, methods of calculation. In: Ingenieure VD. VDI-Richtlinien 4600, Düsseldorf, Germany

51. Weidema BP, Morteson B, Nielsen P, Hauschild M (1996) Elements of an impact assessment of wheat production. Technical University of Denmark, Lyngby, Institute for Product Development

52. WMO (World Meteorological Organization) (1999) Scientific assessment of ozone depletion: 1998. Global Ozone Research and Monitoring Project. Report 44. Geneva, Switzerland

CHAPTER 14

A NEW CONCEPT FOR AQUAPONIC SYSTEMS TO IMPROVE SUSTAINABILITY, INCREASE PRODUCTIVITY, AND REDUCE ENVIRONMENTAL IMPACTS

Werner Kloas, Roman Groß, Daniela Baganz, Johannes Graupner, Henrik Monsees, Uwe Schmidt, Georg Staaks, Johanna Suhl, Martin Tschirner, Bernd Wittstock, Sven Wuertz, Andrea Zikova, Bernhard Rennert

Keywords: Aquaponics concept Tilapia Tomato Sustainability Productivity Minimization of emissions Food security

INTRODUCTION

The growth of the world's population is expected to nearly stabilize at just above 10 billion people after 2062 (UN 2014b) and poses challenges concerning security of water, food, and energy for humans in the 21st century. Water is the primary resource for life and is the main input to produce agricultural goods in the field and throughout the entire agro-food supply chain. Energy is required to distribute water, produce food, and to process and transport agricultural goods. Agriculture is the largest global user of water, accounting for 70% of total withdrawal, and the food production and supply chain consumes ~30% of energy produced globally (UN 2014a). Thus, water, food, and energy are inextricably linked by inter - acting with each other, and the water–food–energy nexus must be managed in a sustainable way concomitant with the protection of the environment to also maintain biodiversity. In order to fulfill global needs for valuable animal protein, aquaculture is by far the fastest-growing sector of agriculture to supply humans with fish (FAO 2014). However, not all aquaculture production technologies for fishes are environmentally sustainable and meet bioeconomical demands. Thus there is an urgent need to integrate aquaculture within the water– food–energy nexus into agricultural production systems using value-added chains concomitant with lowering adverse environmental impacts.

In the early 1970s, recirculating aquaculture systems (RAS) were designed to rear fish in land-based tanks with continuous recycling of water (Bohl 1977, Krüner & Rosenthal 1983) to optimize water use. The major challenge in RAS is the fact that the final product of protein metabolism of fish is excreted as ammonia, which is highly toxic for fish. Therefore,

Citation: Werner Kloas, Roman Groß , Daniela Baganz, Johannes Graupner, Henrik Monsees, Uwe Schmidt, Georg Staaks , Johanna Suhl, Martin Tschirner, Bernd Wittstock, Sven Wuertz, Andrea Zikova, Bernhard Rennert, A new concept for aquaponic systems to improve sustainability, increase productivity, and reduce environmental impacts, doi:10.3354/aei00146.

aquaculture engineering developed bio filters for nitri fication, converting ammonia by aerobic bacteria via nitrite into less toxic nitrate (Collins et al. 1975). However, nitrate accumulation can also have adverse effects on fish (Freitag et al. 2015), and thus, recently, modern RAS focusing solely on fish production often include an anaerobic denitrification unit (Martins et al. 2010), making the whole system more complex. The more conventional ap - proach is to replace daily process water with fresh water in the range of 6 to 12% of the rearing volume to maintain good water quality with regard to nitrogen load. Such fish wastewater contains relatively high concentrations of ni trate and phosphorus, which are essential macro nutrients for plants. Discarding nutrient-rich fish wastewater adversely affects the environment by causing eutrophication of aquatic ecosystems, or it can increase production costs if it has to be passed into sewage treatment plants. Therefore, several attempts have been under - taken in the past to make use of fish wastewater for plant nutrition by aqua ponic systems, a portmanteau for combining freshwater aquaculture with soilless plant production in hydroponic systems (Sneed et al. 1975, Naegel 1977). In principle, the general system design as reviewed by Rakocy et al. (2006) is a single recirculating aqua ponic system (SRAPS) where, in comparison to conventional RAS that only focuses on fish production, the biofilter unit for nitrification is largely reduced or completely reset by a large hydroponic unit of great variety (Rakocy 2012, Love et al. 2014) and can be located in- or outdoors. A conventional aquaponic system is a SRAPS with interconnected aquaculture and hydroponic parts providing value-added chains with regard to the double use of water by fish and plants. Nutrients from fish waste recycled for plant growth in the hydroponic system reduce negative environmental impacts. These designed SRAPS have been a great step forward towards sustainability in regard to resource efficiency for water and nutrients. During the passage through the hydroponic unit, most excreted toxic ammonia is converted by bacteria into much less toxic nitrate that is then consumed by plants, and the cleared water is directly launched further to the fish tanks.

However, although the established SRAPS have already greatly improved the sustainability of aquaculture, there are some principle facts and disadvantages which do not allow optimum production for both fish and plants in such systems. First, all fish and aerobic bacteria used for nitrification, e.g. the genera Nitrobacter and Nitrosomonas, have a pH optimum of ~7 to 9, whereas most hydroponic plant species usually prefer pH levels between 5.8 and 6.2 (Rakocy et al. 2006). Therefore, if the pH is optimal for fish, plant growth is inhibited and vice versa, thereby reducing the productivity of at least 1 of the components of the system. Due to the single circuit, the nutrient concentrations are relatively low in the effluent of fish tanks and cannot accumulate as in a RAS because they are always consistently removed by the hydroponic part. This is why SRAPS are somewhat restricted to produce plants with lower nutrient requirements, such as lettuces and herbs. In order to raise the amount of nitrogen waste, fish are usually reared at relatively high stocking densities and supplied with high amounts of feed, leading to an un - usually high feed conversion ratio (FCR: kg feed / kg biomass gained) as exemplified for tilapia, for which ratios between 1.7 and 1.8 have been reported (Rakocy et al. 2006). Typically, tilapia should have an FCR of ~1.0, which is also the basis for economical production, since fish diets are the biggest cost factor in intense aquaculture (FAO 2006). Finally, due to the large size of the hydroponic units and the evaporation of water by the plants, water use per day is assumed to be relatively high, especially in outdoor systems.

To overcome the disadvantages mentioned above for SRAPS, we developed a new and unique concept for aquaponic systems to improve sustainability, in - crease productivity, and reduce environmental emissions. Within the framework of the project 'aqua - ponic system for (nearly) emission free tomato and fish production in greenhouses' (ASTAF-PRO), a prototype was designed and generated in order to test the practical proof of principle for this new approach.

MATERIALS AND METHODS

System Design

In contrast to SRAPS, the basic design of ASTAFPRO (Fig. 1) consists of a double recirculation aquaponic system (DRAPS) consisting of a RAS for the fish-rearing unit and a recirculating hydroponic unit for plant growth, both being coupled in 2 ways. From the aquaculture to the hydroponic unit, the water flow is unidirectionally connected by a 1-way valve, and thus in the ASTAF-PRO system the water does not directly go back into the fish tanks. The whole system is set up in a greenhouse, including air-conditioning that allows water which has been evapo-transpired via plants and water surfaces to be regained by condensation in cooling traps. The regained process water is then fed back into the aquaculture component. Returning this evaporated water into the RAS minimizes the overall freshwater consumption of the aquaponic system and is an innovative feature of the ASTAF-PRO system. The electrical power for the ASTAF-PRO system was generated by alternative energy using photovoltaics. For the first practical rearing trial in 2009, we selected a combination of Nile tilapia Oreo chro - mis niloticus and tomatoes Solanum lycopersicum.

The greenhouse used is a conventional type made of polycarbonate twin-wall sheet, with floor space of 17.35 × 9.60 m and a total height of 4.75 m. The RAS for fish rearing consists of 9 square fish tanks, of which 7 tanks have a rearing volume of 0.53 m^3 each and the other 2 tanks are 2.1 m^3 each, resulting in a total rearing volume of 7.91 m^3 . One additional 0.53 m^3 tank is used for lime milk addition to stabilize the pH at ~7 for the RAS. The effluent of fish tanks is cleared of solid particles by a lamellae clarifier with a volume of 2.0 m^3 , and the solid-free water is launched into the sump of the biological filter unit (2 × 1.3 × 0.6 m, volume: 1.56 m^3) set up as a trickling filter. Furthermore, an additional 1 m3 water is in the water pipes and trickling filter during the run of the RAS, for a total water volume of the whole RAS of 13 m^3. In the sump, 1 pump (capacity 10 $m^3 h^{-1}$) lifts the water on top of the trickling filter to a rotating irrigation sprinkler, and another pump (capacity 30 m3 h^{-1}) launches the water into the fish tanks to recirculate the process water in the RAS. In addition, 2 heaters, each with a power of 6 kW, are inserted to warm up the water when needed to maintain a constant temperature of ~25°C. The trickling filter consists of a polyethylene cylinder with a width of 1.45 m and a total height of 2.70 m, and the filling material has a volume of ~4 m^3 , for a total surface of ~880 m^2 . The water passing through the trickling filter for nitrification returns degassed and saturated with oxygen. Water flow is regulated to exchange the volume of fish tanks twice per hour. In addition, all fish tanks are aerated via an air blower to maintain the oxygen supply for the fish.

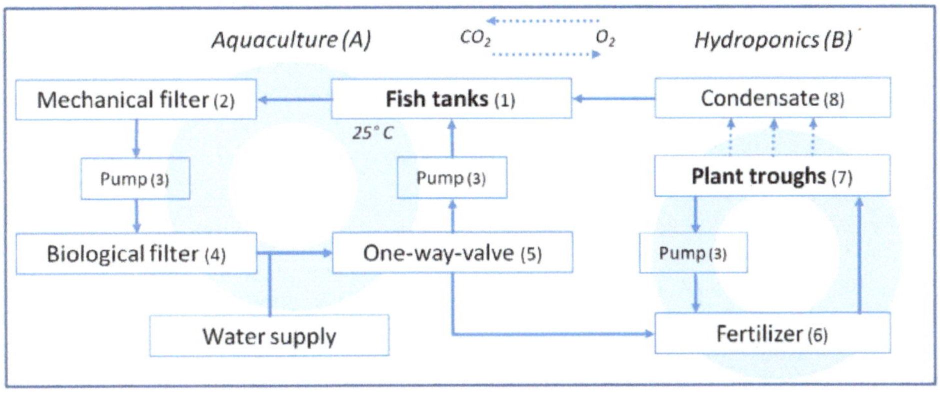

Figure 1. The aquaponic system for (nearly) emission free tomato and fish production in greenhouses (ASTAF-PRO) consists of 2 independent recirculation systems connected unidirectionally from the fish component to the plant component. (A) The aquaculture unit is in principle a recirculating aquaculture system (RAS), and (B) the plant unit resembles a hydroponic system. The RAS consists of (1) fish-rearing tanks, (2) a lamellae clarifier, (3) a recirculation pump system, and (4) a biofilter (trickling filter) for nitrification. (5) A 1-way valve connects the RAS to the (6) fertilizer reservoir of the separate hydroponic unit, where the fish water can be optimized for specific demands of the plant species and a recirculation pump system provides the flow from the reservoir within the hydroponic unit to feed (7) the plant troughs. The plants take up water and nutrients and vaporize water that is mainly gained via condensation by (8) air-conditioning and then re-fed into the aquaculture unit. In addition to value-added chains for water and nutrients, carbon dioxide is used as a nutrient for plants, which in turn provide oxygen via photosynthesis.

The hydroponic unit is based on a nutrient film technique (NFT) to supply water and nutrients via a thin film to the roots and consists of 3 parallel aluminum troughs with lengths of 10.73, 12.51, and 12.96 m, respectively, and a profile of 0.3 × 0.3 m. The troughs are covered by black-and-white foil to prevent light reaching the roots and to inhibit algal growth and evaporation from the water surface. The tomato plants grow soilless on mineral wool cubes placed in the NFT troughs. The fertilizer consists of the clarified fish water from the sump of the trickling filter. It is stored in a reservoir with a volume of 0.8 m3 and is launched via a pump (capacity 9.5 m^3 h^{-1}) into the catchment tank to be evenly distributed into the hydroponic troughs where the nutrient film follows gravity to recirculate back into the fertilizer reservoir.

The specific difference of the ASTAF-PRO system compared to conventional SRAPS is the fact that with the ASTAF-PRO, both recirculation systems run separately from each other and are only connected via a 1-way-valve between the fish RAS and the hydroponic unit. Thus, the major advantage of this DRAPS with separated aquaculture and hydroponic units is the fact that in both parts, optimum conditions can be set up independently for increasing productivity and to prevent any adverse interactions between plant and fish units. For instance, after delivering fish water to the hydroponic reservoir, there is the potential to optimize the fertilizer (i.e. the fish water) with regard to pH and macro- as well as micronutrients for the plant species produced. Therefore, in several aquaponic systems, nutrients like Ca, K, and Fe are added (Rakocy et al. 2004, 2006). The greenhouse is set up with automatic shading and an insulating system for the regulation of light and temperature and, in addition, the hydroponic unit is supplemented with 18 sodium discharge lamps (power 600 W) to bridge periods of low light ex - posure. To regain water vaporized mainly by plant transpiration, an air conditioning system with cooling traps is installed. This system consists of 3 ventilation units, each equipped with a cooler, a heater, and a ventilator, 2 heat pumps (27 kW), and 2 heat ex - changers for heating or cooling. The set value for the greenhouse temperature is 23°C. The water regained by the heat exchanger of the air-conditioning system is collected, measured, and reintroduced into the aquaculture component.

As described above, it is feasible to supplement the fertilizer in the reservoir of the hydroponic unit to optimize nutrient contents. Thus, at the beginning of the practical test, 2 nutrient solutions were dissolved in water, first CalciNitTM (20 g l^{-1}) and second Kristalon™Spezial (10 g l^{-1}) stock solution (both obtained from Yara). CalciNitTM contains 15.5% total N (14.4% N-NO_3 and 1.1% N-NH_4) and 26.3% calcium oxide (CaO), while Kristalon™Spezial is a nitrogen–phosphorus–potassium fertilizer with 18% total N (4.9% N-NO3, 3.3% N-NH_4, and 9.8% NCH_4N_2O), 18% P2O5, 18% K_2O, 3.0% MgO, 2.0% S, and trace elements including Fe, Mn, Zn, B, Cu, and Mo. Kristalon™Spezial was replaced after 3 mo by a solution mixture of $KHCO_3$ and K_2HPO_4.

Evaluation and Testing of the ASTAF-PRO Prototype

For the initial set up (3 March 2009) of the testing period, the aquaculture part was stocked with 1280 Nile tilapia with a total weight of 202 kg. Six out of 7 fish tanks with a volume of 0.53 m^3 and only 1 of the 2 fish tanks with a volume of 2.1 m^3 were stocked, with average weights ranging from 56.0 to 287.7 g $ind.^{-1}$. Stocking densities were in the range of 32.5 to 65.1 kg m^{-3}. Due to growth of fishes the 2 empty fish tanks were stocked until 30 April 2009. The RAS had already been running with fish since 2008 and thus the trickling filter system had already been equipped for the nitrification of nitrogen waste. Three days later, 116 commercially produced tomato plants (strain 'Admiro') with a height of ~40 cm were introduced into the hydroponic unit, in Troughs 1, 2, and 3 with 36, 40, and 40 plants, respectively. The first 4 wk were used to adapt the system for both parts. The lime tank (volume 0.53 m3) was supplied each day with 300 g CaO and set to an exchange rate of 0.5 m^3 d^{-1} in order to stabilize the pH to ~7. The test period started on 3 March with fish stocking adapted to fish weight and tank volume, and the whole testing duration with feeding lasted from 6 March to 26 November 2009. Every 28 d, tilapias were sampled to remove individuals grown to ≥500 g and to retain at least 300 kg total biomass in fish of various sizes in respective tanks. Between each sampling, tilapia were fed with commercial pellets ('Aller Performa 1 gr', Emsland-Aller Aqua, containing 56% protein, 11% fat, 15.5% nitrogen-free extract, 11.2% ash, 1.6% phosphate) at an average feeding rate of 1.25% for 26 d because the day before and during the sampling, the tilapia were not fed. After sampling, the amount of feed given was adapted for each individual fish tank according to the total weight of the reared fish. The lamellae clarifier was cleaned once per week to remove sludge, and water was collected and measured but was not subsequently used for tomato plants due to space limitations for the hydroponic unit. Harvesting of ripe tomatoes started in May, and nutrient concentrations of N-NO_3, N-NO_2, N-NH_4, P-PO_4, K, SO_4, Ca, Mg, Fe, Mo, Cl, Na, and Cu were determined in the aquaculture RAS as well as in the hydroponic recirculation system each week in the sump of the trickling filter and in the catch tank of the hydroponic unit. Electrical conductivity and pH were measured daily

RESULTS

The start of the aquaponictrial with ASTAF-PRO was successful for both components of the system, viz. the aquaculture RAS and the plant hydroponics unit (Fig. 2). Unfortunately, on 1 August 2009, a complete power failure for the whole greenhouse occurred and lasted about half a day, resulting in the loss of >150 kg of tilapia, especially of the smaller ones. In order to replace the lost tilapia, only larger individuals from the last sampling could be used. This affected the entire feeding trial significantly due to the stressed remaining tilapia and the higher FCR of the introduced larger individuals with weights of >500 g. For the tomato plants, no obvious adverse impact was noticed. A flow chart illustrating the movement of nutrients in the system is provided in Fig. 3.

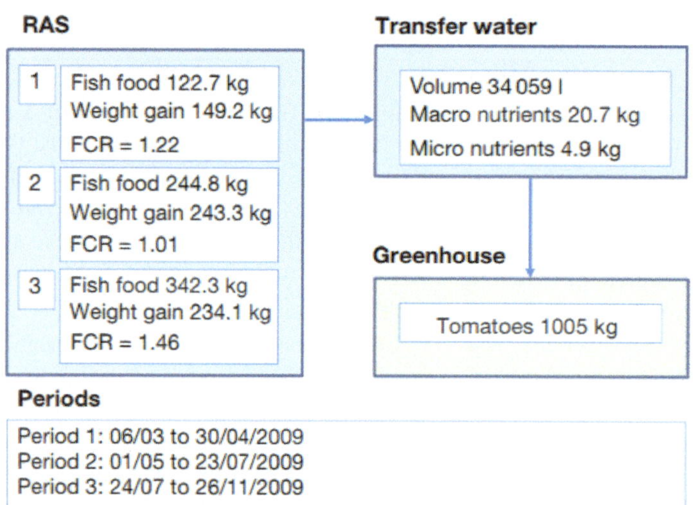

% of optimal plant supply	
Macronutrients	
N-NO_3	128.41%
N-NO_2	-
N-NH_4	2.71%
P-PO_4	17.80%
K	17.02%
SO_4	239.13%
Ca	325.00%
Mg	678.68%
Micronutrients	
Fe	3.75%
Mo	140.00%
Cl	-
Na	-
Cu	60.00%

Figure 3. Nutrient flow through the ASTAF-PRO system (see Fig. 2). The total amounts of nutrients coming from the re - circulating aquaculture system (RAS) are given along with the corresponding percentages of optimum fertilizer supply (table to the right). Fish feed weight, fish weight gain and feed conversion ratio (FCR; kg feed / kg biomass gained) are given for each of the 3 periods; volume of transferred water and tomato biomass are total values over the whole testing period.

Tilapia Production

Until the end of April, a total biomass of >300 kg was reached in the RAS. During that period, fish were also stocked in parallel into the 2 initially fishfree tanks to maintain mean stocking densities of ~38 to 45 kg m^{-3} during each sampling period. Over the whole testing period, the weight gain of tilapia was 600.1 kg using 736.2 kg feed (Fig. 4). Thus, the overall FCR (kg feed / kg biomass gained) over the whole testing period was 1.23. However, after stocking densities were accomplished for all fish tanks containing in total at least 300 kg tilapia, during the first period (1 May to 23 July) without dis turbance by the electrical failure (1 August 2009), the RAS revealed a gain of tilapia of 243.3 kg using 244.8 kg feed, indicating a near optimum FCR of 1.01. After the stress event and the subsequent replacement of dead tilapia by individuals with weights >500 g, the total gain of tilapia was 234.1 kg using 342.3 kg feed, resulting in a higher FCR of 1.46. In order to prevent any further problems concerning continuous energy supply, an emergency power system was implemented in August.

Tomato Production

The tomato plants grew very well, and in all 3 troughs, 1 plant was accidentally lost, resulting in a final total number of 113 flowering tomato plants. Harvesting tomatoes started in May with a total yield of 210.16 kg that increased in June (254.41 kg maximum yield), followed by decreasing yields until November (49.82 kg; Fig. 5). The water use of the tomato plants showed a parallel pattern to the tomato production (Fig. 6). The total harvesting mass of tomatoes was 1005.62 kg, i.e. an average yield of 8.89 kg tomatoes plant−1. It is noteworthy that only a part of the nutrient rich fish water, i.e. 34.06 m^3 , was used by the hydroponic unit for the tomato plants, resulting in a water consumption of 34 l kg−1 tomatoes. The biomass of tomato plants at the end of the experiment was 551 kg in total. As expected, the volume of fish water was larger than what could be consumed by the tomato plants in our facility. However, as we could not establish a larger greenhouse due to space limitations, the surplus fish waste water had to be discharged during the weekly cleaning of the clarifier (2 m^3 , in total ~74 m^3). This volume could have been used as additional fertilizer for another 250 tomato plants, which could have resulted in a total yield of ~3000 kg tomatoes and a total tomato plant biomass of ~1800 kg.

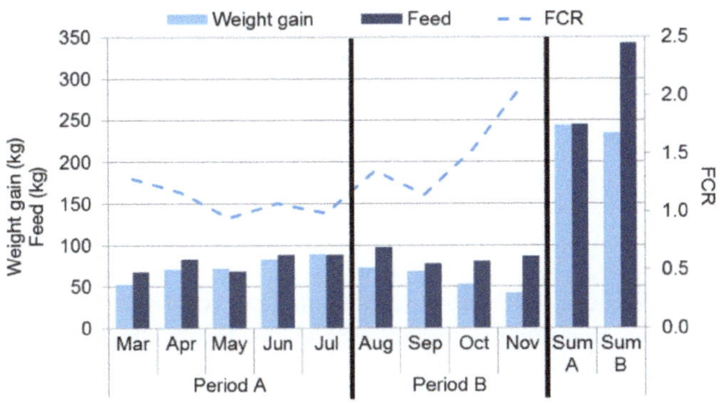

Figure 4. Weight gain of Nile tilapia Oreochromis niloticus, amount of feed used, and feed conversion ratio (FCR: kg feed / kg gained biomass) during the course of the testing period, separated into periods between 2 sampling events. For comparability of stocking densities, the first part of Period A from March to April until completion of fish stocking (i.e. Period 1 in Fig. 3) has been omitted in Sum A so that it represents the weight gain and feed in fully stocked tanks from May to July. On 1 August, fish were stressed by oxygen depletion following a long-lasting power failure, and ~150 kg of fish died and were replaced by older individuals weighing at least 500 g, which notably affected the FCR.

Use of Water

The daily use and introduction of fresh water into the RAS with a total volume of 13 m^3 was 3.83% of the total volume of the RAS. In addition to a small pro - portion of water lost by evaporation in the RAS and the proportion used to water and fertilize the tomato plants, the largest proportion of introduced fresh water was used in the weekly cleaning of the lamellae clarifier by discharging 2 m^3 fish water (see previous section) and sediment into another reservoir and refilling the RAS with 2 m^3 of fresh water. The absolute value of added fresh water was 132.4 m^3 , of which 94.0 m3 were tap water and 38.4 m^3 were condensate gained by the climate control unit and added to the RAS. Thus the daily use and net introduction of tap water into the RAS was remarkably low, with only 2.72%. Considering that we could have supplied additional 250 (instead of only 113) tomato plants with fish water in a larger greenhouse resulting in a total of >360 tomato plants, the regained evapotranspired water might have increased from 38.4 m^3 to ~115 m^3 . Thus, theoretically, the ratio of added fresh water in terms of tap water: condensed water could have changed from 94.0:38.4 m^3 to 17.4:115 m^3 , meaning that only 17.4 m^3 of tap water would have been needed for the whole period be cause most water would be regained by the climate control unit. That theoretical volume of 17.4 m^3 of tap water would result in a daily use and introduction into the RAS of only 0.50% tap water. During our experimental period, on average 220.6 l of fresh water were used (156.7 l of tap water and 63.9 l of condensed water) to produce 1 kg of tilapia and 1.67 kg of tomatoes. As with the FCR, differences be tween testing periods also occurred in terms of water use. During the first undisturbed period (6 March to 23 July) only 190 l of fresh water (134.9 l tap water and 55.1 l condensed water) had to be introduced into the RAS for production of 1 kg of tilapia. In contrast, because of the stress imposed on the fish by the electrical power failure during the second period (24 July to 26 No vember), 267 l of fresh water (189.6 l of tap water and 77.4 l of condensed water) were needed to produce 1 kg of tilapia.

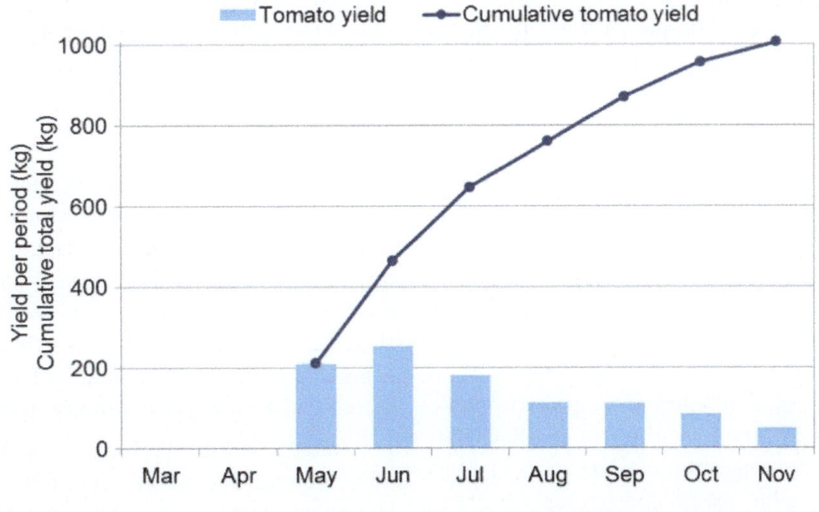

Figure 5. Tomato Solanum lycopersicum production over the whole testing period. The tomato plants grew normally, and harvesting from a total of 113 plants started in May.

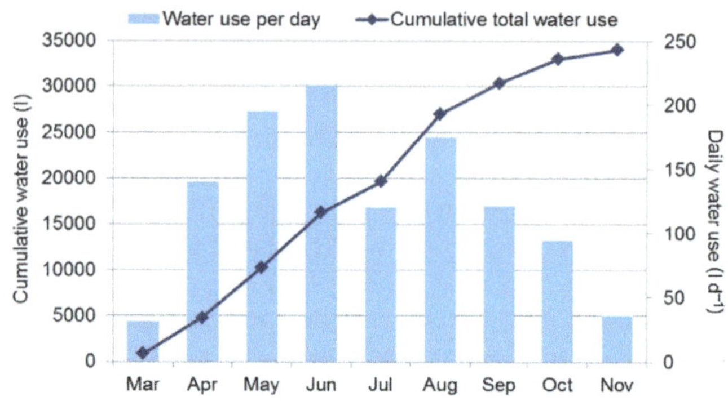

Figure 6. Use of fish water by tomato plants Solanum lycopersicum during the whole testing period.

Composition of the fish water used as Fertilizer

The pH of RAS over the whole testing period was 6.63 ± 0.38 (mean ± SD) with a minimum of 5.43 and a maximum of 7.65. Electrical conductivity was 1.65 ± 0.27 mS cm^{-1}, with a minimum of 0.91 and a maximum of 2.25 mS cm^{-1}. In general, the amounts of the various macro- and micronutrients in the RAS varied during the whole testing period but remained relatively constant in their concentrations, with a tendency to accumulate with decreasing tomato productivity towards the end of the testing period. The mean (±SD) concentration of total nutrients was 753.53 ± 125.30 mg l^{-1}, with the lowest observed value being 526.73 and the highest 983.52 mg l−1. The respective mean values, as well as minimum and maximum values, for N-NO_3, N-NO_2, N-NH_4, P-PO_4, K, SO_4, Ca, Mg, Fe, Mo, Cl, Na, and Cu are summarized in Fig. 7 and Table 1. During the whole testing pe riod, a total of 34.06 m3 fish water was launched via the 1-way valve from the RAS into the reservoir of the hydroponic unit. Thus a total load of 25.66 kg nutrients (cf. Fig. 7 for details concerning nutrients) was transferred from the fish RAS into the NFT hydroponic unit for the tomato plants.

Composition of the Fertilizer of the Hydroponic Unit

The pH of the hydroponic unit was 6.79 ± 0.51 (mean ± SD), with a minimum of 4.30 and a maximum of 7.81. Electrical conductivity was 3.20 ± 0.99 mS cm^{-1} (minimum and maximum: 0.82 and 5.33 mS cm^{-1}), respectively.

In addition to the 25.66 kg nutrients from the fish water (see previous section) to optimize growth conditions, some mineral fertilizer was also introduced into the hydroponic unit (in total: 0.80 kg Kristalon™ Spezial, 0.75 kg CalciNit™, 0.08 kg $KHCO_3$, and 2.80 kg K_2HPO_4). In general, the nutrients in the hydroponic unit seemed to accumulate during the end of the testing period, which was concomitant with reduced tomato productivity due to decreasing light exposure.

As expected, the mean values of nutrients in the hydroponic unit were higher compared to the fish water of the RAS (cf. Tables 1 & 2). The mean concentration of total nutrients was 1516.66 ± 573.40 mg l^{-1}, with lowest and highest observed values of 668.37 and 2709.24 mS cm^{-1}, respectively.

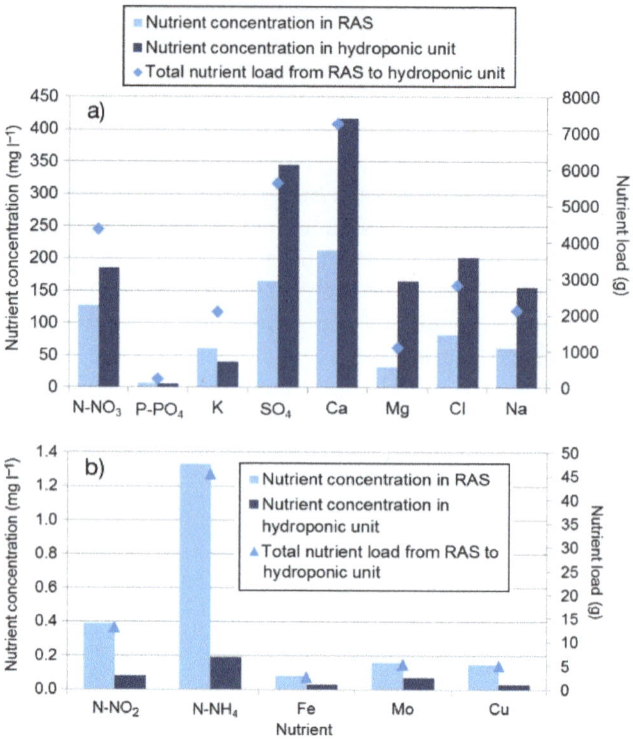

Figure 7. Means of macro- and micronutrients (cf. Table 1) in the recirculating aquaculture system (RAS) and the hydroponic unit and total load of macroand micronutrients launched from aquaculture RAS into the nutrient film technique (NFT) hydroponic unit of ASTAF-PRO during the testing period from 6 March to 26 November 2009. The total volume of fish water transferred into the hydroponic component was 34.06 m^3 . Note the different scale between panels (a,b).

The respective mean values in cluding minimum and maximum values for N-NO_3, N-NO_2, N-NH4, P-PO_4, K, SO_4, Ca, Mg, Fe, Mo, Cl, Na, and Cu are summarized in Tables 1 & 2. Fig. 8 summarizes the means of all nutrients determined over the whole testing period in comparison to the optimum supply for tomato plants according to Lattauschke (2004) and De Kreij et al. (1997). Thus, except for N-NO_3, Ca, Mg, Mo, and SO_4, all other nutrients were more or less lower than optimal (only 2.71 to 60.0% of optimum). It is noteworthy that the supply for the most abundant nutrients like N-NO3 reached ~128.4% of optimum (and ~239.1% for SO_4), but with only 17.0%, 17.8%, and 2.71%, K, PO_4, and NH_4, respectively, were far lower than required for optimum growth.

CO_2 Balance

To further sustainability by reduction of emissions with our ASTAF-PRO system, the co-culture of fish and plants within a nearly closed greenhouse in principle allows the CO_2 released by the fish to be fixed by the tomato plants. However, due to restrictions in our equipment, we did not measure real CO_2 fluxes between fish and plants directly, but we can provide a good conservative estimate of the CO_2 balance based on theoretical assumptions of nutrient metabolism.

In total we used 736.2 kg of diet (Aller Performa 1 gr.; 56% protein, 11% fat) to gain 600.1 kg of fish biomass. Neg lecting any loss with regard to the digestibility of the diet, the total content for metabolization of the diet fed was 412.3 kg protein and 81.0 kg fat. On the other hand, the total gain of tilapia was 600.1 kg, and assuming that tilapia biomass (in terms of whole body content) consists of 15% protein and 5% fat, with negligible carbohydrates (AbdelTawwab et al. 2010), this accounts in total for 90.0 kg protein and 30.0 kg fat; thus, the net values for potential metabolism in fish and/or for further microbial de gradation are 322.3 kg protein and 51.0 kg fat. The respiratory quotient (RQ; CO_2 liberated divided by O_2 consumed) in tilapia is ~1.0 (Peer & Kutty 1981), taking into account that protein is the main energy source and in fish is fully degraded to ammonia (NH2CHCH3COOH + 3 O_2 à 3 CO_2 + 2 H_2O + NH_3), whereas fat has a respiratory quotient of ~0.7. The physical caloric equivalent of protein de graded to NH3 per gram is 22.2 kJ, consumes nearly 1 l of O_2, and according to the RQ of ~1.0, releases ~1 l of CO_2. Thus concerning the complete metabolism of 322.3 kg protein with an RQ of 1.0, the total amount of CO2 that could be released after metabolization is 322 300 l. The density of CO_2 is 1.977 kg m^{-3} (at 1 atm and 0°C) and thus 322 300 l (322.3 m^3) CO_2 equals 637.2 kg CO_2. The caloric equivalent per gram of fat is 38.9 kJ, consumes ~2 l of O_2, and releases ~1.4 l of CO_2. Therefore, the total amount of CO_2 that could be liberated by fat metabolization is 71 400 l (71.4 m^3) of CO_2, which equals 141.2 kg CO_2. In total, complete meta - bolization of protein and fat by fish and microbial metabolism could re sult in a total release of 778.4 kg CO_2.

Table 1. Mean, minimum, and maximum concentrations (mg l^{-1}) of macro- and micronutrients in the aquaculture recirculation system during the course of the testing period

	Mean	Min. - Max.
Macronutrients		
N-NO_3	127.69	86–168
N-NO_2	0.39	0.09–0.78
N-NH_4	1.33	0.19–5.79
P-PO_4	6.89	3.2–16.3
K	61.21	32.6–129
SO_4	165.37	99–238
Ca	212.90	142–311
Mg	32.33	10–60.8
Total	**608.11**	**373.08–929.67**
Micronutrients		
Fe	0.08	0–0.46
Mo	0.16	0–0.52
Cu	0.15	0–2.11
Cl	82.63	41–116
Na	62.40	36–98
Total	**145.42**	**77–217.09**

On the other hand, the tomato plants fix CO_2 by photosynthesis during plant growth in biomass and, therefore, the plant component of the ASTAF-PRO system is a sink for CO2. Mota et al. (2010) revealed that in 1 kg biomass of tomato plants, 0.279 kg CO_2 is fixed. We harvested 1005.62 kg tomatoes, and in addition, we determined that 551 kg biomass were produced at the end of the plant period, not including the biomass of old leaves removed during the growing period; thus, the total biomass yield was ~1556 kg of tomato plants. However, as mentioned above with the discharged cleaning water of 74 m^3 , an additional 250 tomato plants could have been reared, which would result in a total harvest of 363 tomato plants producing ~3000 kg of tomatoes and ~1800 kg of additional biomass. The total tomato plant biomass could therefore have reached 4800 kg. Considering that 0.279 kg CO_2 is fixed in 1 kg tomato plant biomass, the potential total biomass of tomato plants would have fixed ~1339 kg CO_2.

Table 2. Mean, minimum, and and maximum concentrations (mg l^{-1}) of macro- and micronutrients in the hydroponic unit (nutrient film technique, NFT) during the course of the testing period

	Mean	Min. - Max.
Macronutrients		
N-NO_3	185.17	104–316
N-NO_2	0.08	0–0.84
N-NH_4	0.19	0–0.99
P-PO_4	6.62	0–23.9
K	39.92	2.2–170
SO_4	344.83	167–917
Ca	416.97	158–774
Mg	164.92	21–852
Total	**1158.7**	**452.2–3054.73**
Micronutrients		
Fe	0.03	0–0.15
Mo	0.07	0–0.28
Cu	0.03	0–0.33
Cl	202.43	41–422
Na	155.40	36–237
Total	**357.96**	**77–659.76**

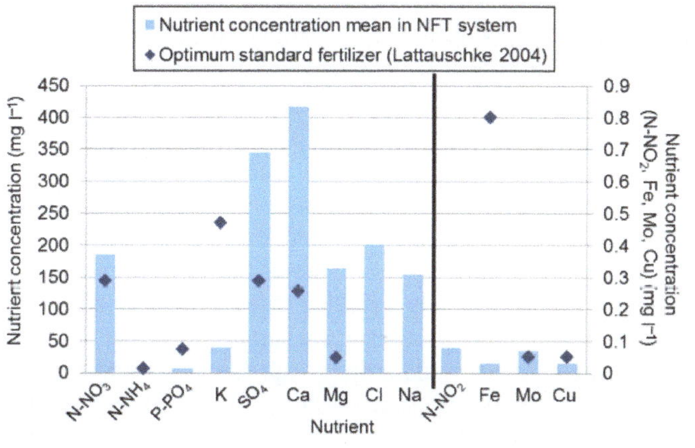

Figure 8. Optimum standardized fertilizer for tomatoes Solanum lycopersicum according to Lattauschke (2004) (for nitrogen compounds, P, K, SO_4, Ca, Mg, Fe, Mo, Cu) in comparison to means of macro- and micronutrients in the hydroponic unit (nutrient film technique, NFT) during the course of the experimental testing period revealing the grade of the optimum supply for tomato plants. The critical values for Na and Cl according to De Kreij et al. (1997) are 276 and 530 mg l^{-1}, respectively. Left (right) of vertical line: left-hand (right-hand) scale applies.

Thus, by a very conservative estimation, our experimental trial indicates that if all of the fish water removed from the RAS could be used for the hydroponic unit to grow tomato plants, this would result in a net sink for CO_2. Complete metabolization of protein and fat not transferred into fish biomass could lead to a release of ~778.4 kg CO_2 (1.30 kg CO_2 kg^{-1} fish biomass), while the complete use of fish water for potential growing of tomato plants would yield a total biomass of ~4800 kg, which could fix ~1339 kg CO_2. Thus in summary, concerning CO_2 balance within our ASTAF-PRO system, theoretically no emission of CO_2 would arise, and a net sink for CO_2 of ~560.6 kg might even occur.

Patented System

The innovative principle proposed by the ASTAFPRO system and validated practically herein was patented (Kloas et al. 2008) as a DRAPS characterized by 2 independent recirculation systems for each component, i.e. aquaculture and hydroponic, connected unidirectionally from the fish to the plant unit and gaining vaporized water for the aquaculture unit by condensation via air conditioning, thereby conserving water (Fig. 1). Value-added chains concerning water, nutrients, and carbon dioxide released from fish metabolism supplementing plant growth, which in turn produces oxygen by photosynthesis, foster sustainability of aquaculture RAS by integration with horticulture.

DISCUSSION

The test evaluation of the ASTAF-PRO system was successful, and despite the electrical power failure in August that adversely impacted at least the tilapia in the RAS, ongoing production of fish and tomatoes was achieved. To prevent similar problems in the future, an emergency power system was implemented within the same month.

Tilapia Production

The total gain in tilapia biomass over the whole feeding trial (264 d) was 600.1 kg, with a productivity rate of about 75.9 kg tilapia m^{-3} in a rearing volume of 7.91 m3 . However, considering that the fish tanks could all be stocked with a total mass of 300 kg for the first time at the end of April, the productivity between 1 May and 23 July was considerably higher, with 30.8 kg m^{-3} in 84 d. The adverse effect of the electrical power failure on 1 August that stressed the fish, which was balanced by restocking with indi - viduals already weighing at least 500 g, also became evident as the productivity rate reached only 29.6 kg m^{-3} after 124 d. In testing an outdoor ap proach using SRAPS, Rakocy et al. (2004, 2006) revealed a productivity rate for Nile tilapia of 61.5 kg m^{-3} in 168 d using stocking densities of ~60 kg m^{-3}, and Al–Hafedh et al. (2008) reported a gain of ~43.8 kg m^{-3} within 13 mo. In another SRAPS study in greenhouses, the productivity rate of tilapia was only 15 kg m^{-3} after 105 d using a very small (0.5 m^3) aquaculture unit (McMurtry et al. 1997). Thus, together with the wellestablished SRAPS by Rakocy et al. (2006), our DRAPS ASTAF-PRO system demonstrates the highest productivity rate for tilapia reared in aquaponic systems reported to date. However, the stocking densities used in the recent testing were only between 38 and 45 kg m^{-3} instead of 60 kg m^{-3} used in the SRAPS. In addition, we used a mixed-sex population, whereas Rakocy et al. (2006) reared a faster-growing mono-sex Nile tilapia culture (Popma & Masser 1999), which is also a parameter to increase productivity.

The FCR (1.01) calculated under undisturbed regular practical conditions (1 May to 23 July) was similar to our conventional RAS unit. After the power failure on 1 August, the ratio increased to 1.46, demonstrating that stressed fish and the restocked individuals that were already >500 g drastically diminished feed conversion; nevertheless, the results demonstrate that in DRAPS with a separated RAS, a nearly optimum FCR for tilapia of ~1.0 can be obtained, whereas previous aquaponic studies using SRAPS reported FCRs of 1.7 (Rakocy et al. 2004), 1.59 (Watten & Busch 1984), and 1.4 (Al–Hafedh et al. 2008). Thus the ASTAF-PRO system demonstrates that with regard to productivity and FCRs, very good yields can be obtained in comparison to separated RAS. However, in the presented testing approach, only very moderate mean stocking densities of tilapia between 38 and 45 kg m^{-3} were used. Thus it is very likely that higher stocking densities of up to 70 to 80 kg m−3 (W. Kloas & B. Rennert un publ. results for our conventional RAS), which still do not affect FCR, can further increase the productivity of tilapia. In addition, the potential use of mono-sex cultures of Nile tilapia as shown by Rakocy et al. (2006) might lead to significant increases in weight gain of tilapia and surpass the yield of established SRAPS.

Tomato Production

The total tomato production of our ASTAF-PRO system with 113 plants resulted in 1005.6 kg over a growing period of less than 9 mo (6 March to 26 November), for a yield of 8.89 kg $plant^{-1}$. In outdoor SRAPS, tomato productivities under different conditions varied greatly, from 0.48 to 3.69 kg $plant^{-1}$ (Watten & Busch 1984) or 4.6 kg $plant^{-1}$ (Lewis et al. 1978). In a thorough review, Rakocy et al. (1992) summarized all vegetables grown in aquaponic systems and reported a maximum productivity of 9.07 to 9.98 kg $plant^{-1}$ yr^{-1} for tomato plants, depending on the strain used. McMurtry et al. (1997) obtained values be tween 2.4 and 5.0 kg $plant^{-1}$, depending on the relationship of the volumina between the number of biofilters and the mass of fish reared.

Thus, the tomato production by ASTAF-PRO of 8.89 kg plant−1 achieved in the present study within less than 9 mo is higher in comparison to previously reported SRAPS (outdoor and indoor) yields, because in conventional greenhouse hydroponics using NFT, the introduction of produced plants usually takes place in January and thus our production period could have been 2 mo longer. In specialized greenhouses for tomato production using comparable NFT systems, a productivity of 15.88 to 20.41 kg $plant^{-1}$ has been documented (Anderson 1995). However, in such specialized green houses, plant production is increased not only due to optimum fertilizer but also due to optimum artificial illumination which we performed only occasionally (during October and November) and never exceeding the 12 h day period from 06:00 to 18:00 h. Artificial illumination maintaining optimum light conditions during the natural day and/or extending the daily light period up to 16 or 18 h might significantly improve tomato yield ~50%. Thus, the hydroponic unit of ASTAF-PRO has the potential to reach the high performance reported for specialized greenhouses using NFT by increasing the growth period by about 2 mo and by optimizing the illumination conditions. In addition, optimization of the hydroponic fertilizer can be established by increasing stocking densities of fish, leading to higher nutrient concentrations in the RAS, and by adding commercial mineral fertilizer to reach the optimum for otherwise deficient nutrients.

As mentioned in the 'Results' above, we were forced to discharge the fish water during the weekly cleaning of the clarifier due to space limitations. Due to budget restrictions, it was not feasible to enlarge the greenhouse facilities for the hydroponic unit to add another 250 tomato plants to make complete use of the fish water and its nutrients for the hydroponic unit. Considering the potential for a larger hydroponic unit under similar conditions and adapting the corresponding ratio between aquaculture and hydroponics, the total yield of tomatoes should become significantly >3000 kg.

Use of Water

The freshwater use in the recent ASTAF-PRO performance was 3.83% d^{-1} of the total RAS volume. However, only 2.72% of the fresh water used was tap water. The rest was introduced as condensed water from the cold traps of the air-conditioning that originated from plant transpiration and water surface evaporation. However, it is hard to compare such values to outdoor aquaponic systems because they gain additional water partially via rain and drainage. Thus, it is more suitable to compare the freshwater use of ASTAF-PRO with indoor RAS in specialized aquaculture units. Optimized RAS systems usually need between 30 and 300 l of fresh water to produce 1 kg fish biomass (Martins et al. 2010). Even a daily exchange of 10% is acceptable for practitioners. However, values < 300 l kg^{-1} fish can only be ob - tained if the RAS includes an additional denitrification unit, which would, besides the ad ditional in - troduction of a complex system unit, deteriorate sus tainability concerning the further use of fish water as fertilizer for plants. During the testing period of ASTAF-PRO, 132.4 m^3 fresh water, 94 m^3 tap water, and 38.4 m3 condensed water gained by the climate control unit were added to the RAS. As a result of the water gained by the climate control unit feeding the aquaculture unit, the present study revealed the lowest freshwater use (220.6 l of fresh water comprising 156.7 l of tap water and 63.9 l of condensed water) per kg fish produced to our knowledge in comparison to established RAS without a de nitrification unit. Assuming that an optimized ASTAF-PRO system would have a 3 times larger hydroponic component making use of the complete fish water including the discharged water for clarifier cleaning, it is highly probable that the proportion of water gained by air conditioning would accordingly become greater, which in turn would markedly decrease the need to add fresh water. Theoretically (see 'Results: Use of water') in a larger greenhouse, by transferring all fish water, including water from cleaning the clarifier, to supplement 250 additional tomato plants with fish water, the relationship between tap water and condensed water could shift to 29.0 l of tap water and 191.6 l of condensed water per kg tilapia and would result in a daily introduction rate for tap water of only 0.5%. Depending on the geographic region, additional collection of rain water in the green house area could lead to a complete balance concerning freshwater use, at least in central Europe. However, the energy consumed for condensation, especially during the summer, is high. Thus especially in moderate climate zones with a good water supply, there is a need for an economical evaluation of the optimum regime concerning water conden sation via climate control, which is the subject of ongoing research.

Composition of the Fish Water

The pH values of the aquaculture unit were between 5.43 and 7.65 (mean ± SD: 6.63 ± 0.38). For tilapia, the optimum pH range is between 6 and 9 (Popma & Masser 1999) and thus during the testing period, the tilapia experienced pH values mostly within their lower pH optimum. However, nitrification by aerobic bacteria in biofilters requires an optimum pH between 7 and 8 (Masser et al. 1999) and thus, because of the low buffering capacity of the fresh water used, there was a steady need to stabilize the pH of the RAS by adding CaO through the lime tank. At 1.65 ± 0.27 mS cm^{-1}, the electrical conductivity of the fish water was in an acceptable range for the euryhaline tilapia. Furthermore, and most importantly for the wellbeing of fish, ammonium, nitrite, and nitrate values within the RAS remained well below any concentration that would adversely affect the fish. In summary, the conditions of the aquaculture part of ASTAFPRO, i.e. the RAS, were always in a very good range for tilapia, and the freshwater use seemed to be advantageous compared to conventional RAS without denitrification.

Composition of the Hydroponic Fertilizer

The aquaponic principle, i.e. the transfer of nutrients excreted by fish into plants, was managed successfully in the ASTAF-PRO. Even the productivity of tomato plants was at the high end for aquaponic systems but showed the potential to become even more improved by starting the rearing period earlier and using intensive artificial illumination during sunless periods to reach the same productivity as conventional NFT hydroponic units. In addition, the com - position of the fertilizer can be optimized to further increase the tomato yield. Although the concentrations of nutrients could be maintained within the same order of magnitude during the most prominent vegetative period from May to September, the nutrient contents of the fish component were lower due to the fact that tomato plants can take up nutrients selectively, which led to an accumulation of some nutrients in the solution (N, SO4, Ca, Mg, Cl, Na; Fig. 7). Directly related to the nutrient concentrations is the electrical conductivity, which had a mean of 3.20 mS cm^{-1}, indicating a very good value for hydroponic tomato growth. Electrical conductivity in hydro - ponic systems used to grow tomatoes in NFT has been reported to be optimal at 3.7 mS cm^{-1}, with a range from 2.5 to 5.5 mS cm^{-1} (De Kreij et al. 1997). Another experiment reported an optimum of 2.5 mS cm^{-1} (Sonneveld & Voogt 1990). Lattauschke (2004) suggested an electrical conductivity of 2.5 to 5.0 mS cm^{-1} in the root zone of tomatoes growing in rock wool. Thus the electrical conductivity was always near optimum conditions in our study.

The mean pH during the testing period was 6.79, which is slightly above optimum levels according to the hydroponics literature, where optimum limits have been reported to be between 5.0 and 6.0 (De Kreij et al. 1997), 5.2 and 6.2 (Lattauschke 2004), and 5.5 and 6.5 (Hochmuth 2013). Thus, stabilizing the pH between 6 and 7 occurred after half of the testing period but should be improved for future experiments. Fig. 8 revealed another more complex pattern regarding which nutrients should be supplemented to optimize the fertilizer (Lattauschke 2004) derived from the aquaculture component. Therefore, nutrient composition of the NFT component documented here is clearly a topic for further optimization of tomato production by adjusting the aquaculture unit to get more of the deficient nutrients or by adding additional nutrients such as NO_3, NH_4, PO_4, K, SO_4, Fe, and Cu in a complementary way. The concentrations of Cl and Na were below adverse levels (De Kreij et al. 1997), but future adaptations must ensure that these ions do not rise beyond optimum concentrations (Fig. 8).

CO_2 Balance

One of the prominent emissions of animal production sites is the release of CO_2 caused by the metabolism of animals, which contributes markedly to the greenhouse gas impact. Unlike birds and mammals, fish do not have to waste energy for body temperature regulation because of their physiological nature as poikilotherms, which is a reason why the production of fish biomass releases relatively low amounts of CO_2 by metabolism compared to homoeothermic animals. However, aquaculture fish production sites do release CO_2 and thus, in order to foster sustainability by reducing emissions with our ASTAF-PRO system, the co-culture of fish and plants within a nearly closed greenhouse in principle allows the CO_2 released by the fish to be fixed by the tomato plants. The conservative estimation for CO_2 fluxes within our aquaponics facility reveals an overall theoretical total CO_2 release of 778.4 kg by fish and further microbial metabolization of the feed used and not integrated into harvested fish. On the other hand, the total assumed tomato plant biomass can fix 1339 kg CO_2, meaning that the CO_2 balance for ASTAF-PRO reveals a potential net sink for 560.6 kg CO_2 in ASTAF-PRO DRAPS, thereby fostering sustainability. The plant biomass not used for consumption can be further processed for use in biogas plants.

CONCLUSIONS

In summary, the proof of principle by practical testing of the prototype ASTAF-PRO revealed that such an innovative DRAPS might provide the potential to significantly increase not only the sustainability of aquaculture but also the productivities of both components, viz. aquaculture represented by a RAS and the hydroponic unit using NFT. The innovation is due to a new combination of systemic parts, whereby instead of SRAPS there is a DRAPS composed of 2 independent recirculating units, i.e. the aquaculture RAS rearing fish and the hydroponic unit producing tomatoes. Both systems are connected unidirectionally by a 1-way valve to launch nutrient-containing fish water into the hydroponic reservoir for optimization. Furthermore, the air-conditioning of the greenhouse recycles the transpired water from plants and water evaporation of both systems by condensation via a heat exchanger. The system allows integration of value-added chains for water, nutrients, and meta - bolized carbon dioxide in a sustainable way and for optimum productivities of both fish and vegetables. Under these conditions it is possible that with 1 kg feed, ~200 l of fresh water (minimizing the proportion of tap water use by condensation of evapo-transpired water), and a small amount of mineral fertilizer, 1 kg tilapia and 5 kg tomatoes can be produced in a sustainable way, since the power is supplied by alternative energy and the ASTAF-PRO might even provide a net sink for CO_2. Thus a new concept for aquaponic systems has been implemented and validated to im - prove sustainability, increase productivity, and re - duce environmental impacts. In the future, this innovative DRAPS will allow to optimize each part, the RAS and the hydroponic unit, independently. For example, it seems worthwhile to test other promising fish species for their adaptability to the RAS aquaculture part, and for the hydroponic plant part, in addition to using other plant species, to integrate more recent technical developments such as trickling irrigation systems. Recently, the optimization and im - plementation of the ASTAF-PRO system has been the focus of the ongoing EU-project 'INAPRO - Innovative model & demonstration based water management for resource efficiency in integrated multitrophic agriculture and aquaculture systems' (www. inapro-project.eu/) to develop further improvements concerning practicability and efficiency for a yeararound production, especially of plants, to facilitate commercial applications.

Acknowledgements

We gratefully acknowledge funding support from the Federal Ministry of Education and Research, Germany (BMBF), via grants 01SF0719, 033L059, and 03IO1304 and from the European Commission via INAPRO (grant agreement number 619137). We also thank Ingo Cuppok, Matthias Kunow, and Jessica Andrä for excellent technical support.

REFRENCES

1. Abdel-Tawwab M, Ahmad MH, Khattab YA, Shalaby AM (2010) Effect of dietary protein level, initial body weight, and their interaction on the growth, feed utilization, and physiological alterations of Nile tilapia, Oreochromis niloticus (L.). Aquaculture 298: 267–274
2. Al-Hafedh YS, Alam A, Beltagi MS (2008) Food production and water conservation in a recirculating aquaponic system in Saudi Arabia at different ratios of fish feed to plants. J World Aquacult Soc 39: 510–520
3. Bohl M (1977) Some initial aquaculture experiments in recirculating water systems. Aquaculture 11: 323–328
4. Collins M, Gratzek J, Shotts E Jr, Dawe D, Campbell LM, Senn D (1975) Nitrification in an aquatic recirculating system. J Fish Res Board Can 32: 2025–2031
5. De Kreij C, Voogt W, Van den Bos A, Baas R (1997) Voedingsoplossingen voor de teelt van tomaat in gesloten teeltsystemen. Proefstation voor Bloemisterij en Glasgroente, Naaldwijk
6. Freitag AR, Thayer LR, Leonetti C, Stapleton HM, Hamlin HJ (2015) Effects of elevated nitrate on endocrine function in Atlantic salmon, Salmo salar. Aquaculture 436: 8–12
7. Hochmuth G (2013) Fertilizer management for greenhouse vegetables. Florida greenhouse vegetable production handbook, Vol 3. Publication no. HS787. IFAS Extension, University of Florida, Gainesville, FL
8. Krüner G, Rosenthal H (1983) Efficiency of nitrification in trickling filters using different substrates. Aquacult Eng 2: 49–67
9. Lewis WM, Yopp JH, Schramm HL Jr, Brandenburg AM (1978) Use of hydroponics to maintain quality of recirculated water in a fish culture system. Trans Am Fish Soc 107: 92–99
10. Love DC, Fry JP, Genello L, Hill ES, Frederick JA, Li X, Semmens K (2014) An international survey of aquaponics practitioners. PLoS ONE 9: e102662
11. Martins C, Eding E, Verdegem M, Heinsbroek L and others (2010) New developments in recirculating aquaculture systems in Europe:a perspective on environmental sustainability. Aquacult Eng 43: 83–93
12. McMurtry M, Sanders D, Cure J, Hodson R, Haning B, St

Amand E (1997) Efficiency of water use of an integrated fish/vegetable co–culture system. J World Aquacult Soc 28: 420–428

13. Naegel LC (1977) Combined production of fish and plants in recirculating water. Aquaculture 10: 17–24
14. Peer MM, Kutty M (1981) Respiratory quotient and ammonia quotient in Tilapia mossambica (Peters) with special reference to hypoxia and recovery. Hydrobiologia 76: 3–9
15. Rakocy JE (2012) Aquaponics—integrating fish and plant culture. In:Tidwell JH (ed) Aquaculture production systems. Wiley-Blackwell, Oxford, p 344–386
16. Rakocy JE, Masser MP, Losordo TM (2006) Recirculating aquaculture tank production systems:aquaponics—integrating fish and plant culture. Publication no. 454. Southern Regional Aquaculture Center, Stoneville, MS
17. Sonneveld C, Voogt W (1990) Response of tomatoes (Lycopersicon esculentum) to an unequal distribution of nutrients in the root environment. Plant Soil 124: 251–256
18. Watten BJ, Busch RL (1984) Tropical production of tilapia (Sarotherodon aurea) and tomatoes (Lycopersicon esculentum) in a small-scale recirculating water system. Aquaculture 41: 271–283
19. Xu J, Mancl KM, Tuovinen OH (2014) Using a hydroponic system with tall fescue to remove nitrogen and phosphorus from renovated turkey processing wastewater. Appl Eng Agric 30: 435–441

CHAPTER 15 PRODUCTIVITY IN COASTAL AQUACULTURE

Peggy Schrobback, Sean Pascoe, Louisa Coglan

INTRODUCTION

Sustainable management of coastal resources is an important policy goal for all governments of countries with coastlines [1]. This is particularly so in coastal waters adjacent to major population centers, where recreational, conservation and economic uses of the areas often co-exist and compete for space. The latter is particularly vulnerable, as non-market values associated with recreational use (such as recreational fishing or scuba diving) and marine conservation are often high, requiring areas designated for commercial purposes to either be efficiently utilized or given over to the non-commercial activities. Increasingly, marine conservation spatial planning tools are considering the opportunity cost of commercial activities forgone when aiming to achieve conservation objectives [2], and commercial activities that are not operating fully efficiently are most likely to be first displaced. Conversely, where commercial industries are maintained within coastal waters, it is important that management of these industries ensures that they operate at their most productive level.

Ensuring sustainable use of marine resources in coastal waters requires an understanding of drivers of productivity. Productivity analysis is a well established technique in applied economics, and has been applied to a wide variety of industries [3]. In this study, by way of example to illustrate its use in coastal aquaculture, we focus on the potential to enhance productivity in the Sydney rock oyster (*Saccostrea glomerata*) (SRO) industry in Moreton Bay, Queensland. Much of the area of Moreton Bay is designated as a multiple use marine park, with some areas designated as no-take zones, some for recreational fishing only and others for low-impact commercial fishing and aquaculture. The Bay is adjacent to Brisbane, the capital city of Queensland and the third largest city in Australia. Much of the Brisbane population use the Bay area for recreational purposes, and recreational use value is believed to far exceed the value of commercial activities in the Bay [4], [5].

Oyster farming in Moreton Bay has been particularly challenged over the past decades, with current production well below historic levels. At one point, Moreton Bay was the largest oyster producing region in Australia, supplying the Sydney and Melbourne markets as well as Brisbane [6]. Overfishing, disease and changes in market conditions have all contributed to the decline in the Moreton Bay industry [6]. However, the area available for production (i.e., under commercial leases) and number of farmers is still one of largest of all estuaries along the Australian east coast, yet total production is relatively low in comparison to other areas [7], [8]. Roughly 17 per cent of all Australian SRO lease holders are located in Moreton Bay,

Citation: Schrobback P, Pascoe S, Coglan L (2014) Productivity in Coastal Aquaculture. doi:10.1371/journal.pone.0115912.

yet they produce less than 2 per cent (by value) of Australian SRO production[7].

In response to this declining production and the opportunity cost this creates in terms of alternative activities in the Bay this creates, the Oyster Industry Management Plan for Moreton Bay Marine Park [9] includes the objectives of increasing production from the existing leases, to promote the commercial industry development and to improve the image of the industry. The related policy of non-productive oyster leases [10] includes a provision for minimum production levels in oyster leases and the requirement to "show cause" for non-productive farmers as to why they should retain their lease.

Increasing production from a given set of leases can only occur through increased efficiency and/or capacity utilisation. To determine the extent of any potential production increases, the existence and causes of any inefficiency and/or capacity underutilisation needs to be determined. It is unclear whether the current situation of the SRO industry in Moreton Bay is due to oyster farmers' business choices, farmers' personal characteristics or whether environmental conditions in the Moreton Bay limit the capacity of the oyster industry in this region.

Coastal fishery and aquaculture production can be influenced by the condition of coastal waters, which can be affected by human activities that occur along shorelines (e.g., increasing urbanisation, industrial development, and run-off from agriculture) [11]. Furthermore, the personal traits of fishermen (e.g., experience, educational level) can also have an effect on the resourceful operation of coastal fisheries. The understanding of the relationship between production factors (e.g., degree of government regulation, age of labour force, weather, and coastal water quality) and the productive performance of fisheries helps determine the extent and means by which productivity can be enhanced.

In this study, the productivity of the Moreton Bay SRO industry was assessed through measures such as technical efficiency, scale efficiency, allocative efficiency and capacity utilisation. Factors that drive these measures were also examined. These approaches are being increasingly applied in aquaculture as a means of providing information to policy makers on how to improve productivity in these industries [12]. The method was based on a two-stage analysis approach, with productivity measures derived in the first stage using multi-output data envelopment analysis (DEA). In the second stage of the analysis, the influence of oyster farmers' personal characteristics and environmental conditions at different production sites on the derived efficiency and capacity scores was examined.

Moreton Bay Sydney Rock Oyster Industry

The Sydney rock oyster industry which is one of Australia's oldest industries, dating back to European settlement [13]. This aquaculture industry is located in estuaries on Australia's east coast ranging from the border between Victoria and New South Wales in the south to Moreton Bay in Queensland in the north (Fig. 1). The industry is based on a native species that naturally occurs in these waters. The SRO industry has been affected by a range of challenges over the past decades, including reoccurring disease outbreaks; the management of food security, biodiversity and environmental degradation risks; severe weather events; and market competition from the increasing production of the introduced Pacific oyster species [14].

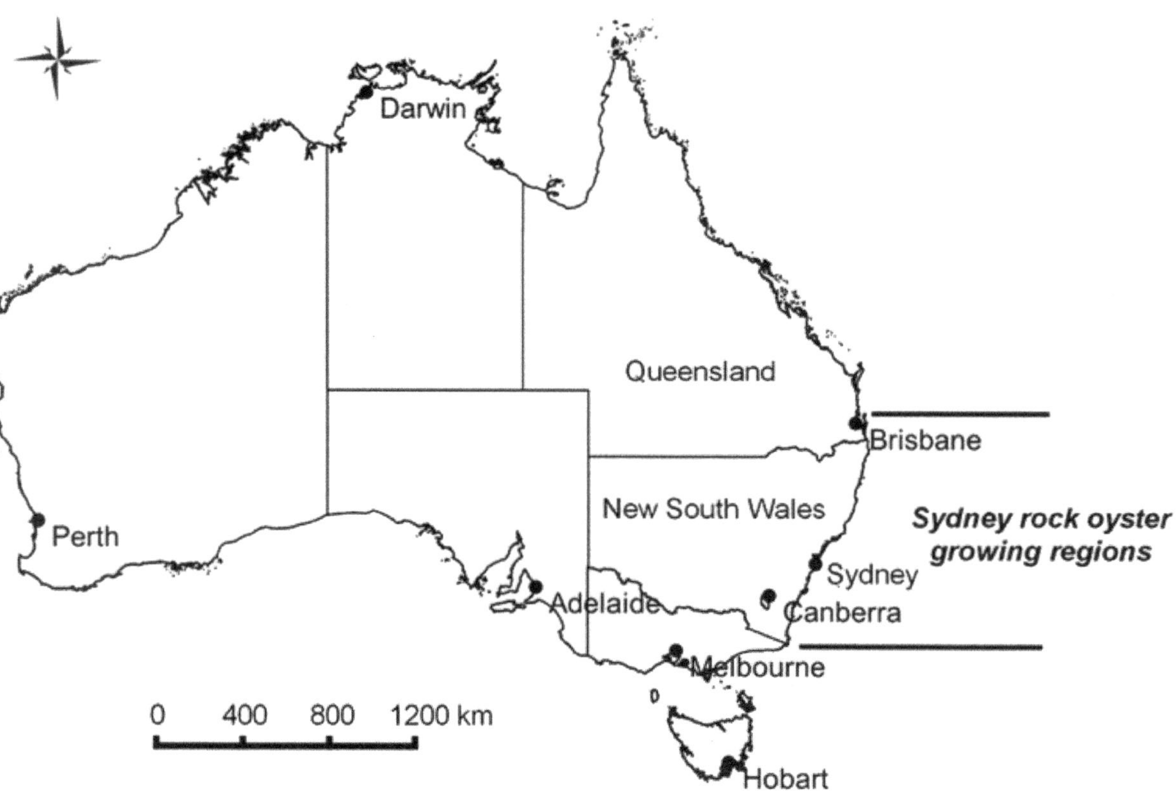

Figure 1. SRO growing regions in Australia.

The history of the SRO industry in Moreton Bay, the northern most cultivation area at the mouth of the Brisbane River, dates back to European settlement in Australia in the early 1800s [15]. The production of oysters in Moreton Bay peaked in the early 1900s [15], [16]. The decline in oyster production in Moreton Bay at that time was linked to mudworm infestation and severe depletion of natural oyster banks [6], [15], [16]. Since then oystering in Moreton Bay continued to be undertaken on a smaller scale in comparison to a relatively large industry in New South Wales. Currently there are 67 oyster farming businesses that take up a total of 97 approved leases in this estuary. In 2011–12 the total annual production volume of oysters in Moreton Bay was about 132,294 dozen valued at approximately $513,400 Australian Dollars [17]. The major market for SROs from the Moreton Bay is Brisbane, a metropolis with a population of 1.8 million people.

Moreton Bay is one of currently 65 Ramsar sites in Australia (which are unique wetlands that are of particular biological importance) [18] and the Bay is designated as a multiple use marine park. The Oyster Industry Management Plan provides an administrative framework for managing the oyster industry within the marine park. The plan is accredited under Marine Parks Regulations 2006 and oyster growers who conduct their operations within the framework of the plan do not require a marine parks permit.

There are currently four areas allocated for oyster growing in Moreton Bay: Moreton Island (hereafter referred to as Eastern Banks), North Stradbroke Island (includes Myora and Canaipa, hereafter referred to as Eastern Bay), Pimpama River and Pumicestone Passage (Fig. 2).

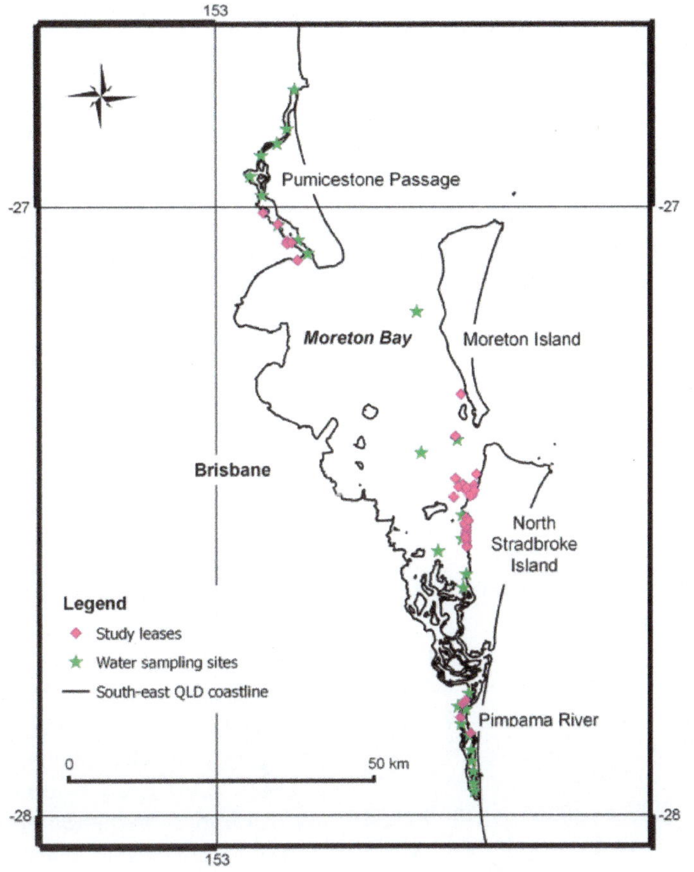

Figure 2. Moreton Bay oyster growing areas. Note: Moreton Island production area is referred to as Eastern Banks, North Stradbroke Island production area is referred to as Eastern Bay.

The oyster areas are located within General Use, Habitat and Conservation zones of the Moreton Bay Marine Park [19]. Both (approved) commercial and recreational fishing activity can occur within oyster areas as long as the activity does not interfere with the aquaculture operation. The total area allocated to oyster leases covers about 435 hectares, which is less than one per cent of the total area of the marine park [19].

The Moreton Bay oyster industry is managed by the Queensland Department of Agriculture, Fisheries and Forestry (QLD DAFF). Resource allocation authorities issued under the Fisheries Act 1994 provide the holders the exclusive right to cultivate and to take oysters from the designated lease areas. Resource allocation authorities are issued for a period of up to 30 years.

SROs are filter feeding organisms and naturally occur in estuaries where the intertidal change provides a suitable habitat. This native Australian oyster species takes about 2.5 to 3.5 years to grow the smallest and largest marketable size, respectively. SROs are typically harvested in the warmer summer months ranging from October to March.

Being filter feeders, they can accumulate any type of pollution present in the water. The monitoring of the safety of oyster for human consumption includes regular water sampling at oyster areas and oyster meat sampling, with supply of oysters from particular leases stopped should water and meat samples not comply with food safety standards. Run-off from agricultural production in nearby river catchments can carry sediments into the estuary which may negatively affect water quality, oyster growth and also food safety. This is particularly problematic after high rainfall events, which also has the effect of reducing salinity further affecting the health of the oysters. The presence of high *E. coli* levels in meat samples, caused, for example, by sewage spills, is also observed occasionally, and leads to ceasing the supply of oysters from affected areas.

METHODS

In this study, a two-stage analysis procedure to analyse and assess inefficiency and capacity utilisation for two reasons was used. First, different producers harvest their oysters at different grow-out periods, resulting in a mix of size grades which requires a multi-output method of assessment. Second, anecdotal evidence suggested that there were a range of different approaches to production, ranging from effectively hobby farm to commercial enterprise. Statistical approaches, such as stochastic distance function approaches [20],[21], [22], [23], [24] effectively assume a common underlying production technology. The limited number of observations also makes parametric estimation of distance function models difficult. Consequently, data envelopment analysis (DEA) was undertaken in order to assess the level of efficiency and capacity utilisation for the Moreton Bay SRO industry. DEA is commonly applied in studies in the context of food production, fisheries and aquaculture[25], [26], [27], [28], [29], [30], [31] and is more commonly applied in general than parametric approaches for productivity analysis [3].

DEA (First Stage)

DEA is a non-paramedic, linear programming method for measuring the relative efficiencies of individual decision making units (DMUs) within a group of individual DMUs, given a set of inputs and produced outputs [32]. A DMU is a term that is frequently used in economics to refer to an individual or entity (e.g., firms, industries, countries) that are responsible for making production decisions [33], [34], [35]. In this study, the term DMU refers to individual oyster farmers who operate within the industry under given industry management plans and regulatory settings. Individual oyster farmers make decisions about how they will undertake oyster production, and that these decisions are reflected in the chosen production inputs and the obtained production outputs. For example, oyster farmers make decisions about the quantity and allocation of production inputs (e.g., labour input). Farmers also make decisions about the use of the total lease area (fixed inputs) they hold, and about their output production mix (e.g., harvest of small, medium or large sized oysters). The outcome of these decisions is the quantity harvested.

Given this notion, DEA can be used as a benchmarking tool to assess the performance of individual DMUs against the efficient frontier of the group which is defined by the most efficient DMUs within the group [36]. However, the frontier approach does not assume that most efficient DMUs within a group are fully efficient [36]; it rather provides a benchmark derived from observed efficient (or best-practice) DMUs.

A key feature of DEA is that it is readily able to incorporate multiple outputs into the analysis. This is particularly relevant since oysters are typically produced in three market sizes, that are small (bottle), medium (bistro) and large (plate). DEA does not impose any assumption about the functional form of the production function and thus is less prone to mis-specifications. However, as a non-parametric method, DEA cannot account for statistical noise and hence efficiency estimates may be biased if the production process is characterised by stochastic elements [36]. This is less problematic for capacity utilisation estimation, as the process for deriving unbiased estimates of capacity utilisation (shown below) has a benefit in that much of the effects of random error are cancelled out [37].

In this study, an output-oriented DEA model was used as the aim was to determine the maximum output of the *j*th DMU given observed inputs. The basic assumption of the output-oriented DEA is that output vector of the *j*th DMU is expanded radially until the combination of inputs of the respective DMU reached the efficient output frontier of the production possibility set for the group of DMUs. The form of the output-oriented DEA model can be given as:

$$\max \mathbf{\Phi}_1$$

$$\text{s.t.} \quad \mathbf{\Phi}_1\, y_{j,m} \leq \sum_j z_j y_{j,m} \qquad m \in M$$

$$x_{1,n} \geq \sum_j z_j x_{j,n} \qquad n \in N \tag{1}$$

where, Φ_1 is a scalar indicating by how much the output of each DMU can be increased relative to the efficient frontier of a group of DMUs; $y_{j,m}$ is the amount of output m by DMU j;$x_{j,n}$ is the amount of input n used by DMU j; and z_j are weighting factors. The input set can be separated into variable inputs (e.g., labour), where values of the variable may change in short-run and fixed inputs (e.g., the area of the lease), where values can only change in long-run. In order to account for the changes in the relationship between fixed inputs and outputs we can impose variable returns of scale (VRS), $\sum_j z_j = 1$ which allows for increasing, constant and decreasing returns within the production process. Various authors[38], [39] suggested the use of VRS in DEA models to account for situations such as imperfect competition and government regulation that may cause a firm to be unable to operate at optimal scale [36]. Without such a restriction, constant returns to scale (CRS) are assumed.

Technical efficiency (TE) is a measure that reflects the ability of DMUs to obtain maximum output from a given input set. The general form of TE is given by:

$$TE = \mathbf{\Phi}_1^{-1} \tag{2}$$

The value obtained for TE is the efficiency score for the *j*th DMU. The derived efficiency scores lie in the interval [0,1], with a value of 1 indicating a point on the frontier and hence a technically efficient firm.

Capacity represents the potential output given a set of fixed input, assuming that these are all fully utilised [40]. Most applied studies are concerned with the level of capacity utilisation, which measures the extent to which fixed inputs are being fully utilised [29], [31], [41], [42]. Capacity utilisation is derived by solving the above model (1) using fixed inputs only. The resultant technical efficiency measure, Φ_2, can be used to derive a capacity utilisation score by:

$$CU(observed) = \frac{y}{\Phi_2 y} = \frac{1}{\Phi_2} = \Phi_2^{-1} \tag{3.1}$$

Färe et al. [42] argued that this CU measure may be biased downward, since it captures both capacity utilisation and technical efficiency. Consequently, an adjustment is required to separate out the CU component to correct for the bias. Färe et al. [42] suggest that an unbiased measure of CU may be calculated as:

$$CU(unbiased) = \frac{\Phi_1 y}{\Phi_2 y} = \frac{\Phi_1}{\Phi_2} \tag{3.2}$$

As noted above, this measure is also less susceptible to random error [37].

The scale efficiency measure provides information about the production scale or level of a DMU compared to other DMUs in a group. The CRS assumption is appropriate when DMUs are operating at an optimal scale [36]. However, the use of VRS imposes the possibility that the scale of production could affect the efficiency of DMUs. The scale efficiency measure is estimated as the ratio of technical efficiency with constant returns to scale (TE(CRS)) to technical efficiency with variable returns to scale (TE(VRS)), a TE(CRS) and a TE(VRS). The relationship can be described as:

$$SE = TE(CRS)/TE(VRS) \tag{4}$$

If the results for TE(CRS) and TE(VRS) scores for a DMU differ, it indicates that this DMU is operating at a scale that is less than efficient. Hence, the results provide an indication as to how close a DMU is to its (technically) optimal scale.

The allocative efficiency (AE) measures is used to identify the degree to which DMUs are adopting strategies that lead to optimisation of revenue from the production process, given the relative prices of each output. The estimation of the revenue efficiency with VRS and TE(VRS) are required for the estimation of allocative efficiency scores.

Revenue efficiency can be obtained by solving the following revenue maximization DEA problem:

$$\max \sum_m p_m \varphi_1 y_{1,m}$$

$$\text{s.t.} \quad \varphi_1 y_{1,m} \leq \sum_j z_j y_{j,m} \quad m \in M$$

$$x_{1,n} \geq \sum_j z_j x_{j,n} \quad n \in N$$

$$\sum_j z_j = 1 \tag{5}$$

Where φ represents the linear expansion factor to the revenue frontier. The revenue efficiency is given by:

$$RE(VRS) = \varphi^{-1} \tag{6}$$

The output-mix allocative efficiency measure is then obtained as:

$$AE = RE(VRS)/TE(VRS) \tag{7}$$

Allocative efficiency scores provide information about the degree to which changes in the production mix (i.e., production of small, medium, large and other sized oysters) could enhance the DMU's revenues.

SECOND STAGE

The ability of DMU's to convert input into outputs can be influenced by exogenous variables that characterise the environment in which production takes place [36]. These exogenous factors can be observable (e.g., government regulation, age of labour) and unforeseen (e.g., disease, weather) [36]. Previous studies that assessed the effect of drivers of efficiency derived in a DEA most commonly use Tobit analysis [29], [31], [43]. Efficiency scores, defined as ratios of actual output to the frontier value of the output, must lie between 0 and 1 or equal 0 or 1. Thus, the application of Tobit analysis is frequently used in the second stage analysis for [0,1] limited and censored distribution of the dependent variable [32], [44],[45]. Hoff [32] compares Tobit and OLS methods and shows that while the Tobit approach may be adequate, the OLS approach may in many cases replace Tobit as a sufficient second stage DEA model. McDonald [45] argues that DEA efficiency measures are not censored but rather fractional or normalized with a heteroskedastic distribution. In this case, Tobit analysis may produce mis-specified estimates, and OLS may be a more appropriate approach [45].

In this study, we apply both Tobit and OLS approaches to estimate how demographic characteristics and environmental conditions may affect productivity measures derived for oyster farmers in Moreton Bay.

Data

Annual cross-sectional oyster production data were made available by the QLD DAFF. The initial production data set includes records covering the period between 1997/98 and 2011/12 for a total of 39 oyster farmers who gave the consent to DAFF for their data to be used for research purposes of this study. Since individual oyster farmers (the DMUs within the oyster industry) take decisions about their individual production process it can be assumed that these choices are reflected in the observed input and output production data.

The production data includes information about production output volume (number of dozens of oysters) and production values for four different product grades (sizes), namely bottle (small), bistro (medium), plate (large) and other. Product prices were derived from the production volume and values. In cases in which production values were unavailable for an individual, average annual prices derived from the available observation for each year were used.

The production data also included information on labour inputs, separated into three categories: lease owner full-time equivalent (FTE, based on 40 hour work per week), permanent FTE and casual FTE. Information about the total leased area size (hectare) per lease owner as well as the geographic location of the leases were used as fixed inputs. For larger scale oyster cultivation, there are commonly a number of leases used for different stage in the cultivation process. For example, the initial phase of catching of oysters spat usually requires areas where oyster spat is available in abundance while grow-out leases are used to fatten the oysters and depuration leases are used for purification prior to the harvest of oysters. Information collected from farmers that held more than one lease did not cover the particular use of each oyster leases during the cultivation process. Thus, the available production data only reported total production volumes and values per lease owner.

The total number of observations initially used for the first stage DEA was 288 (i.e., all observations). However, due to limitations in environmental and demographic data used in the second stage (see below), we also performed the first stage DEA on a sub-sample of 113 observations for which complete data were available. Descriptive statistics of the full and sub-sample of the data are shown in Table 1.

Table 1. Descriptive statistics of production data.

Variable	Mean Full-sample [sub-sample]	Coeff. of variation Full-sample [sub-sample]
Outputs quantity (dozen)		
Bottle grade	1,418 [1,925]	272% [201%]
Bistro grade	1,198 [1,314]	225% [222%]
Plate grade	612 [523]	221% [166%]
Other grade	513 [720]	288% [269%]
Output price per dozen		
Bottle grade	3.48 [3.87]	38% [32%]
Bistro grade	4.92 [5.69]	40% [39%]
Plate grade	6.57 [7.42]	37% [30%]
Other grade	3.94 [4.47]	58% [61%]
Inputs		
Hectare size	3.29 [10.23]	66% [188%]
Total labour (FTE)	0.13 [0.17]	218% [213%]

doi:10.1371/journal.pone.0115912.t001

For the second stage analysis, we use demographic characteristics of oyster farmers and data of environmental parameters in proximity of the respective oyster leases. An oyster farm survey was undertaken by the authors of this study in 2012, which provided information about the socio-economic characteristics of Moreton Bay oyster farmers (see Table 2). The collection of this primary data set has been approved by the Queensland University of Technology's Human Research Ethics Committee (approval number: 1200000303). The participants of the survey were made aware in written form about the confidential use of the data for research purposes and that the consent to use the data for this purpose is provided by the participants by completing the survey and returning it to the authors of this study. In order to comply with ethical research standards, none of the participants or their business are identifiable in this study.

Table 2. Socio-economic characteristics of the Moreton Bay oyster industry.

Socio-economic characteristics	Results	Socio-economic characteristics	Results
Gender(per cent of all farmers)		***Household***	
Female	17%	Number of people living in household	2.1
Male	83%	Number of children	2.2
Age (years)		***Annual income* (weekly disposable income)**[1,#]	
Minimum age	29	$0–$40,000 ($0–$669)	48%
Q1 age	51	$40,001–$60,000 ($670–$922)	5%
Average age	57.5	$60,001–$80,000 ($923–$1,174)	10%
Median age	56.5	$80,001–$100,000 ($1,175–$1,411)	29%
Q3 age	65	$100,001–$120,000 ($1,412–$1,646)	0%
Maximum age	76	Over $120,000 (over $1,646)	10%
Farmers younger than 35 years	4%	Off-farm income[#]	73%
National origin[#]		Proportion of total income from oyster farming (average)[#]	14%
Australian born	96%	**Other**	
Experience in oystering (years)		First Generation is oyster farming[#]	83%
Q1 experience	4	Average number of generation in oystering if not first generation	2.5
Minimum experience	0	Member in farming association[#]	100%
Q1 experience	4	Experience with other fish/shellfish species[#]	13%
Average experience	14		
Median experience	10		
Q3 experience	28		
Maximum experience	50		
Educational attainment[#]			
Year 10 certificate & below	30%		
Year 12 certificate	39%		
TAFE degree/Apprenticeship	4%		
University degree	26%		

Data collected in a farm survey in 2012. Weekly disposable income (net income) estimates for income brackets derived from Australian Taxation Office [71].
[#]Per cent means, data represent as proportion on all farmers. All income values are in Australian Dollars.

doi:10.1371/journal.pone.0115912.t002

The socio-economic characteristics of Moreton Bay oyster farmers were matched with the production data where available. The second stage analysis was undertaken on a sub-sample of the production data as demographic information was not available for all farmers. For observations that included demographic information, we augmented records of continuous variables (e.g., age, years of experience) to account for the continuous involvement of farmers in the industry. Categorical data (e.g., level of farmer education) was assumed to be constant over time, with dummy variables also included to capture any effects of gender (Male=1), education (tertiary educational=1) and generation (1=more than one generations of experience in oyster farming within the family) on productivity.

Environmental data for Moreton Bay (see Table 3) were obtained from Healthy Waterways Ltd. [46]. This data set contains monthly records for water quality indicators collected at estuarine zones within Moreton Bay. The environmental data includes records ranging from 2000 to 2012. Although earlier records are available, they were collected by different agencies and contain less frequent and spatially distributed information and were therefore excluded from this analysis. We mapped oyster production areas against water collection sites and only used data for sites that were in close proximity to the production areas. Details are provided in Fig. 2.

Table 3. Environmental variables used in the analysis.

Environmental variable (unit)	Mean	Coeff. of variation
Salinity (ppt)	31.98	16%
Temperature (℃)	22.47	4%
Dissolved oxygen (%)	94.52	12%
Light penetration	3.34	57%
Turbidity (NTU)	5.92	116%
Dissolved total nitrogen (mg/L)	0.25	76%
Dissolved total phosphorus (mg/L)	0.02	52%
Chlorophyll-a (μg/L)	2.46	147%
pH	7.96	6%

The key variables considered were salinity, water temperature, dissolved oxygen, light penetration, turbidity, dissolved nitrogen and phosphorus, chlorophyll-a levels, and acidity (pH). The relationship between oyster shell and flesh growth and environmental factors is very complex, depending on average as well as extreme levels and their duration.

Several of these variables are believed to have a direct impact on the growth and survival of the oysters. A low level of salinity may compromise the development and growth as oysters close their valves and stop feeding at low salinity levels [47]. Prolonged rainfall periods typically lead to low salinity levels in estuaries. Optimal salinity levels range from 25 to 35 parts per thousand (ppt) [48], [49]. However, the salinity tolerance varies significantly depending on the life stage of oysters, with younger oyster tolerating 15–39 ppt and adult oysters tolerating 0–50 ppt for limited period of time [49]. The optimal water temperature for SRO development and growth ranges from 14–28°C with a tolerance of 11–30°C [48], [49]. Low levels of dissolved oxygen also affect the metabolism of oysters [50]. High acidity in water affects shell formation of oysters and, thus, their growth [51], with the optimal pH range for SRO being 6.5–8.5 [52]. High level of turbidity, in particular of inorganic particles, may lead to congested gills affecting their ability to filter water and extract food [53]. Turbidity typically increases after rain events.

Other environmental variables affect the supply of the food source for oysters, indirectly affecting their growth. The depth to which light penetrates the water affects the presence of phytoplankton/microalgae biomass, an energy source for oysters. High level of turbidity also reduce the amount of light and therefore the food supply [53]. Similarly, a low level of oxygen may affect phytoplankton/microalgae biomass production. Chlorophyll-a is a direct measure of the presence of phytoplankton.

The level of dissolved nutrients can reduce the food safety of the oyster, with too high levels leading to harvesting being delayed. Nutrient levels also may affect the production of the food supply, with excessive levels leading to algal blooms, and in extreme cases eutrophication. Dissolved total nitrogen measures the presences of all forms of nitrogen (e.g., nitrate, nitrite, ammonia) in water. Urban and agricultural runoff, industrial wastes, and sewage effluents typically lead to high nitrogen concentrations in estuaries. Oysters are able to assimilate nitrogen from the water in their soft tissue and shells [54]. Dissolved total phosphorus is a measure for the presence of all forms of phosphorus present in water. The presence of high levels of phosphorous in estuaries can be attributed to similar sources as for nitrogen (see above). Oysters are able to assimilate phosphorus from the water in their soft tissue and shells [55].

Monthly records were used to obtain an annual average value for each environmental variable at each production area. Extreme values, such as annual minima or maxima, were not considered appropriate for our analysis as the data set does not provide information about the frequency and duration of extreme values within a month. Such information would have been vital for estimating the magnitude and significance of extreme environmental values on productivity [49], [52]. Since demographic observations were unavailable for leases in the Pumicestone Passage we could only include three for the four oyster production areas in Moreton Bay in the second stage analysis. Dummy variables for Eastern Bay and the Eastern Banks oyster growing areas (Fig. 2) were included to account for any spatial effects that are not picked up by environmental variables.

RESULTS

The DEA analysis revealed that a high proportion of oyster businesses in Moreton Bay were relatively inefficient (Fig.

3.a, Table 4). In contrast, most of the oyster businesses operated at a high or full capacity utilisation rate (UCU median of 0.85, Table 4). That is, the businesses are mostly providing an appropriate amount of variable inputs (labour), but not are using it efficiently. The majority of oyster businesses operate close to or at the technical optimal scale (Fig. 3.c, Table 4) with a median scale efficiency scores of 0.81. Given this, we can conclude that most businesses in this industry would not be able to significantly increase their productivity by changing their level of activity (labour) or the scale of their operations.

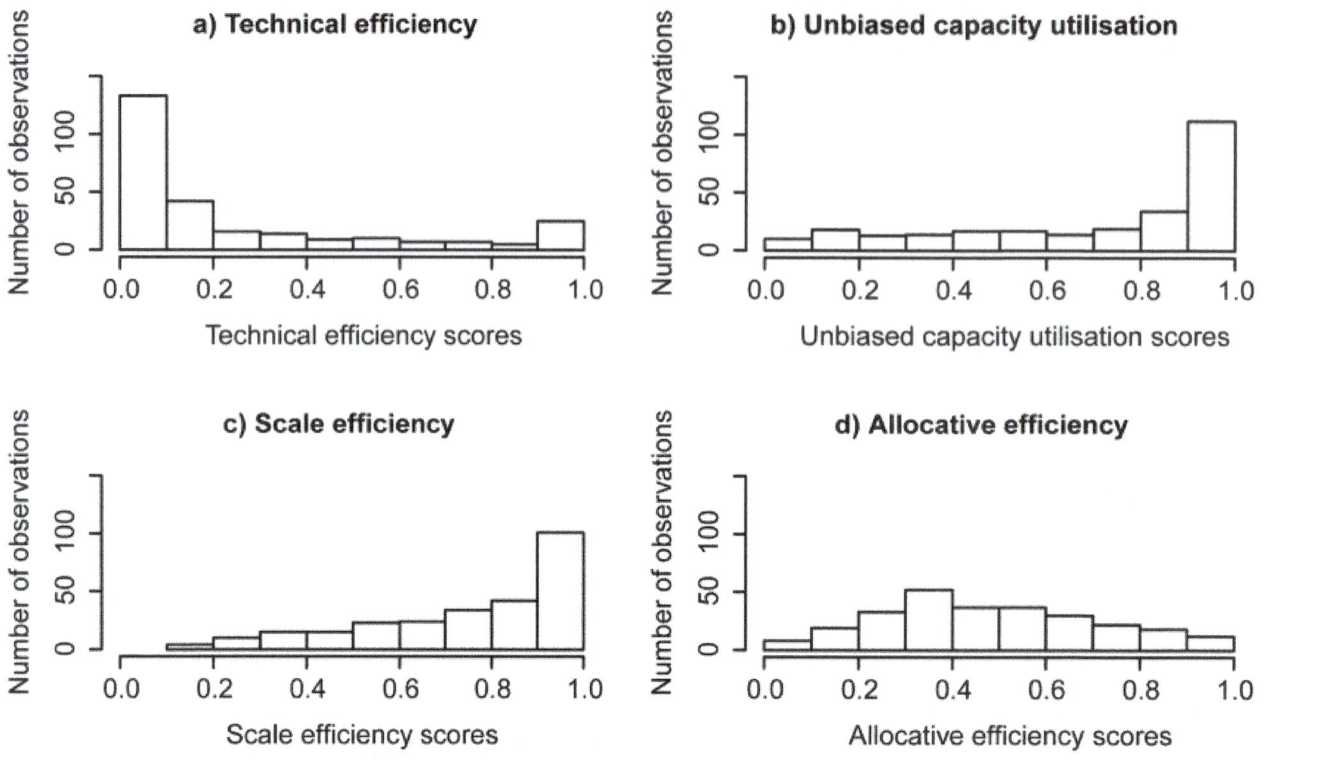

Figure 3. Distribution for capacity utilisation and efficiency scores (all observations).

Table 4. Summary of the key DEA results (all observations).

Capacity utilisation/efficiency measure	Min.	Median	Mean	Max.	Standard deviation
Observed CU	0.000	0.059	0.177	1.000	0.267
Unbiased CU	0.018	0.850	0.716	1.000	0.305
TE (VRS)	0.000	0.099	0.249	1.000	0.311
Scale efficiency	0.104	0.808	0.751	1.000	0.232
Allocative efficiency	0.000	0.438	0.477	1.000	0.230

CU for capacity utilisation, TE (VRS) for technical efficiency (variable returns of scale).

doi:10.1371/journal.pone.0115912.t004

Allocative efficiency scores were found to be relatively dispersed (Fig. 3.d, Table 4). Allocative efficiency compares technical efficiency against revenue efficiency and thus, indicates the degree of which changes in the output mix (different grades of oysters sold) could enhance the revenue of businesses in the industry. The wide distribution of allocative efficiency scores indicates that the current product mix is not optimal for a high proportion of the industry.

The derived technical efficiency scores, capacity utilisation scores and scale efficiency scores for the sub-sample used in the first stage analysis follow a very similar distributional pattern with only minor variation in comparison to the results obtained in the analysis using the full data set (Fig. 4, Table 5). The distributions of allocative efficiency scores using the full data set and the sub-sample set show differing patterns (see Fig. 3.d, Fig. 4.d, Table 4,Table 5), which suggests that the results using the sub-sample data set should be interpreted with caution.

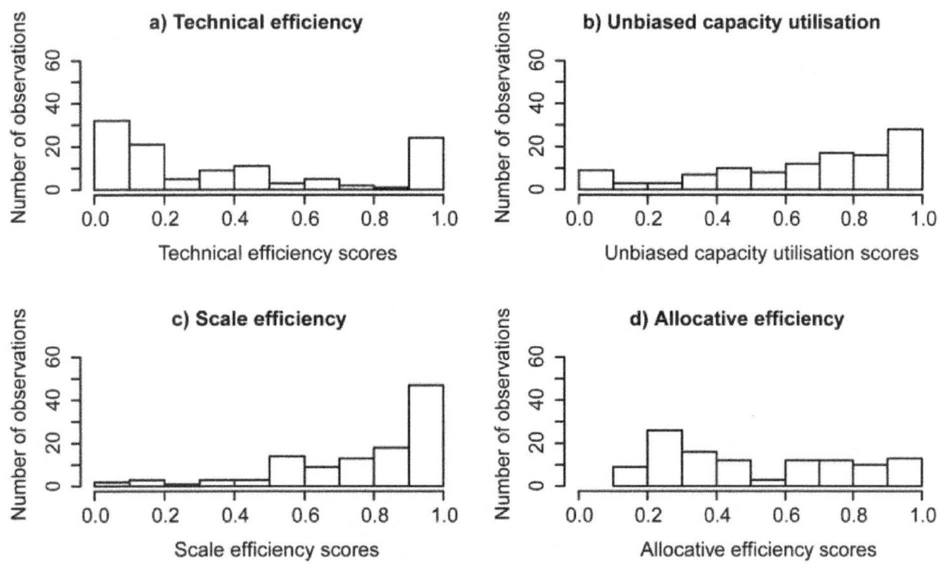

Figure 4. Distribution for capacity utilisation and efficiency scores (sub-sample).

Table 5. Summary of the key DEA results (sub-sample).

Capacity utilisation/efficiency measure	Min.	Median	Mean	Max.	Standard deviation
Observed CU	0.000	0.114	0.262	1.000	0.313
Unbiased CU	0.010	0.752	0.664	1.000	0.293
TE (VRS)	0.001	0.291	0.398	1.000	0.367
Scale efficiency	0.038	0.847	0.778	1.000	1.000
Allocative efficiency	0.113	0.445	0.518	1.000	0.271

CU for capacity utilisation, TE (VRS) for technical efficiency (variable returns of scale).

doi:10.1371/journal.pone.0115912.t005

The results for the second stage OLS and Tobit estimations are shown in Table 6 and Table 7. Both OLS and Tobit estimation methods generate consisted results with only minor differences in significance levels in TE and UCU model results. The TE and UCU models were jointly significant, although the explanatory power of the OLS models were generally low. While the allocative efficiency model was jointly significant using the Tobit approach, variables in the OLS model were not jointly significant. Thus, the second stage analysis results for the allocative model should be interpreted with caution.

Table 6. Tobit analysis results.

TOBIT	TE					UCU					Allocative				
Coefficients	Estim.	Std. Err.	p-val.	Magn.	Rank	Estim.	Std. Err.	p-val.	Magn.	Rank	Estim.	Std. Err.	p-val.	Magn.	Rank
(Intercept)	0.311	0.283	0.273			0.469	0.220	0.033			6.967	7.537	0.355		
Male	0.018	0.172	0.919	0.018		0.462	0.135	0.001	0.462	2	0.186	0.137	0.175	0.186	
Age	−0.291	0.059	0.000	0.291	5	−0.003	0.049	0.950	0.003		0.001	0.004	0.884	0.007	
Experience	0.141	0.044	0.001	0.141	8	0.016	0.035	0.654	0.016		−0.002	0.003	0.626	0.017	
Education_2	−0.219	0.114	0.055	0.219	6	−0.116	0.091	0.202	0.116		0.105	0.090	0.243	0.105	
Generation_2	−0.162	0.101	0.110	0.162		0.166	0.081	0.039	0.166	3	0.135	0.078	0.085	0.135	3
Salinity	0.019	0.058	0.749	0.019		−0.031	0.046	0.504	0.031		0.083	0.043	0.052	0.088	4
Temperature	0.057	0.041	0.172	0.057		0.009	0.033	0.793	0.009		−0.044	0.077	0.567	0.019	
Dissolved Oxygen	−0.008	0.063	0.897	0.008		−0.073	0.050	0.144	0.073		−0.017	0.021	0.421	0.039	
Light penetration	−0.307	0.107	0.004	0.307	4	0.068	0.082	0.412	0.068		−0.063	0.048	0.184	0.109	
Turbidity	0.592	0.172	0.001	0.592	1	−0.036	0.135	0.792	0.036		0.045	0.060	0.454	0.096	
Nitrogen	0.171	0.090	0.056	0.171	7	−0.010	0.072	0.889	0.010		−0.409	1.818	0.822	0.016	
Phosphorous	−0.351	0.134	0.009	0.351	3	−0.064	0.107	0.549	0.064		29.642	27.338	0.278	0.112	
Chlorophyll-a	−0.373	0.124	0.003	0.373	2	0.011	0.097	0.911	0.011		−0.150	0.186	0.419	0.075	
pH	0.128	0.068	0.062	0.128	9	−0.027	0.055	0.621	0.027		−0.881	0.880	0.317	0.053	
Eastern Banks	−0.290	0.454	0.522	0.290		0.657	0.359	0.067	0.657	1	0.773	0.348	0.026	0.773	1
Eastern Bay	0.301	0.307	0.328	0.301		−0.342	0.237	0.149	0.342		0.494	0.234	0.035	0.494	2
Log Sigma	−1.136	0.079	0.000			−1.368	0.075	0.000			−1.363	0.069	0.000		
Model Statistics	**TE**					**UCU**					**Allocative**				
Left-censored:	0					0					0				
Uncensored:	90					95					107				
Right-censored:	23					18					6				
Log-likelihood:	−46.663					−18.723					−13.1-28				
LR chi-squared:	71.116					62.660					23.081				
p-value:	0.000					0.000					0.000				

doi:10.1371/journal.pone.0115912.t006

Table 7. OLS analysis results.

OLS	TE					UCU					Allocative				
Coefficients	Estim.	Std. Err.	p-val.	Magn.	Rank	Estim.	Std. Err.	p-val.	Magn.	Rank	Estim.	Std. Err.	p-val.	Magn.	Rank
(Intercept)	−14.467	8.402	0.088			4.368	7.078	0.539			5.517	7.715	0.476		
Male	0.020	0.153	0.894	0.020		0.458	0.129	0.001	0.458	1	0.207	0.141	0.144	0.207	
Age	−0.020	0.004	0.000	0.258	4	0.002	0.003	0.481	0.031		−0.000	0.004	0.994	0.000	
Experience	0.013	0.004	0.002	0.125	9	−0.001	0.003	0.819	0.008		−0.001	0.004	0.751	0.011	
Education_2	−0.195	0.102	0.058	0.195	6	−0.075	0.086	0.384	0.075		0.100	0.093	0.286	0.100	
Generation_2	−0.185	0.088	0.039	0.185	7	0.143	0.074	0.057	0.143	3	0.135	0.081	0.100	0.135	3
Salinity	0.017	0.047	0.728	0.018		−0.015	0.040	0.717	0.015		0.073	0.044	0.096	0.078	4
Temperature	0.109	0.086	0.211	0.046		−0.001	0.073	0.992	0.000		−0.044	0.079	0.578	0.019	
Dissolved Oxygen	−0.002	0.024	0.923	0.005		−0.018	0.020	0.381	0.040		−0.013	0.022	0.560	0.029	
Light penetration	−0.144	0.053	0.008	0.248	5	0.043	0.045	0.338	0.074		−0.052	0.049	0.293	0.089	
Turbidity	0.194	0.067	0.005	0.415	1	−0.016	0.057	0.777	0.034		0.034	0.062	0.582	0.073	
Nitrogen	4.122	2.050	0.047	0.159	8	0.150	1.727	0.931	0.006		−0.387	1.882	0.838	0.015	
Phosphorous	−76.796	30.719	0.014	0.290	2	−14.330	25.880	0.581	0.054		28.570	28.210	0.314	0.108	
Chlorophyll-a	−0.560	0.209	0.009	0.281	3	0.009	0.176	0.961	0.004		−0.126	0.192	0.513	0.063	
pH	1.751	0.981	0.078	0.105	10	−0.212	0.827	0.799	0.013		−0.714	0.901	0.430	0.043	
Eastern Banks	−0.407	0.389	0.297	0.407		0.540	0.327	0.102	0.540		0.700	0.357	0.053	0.700	1
Eastern Bay	0.170	0.261	0.515	0.170		−0.382	0.220	0.085	0.382	2	0.418	0.239	0.084	0.418	2
Model Statistics		**TE**				**UCU**					**Allocative**				
Df:		96				96					96				
Residual standard error:		0.289				0.243					0.265				
Multiple R-squared:		0.469				0.410					0.180				
Adjusted R-squared:		0.380				0.311					0.044				
F-statistic:		5.292				4.162					1.320				
p-value:		0.000				0.000					0.201				

doi:10.1371/journal.pone.0115912.t007

The OLS estimation of the relationship between the assessed exogenous production factors and the derived TE scores suggest that the age of farmers has negative significant impact on TE scores (Table 7). Tertiary education, and two or more generations within families in oyster farming also had a negative effect on TE scores compared to farmers who have a lower educational level and are first generation in oyster farming. TE is likely to be positively influenced be a higher level of experience as an oyster farmer. Gender did not affect TE in our estimation.

Light penetration, turbidity, nitrogen, phosphorous, chlorophyll and pH were all found to be significant, as expected. The spatial dummy variables were not significant, suggesting any differences between areas were adequately captured by the environmental variables.

An OLS regression analysis using scaled independent variables provided information about the magnitude and rank of the impact that significant exogenous variables have on the TE score (Table 7). Based on this, most demographic and environmental conditions appear to affect the level technical efficiency of oyster businesses, with the latter having a generally greater influence.

In terms of (unbiased) capital utilisation, being male and of more than one family generation in the oyster business has a positive and significant impact. In contrast, none of the environmental variables had a significant effect on UCU. This is not surprising, as the measure reflects to a large extent the degree to which output could be increased by increasing variable inputs, all other things being equal. The dummy variables for Eastern Bay (in OLS model) and Eastern Banks (Tobit model only) were significant (Table 6 andTable 7), suggesting that output of leases in the Eastern Banks (positive coefficient) was more fully utilised than the other areas, while Eastern Bay (negative coefficient) had greatest potential to increase output from increased variable input use.

The results for the allocative efficiency models show that more than one generation in oyster business, average salinity and spatial dummy variables are weakly significant (Table 6 and Table 7). However, the F-statistic in the OLS model indicates that the variables are not jointly significant, and the very low R-squared coefficient suggest that these factors explain very little of the actual variation in allocative efficiency. Thus, we concluded that the level of allocative efficiency observed in the industry is likely explained by factors other than those assessed in this study.

We did not undertake a second stage analysis on the derived scale efficiency scores. Lease sizes are determined exogenously (by the Government), and are not within the control of the farmers.

DISCUSSION AND CONCLUSION

Increasingly, marine protected areas are being designed to achieve ecological objectives with least cost in terms of forgone commercial production through allowing multiple uses of conservation areas [2], [56], [57]. However, with potentially a high opportunity cost in terms of conservation benefits at stake, it is essential that remaining commercial activities are managed as efficiently as possible. Conservation management in marine protected areas should therefore be complemented by effective fisheries and aquaculture management.

Ensuring that the productivity of economic activities is maximised requires an understanding of the potential output of the industry and the level and drivers of efficiency and capacity utilisation within the industry. In this study the case of the Moreton Bay SRO industry was considered as this is an industry with a long history in the region, but has declined substantially over recent decades. Further, it is based on a native species, so preservation of the industry also has conservation as well as commercial value. Finally, it is facing increasing competition from other activities within Moreton Bay for space, particularly recreational and conservation uses.

While production of the industry has declined, the number of active leases in the area, in contrast, has not decreased by the same proportion, suggesting substantial decreases in productivity in the region. The Oyster Industry Management Plan for Moreton Bay Marine Park [9] includes the objectives of increasing production from the existing leases, and measures are available in the related policy on management of non-productive oyster leases [10] to potentially confiscate leases that do not meet minimum performance standards.

Given a fixed number of leases, and given that these leases are mostly operated at the optimal scale (from the scale efficiency measures), production can only be increased through either working the leases harder or through increased efficiency. The distribution of capacity utilisation from the analysis suggests that the potential to increase output through greater utilisation is limited for many leases, although a small number of leases were relatively underutilised. In contrast, a high proportion of the leases were operating inefficiently, and improved efficiency is the only way in which total productivity is likely to increase.

The potential to increase efficiency in the area depends on the factors that drive inefficiency, and the degree to which these can be influenced by policy. From the second stage analysis, the key drivers of efficiency differences between farms were largely environmental, and largely related to water quality. Hence productivity improvements are more likely to be improved through improvements in water quality in the region than through activities of individual farmers. Declining water quality has been attributed to substantial degradation of other components of the Moreton Bay marine ecosystem [58], and there is an active program underway to improve water quality through improved catchment management [59],[60]. Oyster farming has not generally been associated with water quality impairment, with some studies suggesting both beneficial attributes (e.g., nutrient recycling) as well as negative aspects (e.g., oxygen consumption) [61]. An assessment of the effects on the water quality in Moreton Bay and subsequent effect on the productivity of the oyster industry was beyond the scope of this study.

Some farmer specific variables were found to be significant, however, some potential to improve efficiency (and hence production) does exist. The key farmer characteristics that affected the level of efficiency included age, experience, education and family history in the industry. As might be expected, efficiency decreases with the age of farmers but increases with their experience. The fishery is characterised by an older population, many of which enter the industry at a relatively old age (compared with most industries). When comparing these findings with more detailed information about socio-economic characteristics of oyster farmers collected in 2012 (shown in Table 1), we can conclude that there is a high degree of hobbyist or lifestyle oyster farmers present in this industry. This type of oyster farmers may have lower incentives in operating their business efficiently than commercial oriented farmers, and thus, this may explain the observed low technical efficiency.

In such a case, efficiency would be enhanced by recruiting younger farmers to the fishery with a greater dependence on the industry for income, but given generally low earnings from the activity and the higher opportunity cost of labour of younger (potential) farmers, due to the co-location with a major city, encouraging younger farmers to the industry is difficult. Given this, the potential requirement of minimum production volumes over a number of years may be counter-productive. While hobbyist or lifestyle farmers could be forced out of industry as a result of the policy, leases that subsequently became available may not necessarily be taken up by existing or new oyster farmers.

The efficiency increase associated with experience suggests that skills can be learnt through time which improves productivity. Understanding these skills and undertaking training may help expedite these productivity benefits. Experience is a common factor affecting efficiency in both wild caught fisheries [30], [62], [63] and aquaculture [43], [64],[65].

The result that higher levels of education do not necessarily increase efficiency (and may, in fact, decrease efficiency) is not uncommon in studies of aquacultural efficiency [66], [67], although other studies have found that efficiency levels are related to the level of education[68], [69]. In wild caught fisheries, Pascoe and Coglan [62] found that education improved the efficiency of vessels using mobile gear (e.g., trawl), but decreased the efficiency of fishers using static gear (e.g., lobster pots). Oyster farming is a largely passive activity, as there is relatively little ongoing intervention required in their husbandry. One possible explanation then is that more educated farmers may be more prone to unnecessarily employing too much labour trying to improve production with less than proportional results. As many of the farmers came to the industry at a more advanced age, another possibility is that more educated farmers came from occupations that involved a very different skill set to those who were less educated.

The outcome for allocative efficiency scores implies that there is the potential for changes in the production mix to enhance production revenues (Fig. 3 and Fig. 4). However, there was no significant link between observable demographic and environmental factors and the allocative efficiency scores. Key factors that may influence the production mix are the risk of stock loss and the need to maintain of a cash flow. Although oyster farmers would potentially gain a higher price for selling plate (large) sized oysters, this would take a longer than harvesting a smaller size, as SRO take 2.5 and 3.5 years to grown to bistro and plate grade respectively. During this extra year, there is the risk of production loss through diseases, water pollution, extreme weather events or poaching (an ongoing problem in the industry). More risk averse farmers are likely to harvest a higher proportion of their stock earlier than what otherwise might be considered optimal [70]. Maintaining a cash flow during this period may also be important for farmers, especially those who do not receive a sufficient income from off-farm activities.

The aim of this study was to assess productivity measures such as technical efficiency, scale efficiency, allocative efficiency and capacity utilisation measure for the Moreton Bay SRO industry, and to determine the extent to which these measures could be influenced by management to enhance productivity in the industry. We found that there is a relatively low level of technical efficiency in the industry. Some of this can be explained by differences in environmental conditions in Moreton Bay. As such, improvements in water quality in the Bay may result in increased productivity in the industry. However, some demographic traits of the farmers are also significant drivers of efficiency. In particular, the high numbers of pre-retirement hobbyists present in this industry who potentially undertake their oyster business with a low incentive for technical efficient production, and also potentially with the wrong skill set to operate efficiently. Forcing these producers out of the industry through command and control measures (i.e., minimum production requirements) may not be effective in increasing productivity as there are few incentives for younger farmers to enter the industry. Developing appropriate training programs aimed at specific skills may be a more effective means of improving efficiency in the industry.

While the results of this study are not immediately transferable to other coastal aquaculture (or fishing) industries competing in a multi-use environment, the methods we have employed are readily transferable. Multiple use management of marine protected areas requires each of the uses to be optimised. This is particularly so in high population areas where use is correspondingly high, and the opportunity cost in terms of foregone recreational or conservation value from an underperforming commercial activity may be substantial.

ACKNOWLEDGMENTS

The authors of this article are grateful to John Dexter (QLD DAFF) who provided helpful comments and assistance in the data collection process as well as feedback on regulatory aspects of the Sydney rock oyster industry in Moreton Bay. We thank Ana Rubio Zuazo (University of Sydney) for useful remarks on environmental aspects that can affect oyster growth and development. We are particularly grateful to the oyster farmers in Moreton Bay who agreed to share their production data and personal data for the purpose of this study. We also thank two anonymous reviewers for their comments on the original manuscript of this study.

Author Contributions

Conceived and designed the experiments: PS SP. Performed the experiments: PS SP. Analyzed the data: PS SP. Contributed reagents/materials/analysis tools: PS SP LC. Contributed to the writing of the manuscript: PS SP LC.

REFERENCES

1. Byron C, Bengtson D, Costa-Pierce B, Calanni J (2011) Integrating science into management: Ecological carrying capacity of bivalve shellfish aquaculture. Marine Policy 35:363–370. doi: 10. 1016/j. marpol. 2010. 10. 016
2. Mazor T, Possingham HP, Edelist D, Brokovich E, Kark S (2014) The Crowded Sea: Incorporating Multiple Marine Activities in Conservation Plans Can Significantly Alter Spatial Priorities. PLoS ONE 9:e104489. doi: 10. 1371/ journal. pone. 0104489
3. Lampe HW, Hilgers D (2014) Trajectories of efficiency measurement: A bibliometric analysis of DEA and SFA. European Journal of Operational Research.
4. Pascoe S, Doshi A, Dell Q, Tonks M, Kenyon R (2014) Economic value of recreational fishing in Moreton Bay and the potential impact of the marine park rezoning. Tourism Management 41:53–63. doi: 10. 1016/j. tourman. 2013. 08. 015
5. Whitmore M, De Lacy T (2005) Sustainable development and management of tourism in Moreton Bay. Brisbane: Sustainable Tourism CRC. 204 p.
6. Smith G (1981) Southern Queensland's oyster industry. Journal of the Royal Historical Society of Queensland 11:45–58.
7. ABARES (2013) Australian fisheries statistics 2012. Canberra: Australian Bureau of Agricultural and Resource Economics and Sciences (ABARES).
8. NSW DPI (2013) Aquaculture production report 2011–2012. In: Livingstone S, editor. Port Stephens: ISSN 1444-840. New South Wales Department of Primary Industries.
9. QLD DPI&F (2008) Oyster Industry Manamgement Plan for the Moreton Bay Marine Park. Brisbane: Queensland Department of Primary Industries and Fisheries, August 2008.
10. QLD DPI&F (2007) Policy for maximising rock oyster production: Management of non-productive oyster areas. Brisbane: Queensland Department of Primary Industries and Fisheries, May 2007.
11. Shahidul Islam M, Tanaka M (2004) Impacts of pollution on coastal and marine ecosystems including coastal and marine fisheries and approach for management: a review and synthesis. Marine Pollution Bulletin 48:624–649. doi: 10. 1016/j. marpolbul. 2003. 12. 004
12. Iliyasu A, Mohamed ZA, Ismail MM, Abdullah AM, Kamarudin SM, et al. (2014) A review of production frontier research in aquaculture (2001–2011). Aquaculture Economics & Management 18:221–247. doi: 10. 1080/13657305. 2014. 926464
13. Ogburn DM, White I, McPhee DP (2007) The disappearance of oyster reefs from eastern Australian estuaries–Impact of colonial settlement or mudworm Invasion? Coastal Management 35:271–287. doi: 10. 1080/08920750601169618
14. Schrobback P, Pascoe S, Coglan L (2014) Impacts of introduced aquaculture species on markets for native marine aquaculture products: the case of edible oysters in Australia. Aquaculture Economics & Management 18:248–272. doi: 10. 1080/13657305. 2014. 926465
15. Smith G (1985) The Queensland oyster fishery: An illustrated history. Queensland Department of Primary Industries, Brisbane, Australia.
16. Lergessner JG (2006) Oysterers of Moreton Bay. Bribie Island, Queensland.
17. Wingfield M, Heidenreich M (2013) Annual Sydney rock oyster production data. Bribie Island/Brisbane:

Queensland Department of Agriculture, Fisheries and Forestry.

18. Australian Government (2014) Australian Ramsar Wetlands. Department of the Environment and Water Resources.

19. QLD DPI (2008) Oyster Industry Management Plan for Moreton Bay Marine Park. PR08–4137. State of Queensland, Department of Agriculture, Fisheries and Forestry.

20. Coelli TJ, Perelman S (2000) Technical efficiency of European railways: a distance function approach. Applied Economics 32:1967–1976. doi: 10. 1080/00036840050155896

21. Färe R, Grosskopf S (2000) Theory and application of directional distance functions. Journal of Productivity Analysis 13:93–103. doi: 10. 1023/a:1007844628920

22. Grosskopf S, Margaritis D, Valdmanis V (1995) Estimating output substitutability of hospital services: A distance function approach. European Journal of Operational Research 80:575–587. doi: 10. 1016/0377-2217(94)00138-3

23. O'Donnell C, Coelli T (2005) A Bayesian approach to imposing curvature on distance functions. Journal of Econometrics 126:493–523. doi: 10. 1016/j. jeconom. 2004. 05. 011

24. Pascoe S, Punt AE, Dichmont CM (2010) Targeting ability and output controls in Australia's multi-species Northern Prawn Fishery. European Review of Agricultural Economics 37:313–334. doi: 10. 1093/erae/jbq022

25. Alam F (2011) Measuring technical, allocative and cost efficiency of pangas (Pangasius hypophthalmus: Sauvage 1878) fish farmers of Bangladesh. Aquaculture Research 42:1487–1500. doi: 10. 1111/j. 1365-2109. 2010. 02741. x

26. Chambers R, Hailu A, Quiggin J (2011) Event-specific data envelopment models and efficiency analysis. Australian Journal of Agricultural & Resource Economics 55:90–106. doi: 10. 1111/j. 1467-8489. 2010. 00517. x

27. Mugera AW (2013) Measuring technical efficiency of dairy farms with imprecise data: a fuzzy data envelopment analysis approach. Australian Journal of Agricultural & Resource Economics 57:501–520. doi: 10. 1111/1467-8489. 12008

28. Reid C, Squires D, Jeon Y, Rodwell L, Clarke R (2003) An analysis of fishing capacity in the western and central Pacific ocean tuna fishery and management implications. Marine Policy 27:449–469. doi: 10. 1016/s0308-597x(03)00065-4

29. Tingley D, Pascoe S (2005) Factors Affecting Capacity Utilisation in English Channel Fisheries. Journal of Agricultural Economics 56:287–305. doi: 10. 1111/j. 1477-9552. 2005. 00005. x

30. Tingley D, Pascoe S, Coglan L (2005) Factors affecting technical efficiency in fisheries: Stochastic production frontier versus data envelopment analysis approaches. Fisheries Research 73:363–376. doi: 10. 1016/j. fishres. 2005. 01. 008

31. Vestergaard N, Squires D, Kirkley J (2003) Measuring capacity and capacity utilization in fisheries: The case of the Danish gill-net fleet. Fisheries Research 60:357–368. doi: 10. 1016/s0165-7836(02)00141-8

32. Hoff A (2007) Second stage DEA: Comparison of approaches for modelling the DEA score. European Journal of Operational Research 181:425–435. doi: 10. 1016/j. ejor. 2006. 05. 019

33. Charnes A, Cooper WW, Rhodes E (1978) Measuring the efficiency of decision making units. European Journal of Operational Research 2:429–444. doi: 10. 1016/0377-2217(78)90138-8

34. Azizi H, Ajirlu SF (2010) Measurement of overall performances of decision-making units using ideal and anti-ideal decision-making units. Computers & Industrial Engineering 59:411–418. doi: 10. 1016/j. cie. 2010. 05. 013

35. Humphreys J, Herbert RJH, Roberts C, Fletcher S (2014) A reappraisal of the history and economics of the Pacific oyster in Britain. Aquaculture 428–429:117–124. doi: 10. 1016/j. aquaculture. 2014. 02. 034

36. Coelli TJ, Rao DSP, O'Donnell CJ, Battese GE (2005) An introduction to efficiency and productivity analysis. New York: Spinger Science & Business Media.

37. Holland DS, Lee ST (2002) Impacts of random noise and specification on estimates of capacity derived from data envelopment analysis. European Journal of Operational Research 137:10–21. doi: 10. 1016/s0377-2217(01)00087-x

38. Färe R, Grosskopf S, Logan J (1983) The relative efficiency of Illinois electric utilities. Resources and Energy 5:349–367. doi: 10. 1016/0165-0572(83)90033-6

39. Banker RD, Charnes A, Cooper WW (1984) Some models for estimating technical and scale inefficiencies in Data Envelopment Analysis. Management Science (pre-1986) 30:1078. doi: 10. 1287/mnsc. 30. 9. 1078

40. Färe R, Grosskopf S, Kirkley J (2000) Multi-Output Capacity Measures and Their Relevance for Productivity. Bulletin of Economic Research 52:101–113. doi: 10. 1111/1467-8586. t01-1-00097

41. Dupont DP, Grafton RQ, Kirkley J, Squires D (2002) Capacity utilization measures and excess capacity in multi-product privatized fisheries. Resource and Energy Economics 24:193–210. doi: 10. 1016/s0928-7655(01)00050-1

42. Färe R, Grosskopf S, Kokkelenberg EC (1989) Measuring plant capacity, utilization and technical change: A nonparametric approach. International Economic Review 30:655–666. doi: 10. 2307/2526781

43. Ahmed N, Zander KK, Garnett ST (2011) Socioeconomic aspects of rice-fish farming in Bangladesh: opportunities, challenges and production efficiency. Australian Journal of Agricultural & Resource Economics 55:199–219. doi: 10. 1111/j. 1467-8489. 2011. 00535. x

44. Kieschnick R, McCullough BD (2003) Regression analysis of variates observed on (0, 1): percentages, proportions and fractions. Statistical Modelling 3:193–213. doi: 10. 1191/1471082x03st053oa

45. McDonald J (2009) Using least squares and tobit in second stage DEA efficiency analyses. European Journal of Operational Research 197:792–798. doi: 10. 1016/j. ejor. 2008. 07. 039
46. Healthy Waterways Ltd (2012) Estuarine marine data. Ecosystem Health Monitoring Program. Brisbane.
47. Rubio Zuazo AM (2008) Environmental influences on the sustainable production of the Sydney rock oyster, saccostrea glomerata: A study in two southeastern Australian estuaries. PhD Thesis, Centre for Resource and Environmental Studies, Fenner School for Environment and Society. The Australian National University.
48. Dove MC, O'Connor W (2007) Salinity and temperature tolerance of Sydney rock oysters, saccostrea glomerata, during early ontogeny. Journal of Shellfish Research 26:939–947. doi: 10. 2983/0730-8000(2007)26[939:sattos]2. 0. co;2
49. Holliday JE (1995) Nursery culture of Sydney rock oysters, saccostrea commercialis (Iredale and Roughley, 1933) and Pacific oysters, crassostrea gigas (Thunberg, 1793). Unpublished PHD Thesis, Department of Aquaculture, University of Tasmania.
50. Bayne BL, Svensson S, Nell JA (1999) The Physiological Basis for Faster Growth in the Sydney Rock Oyster, Saccostrea commercialis. The Biological Bulletin 197:377–387. doi: 10. 2307/1542792
51. Parker LM, Ross PM, O'Connor WA (2011) Populations of the Sydney rock oyster (saccostrea glomerata) vary in response to ocean acidification. Marine Biology 158:689–697. doi: 10. 1007/s00227-010-1592-4
52. Dove MC (2003) Effects of estuarine acidification on survival and growth of the Sydney rock oyster, saccostrea glomerats. PhD thesis Geography Department. University of New South Wales, Sydney.
53. Grant J, Cranford P, Emerson C (1997) Sediment resuspension rates, organic matter quality and food utilization by sea scallops (Placopecten magellanicus) on Georges Bank. Journal of Marine Research 55:965–994. doi: 10. 1357/0022240973224193
54. Kellogg ML, Cornwell JC, Owens MS, Paynter KT (2013) Denitrification and nutrient assimilation on a restored oyster reef. Marine Ecology Progress Series 480:1–13. doi: 10. 3354/meps10331
55. Higgins CB, Stephenson K, Brown BL (2011) Nutrient bioassimilation capacity of aquacultured oysters: Quantifi cation of an ecosystem service. Journal of Environmental Quality 40:271–277. doi: 10. 2134/jeq2010. 0203
56. Klein CJ, Steinback C, Watts M, Scholz AJ, Possingham HP (2009) Spatial marine zoning for fisheries and conservation. Frontiers in Ecology and the Environment 8:349–353. doi: 10. 1890/090047
57. Stewart R, Possingham H (2005) Efficiency, costs and trade-offs in marine reserve system design. Environmental Modeling and Assessment 10:203–213. doi: 10. 1007/s10666-005-9001-y
58. Jackson JBC, Kirby MX, Berger WH, Bjorndal KA, Botsford LW, et al. (2001) Historical overfishing and the recent collapse of coastal ecosystems. Science 293:629–637. doi: 10. 1126/science. 1059199
59. Pantus F, Dennison W (2005) Quantifying and evaluating ecosystem health: A case study from Moreton Bay, Australia. Environmental Management 36:757–771. doi: 10. 1007/s00267-003-0110-6
60. de la Mare W, Ellis N, Pascual R, Tickell S (2012) An empirical model of water quality for use in rapid management strategy evaluation in Southeast Queensland, Australia. Marine Pollution Bulletin 64:704–711. doi: 10. 1016/j. marpolbul. 2012. 01. 039
61. Chapelle A, Ménesguen A, Deslous-Paoli J-M, Souchu P, Mazouni N, et al. (2000) Modelling nitrogen, primary production and oxygen in a Mediterranean lagoon. Impact of oysters farming and inputs from the watershed. Ecological Modelling 127:161–181. doi: 10. 1016/s0304-3800(99)00206-9
62. Pascoe S, Coglan L (2000) Implications of differences in technical efficiency of fishing boats for capacity measurement and reduction. Marine Policy 24:301–307. doi: 10. 1016/s0308-597x(00)00006-3
63. Sharma KR, Leung PS (1999) Technical efficiency of the longline fishery in Hawaii: an application of a stochastic production frontier. Marine Resource Economics 13:259–274.
64. Irz X, McKenzie V (2003) Profitability and technical efficiency of aquaculture systems in pampaanga, philippines. Aquaculture Economics & Management 7:195–211. doi: 10. 1080/13657300309380340
65. Sharma KR, Leung PS (2000) Technical efficiency of carp production in India: A stochastic frontier production function analysis. Aquaculture Research 31:937–947. doi: 10. 1046/j. 1365-2109. 2000. 00521. x
66. Chiang F-S, Sun C-H, Yu J-M (2004) Technical efficiency analysis of milkfish (chanos chanos) production in Taiwan–an application of the stochastic frontier production function. Aquaculture 230:99–116. doi: 10. 1016/j. aquaculture. 2003. 09. 038
67. Onumah EE, Brümmer B, Hörstgen-Schwark G (2010) Elements which delimitate technical efficiency of fish farms in Ghana. Journal of the World Aquaculture Society 41:506–518. doi: 10. 1111/j. 1749-7345. 2010. 00391. x
68. Dey MM, Paraguas FJ, Bimbao GB, Regaspi PB (2000) Technical efficiency of tilapia growout pond operations in the Philippines. Aquaculture Economics & Management 4:33–47. doi: 10. 1080/13657300009380259
69. Singh K, Dey MM, Rabbani AG, Sudhakaran PO, Thapa G (2009) Technical efficiency of freshwater aquaculture and its determinants in Tripura, India. Agricultural Economics Research Review 22:185–195.
70. Pascoe S, Wattage P, Naik D (2002) Optimal harvesting strategies: Practice versus theory. Aquaculture Economics & Management 6:295–308. doi: 10. 1080/13657300209380320
71. Australian Taxation Office (2013) Tax Withheld Calculator - Individual non-business. Canberra: Commonwealth of Australia.

SECTION V:
FISH POSTHARVEST

CHAPTER 16

APPLICATION OF HIGHLY PURIFIED ELECTROLYZED CHLORINE DIOXIDE FOR TILAPIA FILLET DISINFECTION

Chen-Hsing Yu, Tzou-Chi Huang, Chao-Chin Chung, Hao-Hsun Huang and Ho-Hsien Chen

INTRODUCTION

Chlorine dioxide (ClO_2) is a strong oxidant widely applied for sterilization, disinfection, and waste-water treatment. It is commonly used on drinking water and environmental disinfection. It was also recommended as a commercial sanitizer to replace electrolyzed oxidizing water [1, 2], chlorine (Cl_2), hypochlorous acid (HOCl), and hypochlorite (OCl^-) [3–5]. Contact of chlorine dioxide with organic substances in food or water results in microbial resistance and inactivation, but it also produces four trihalomethane (THM) byproducts, that is, chloroform, bromodichloromethane, dibromochloromethane, and bromoform, which are associated with toxicity and carcinogenesis [6–9]. In Taiwan, tilapia fillets are an important economic product, and it is common practice to use sodium hypochlorite (NaClO) as a disinfecting agent for processing tilapia fillets; however, treatment of this type could lead to serious problems involving residual THMs in treated seafood [4,10–12]. As for its application for vegetable and fruit disinfection, ClO_2 gas has been successfully used to disinfect strawberries, lettuce, cabbage, and cucumbers with continuous methods [4, 13–17]. In this work, the bactericidal efficacy of ClO_2 was evaluated for cleaning tilapia fillets with different cleaning methods.

Commercial ClO_2 is commonly generated using chemical methods that react sodium chloride, sodium hypochlorite, or sodium chlorate with sulfuric acid or hydrochloride acid [18, 19]. The chemical method of producing ClO_2 needs a strong acid (pH 2~3), inhibits Cl_2 hydrolysis, and takes a long time for activation. The yield of ClO_2 depends on the purity of the raw materials, the catalyst, pH, reaction time, and temperature [20–24]. Furthermore, it was discovered that electrolyzing sodium chlorite can produce highly purified ClO_2[25, 26]. Therefore, the objective of this study is to develop novel electrolysis equipment to produce highly purified, low-cost ClO_2 to disinfect with water, while simultaneously monitoring trihalomethane (THM) residuals on tilapia fillets.

MATERIALS AND METHODS

Materials

Tilapia fillets were bought from a local traditional market in Pingtung, Taiwan. The microbiological media used in this study were peptone and tryptic soy agar (TSA) purchased from Difco Laboratories (Detroit, MI, USA); these were prepared according to the manufacturer's specifications. 3 M Coliform and E. coli Petrifilm no. 6414 were purchased from Microbiology Products 3 M Health Care (St. Paul, MN, USA).

Citation: Chen-Hsing Yu, Tzou-Chi Huang, Chao-Chin Chung, Hao-Hsun Huang, and Ho-Hsien Chen, "Application of Highly Purified Electrolyzed Chlorine Dioxide for Tilapia Fillet Disinfection. doi:10.1155/2014/619038.

ClO_2 Electrolysis Equipment (ClO_2 Generator)

The self-designed electrolysis equipment consisted of a raw material tank, an electrolyzer, an air pump, two ClO_2 collecting tanks, and a cooling system (Taiwan Patent, no. 200722557) [27]. Figure 1 shows the designed chlorine dioxide electrolysis equipment. The internal structure and reaction of the electrolyzer are shown in Figure 2. Saturated saline and sodium hypochlorite enter and mix in the electrolyzer system using a direct current (100~110 A, 7~8 V), the electrolyzed temperature was controlled to 55~65°C, and the electrolyzed material supply rate was 10 L/h. The temperature of the ClO_2 collecting tank was maintained at 5~10°C by cooling water from the cooling system. NaCl was electrolyzed into $NaClO_2$. The reaction equation is as follows:

$$NaCl + 2H_2O \longrightarrow NaClO_2 + 2H_2 \uparrow . \qquad (1)$$

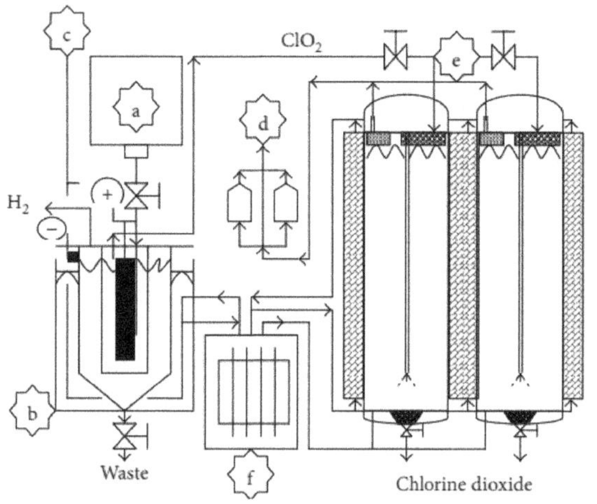

Figure 1: Designed chlorine dioxide electrolysis equipment: (a) material, (b) electrolyzer, (c) electronic control system, (d) air pump, (e) collecting tank, and (f) cooling system.

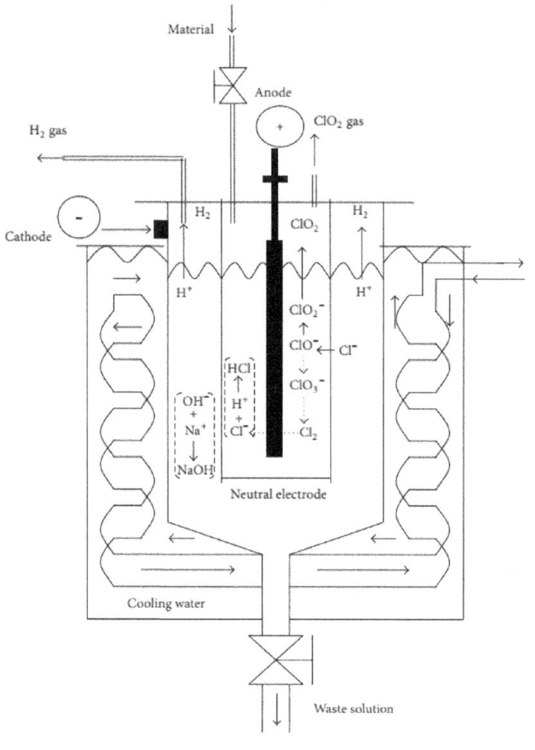

Figure 2: Internal structure and reaction of the electrolysis system.

Meanwhile, the $NaClO_2$ was further electrolyzed, the ClO_2^- was attracted by the cathode, and H_2O was attracted by the anode to release H_2 (Figure 2). The reaction equations are as follows:

$$\text{anode:} \quad ClO_2^- \longrightarrow ClO_2 \uparrow + e^-; \qquad (2)$$

$$\text{cathode:} \quad H_2O + e^- \longrightarrow OH^- + \frac{1}{2}H_2; \qquad (3)$$

$$\text{overall:} \quad 2NaClO_2 + 2H_2O \longrightarrow 2ClO_2 + 2NaOH + H_2 \uparrow \qquad (4)$$

$$NaCl + NaClO_2 + 4H_2O \longrightarrow 2ClO_2 + 2NaOH + 3H_2 \uparrow . \qquad (5)$$

The resultant ClO_2 was aspirated out and collected into 5~10°C pure water in the two collecting tanks. The NaOH solution was collected separately. The oxidation/reduction potential (ORP) and pH of the ClO_2solutions were measured using an ORP/pH meter (Mettler-Toledo Seven Easy ORP/pH meter, Kaohsiung, Taiwan).

ClO_2 analysis: the concentration of ClO_2 was analyzed using the iodine method [28]. The ClO_2 solution at 10 mL was diluted 200 times with pure water. It was adjusted to five pH levels and then titrated with a 0.01 N sodium thiosulfite solution. The titration volumes were A, B, C, D, and E. The following calculation formulas were used to calculate the concentrations of ClO_2, Cl_2, ClO_2^-, and ClO_3^-

$$ClO_2\,(\text{ppm}) = \left(\frac{5}{4}\right) \times (B - D) \times N \times 13,490/\text{sample volume}, \qquad (6)$$

$$Cl_2\,(\text{ppm}) = D \times N \times 16,863/\text{sample volume}, \qquad (7)$$

$$ClO_2^-\,(\text{ppm}) = [E - (A + B)] \times N \times 13,909/\text{sample volume}, \qquad (8)$$

$$ClO_3^-\,(\text{ppm}) = \left[A - \frac{(B - D)}{4}\right] \times N \times 13,909/\text{sample volume}. \qquad (9)$$

Here, N is the concentration of the sodium thiosulfite solution.

Cleaning Methods

Tilapia fillets were incubated at 37°C until the total plate count reached 5~6 log CFU/g. The different purities (45%~99%) of 400 mg/L ClO_2 solutions were used to test the effect of tilapia disinfection. Tilapia fillets were inoculated with coliforms or E. coli at a concentration of 5~6 log CFU/g. At 99% purity, ClO_2 solutions of 50, 100, and 200 ppm were used to wash the fillets by soaking or shaking treatment for 5, 15, and 25 min, and the total plate counts, coliforms, and E. coli of the fillets were determined.

Microbiological Analyses

The total plate count assay followed the China National Standard (CNS 10890 N6186) [29] method: 1 mL of masticated tissue liquid was serially (1 : 10) diluted in 9 mL of 0.1% sterile peptone water, and 0.1 mL portions of appropriate diluents were surface-plated on TSA. The plates were then incubated at 37°C for 48 h in duplicate. CFUs were counted and expressed per gram of sample after logarithmic conversion. The coliform and E. coli assays followed Sasithorn and Sirirat [30] using 1 mL of masticated tissue liquid plated on 3 M Petrifilm no. 6414 (St. Paul, MN), and the plates were incubated at 37°C for 24 h in duplicate.

Residual THMs Analyses

The analytical procedure was modified from Stack et al.'s [31] gas chromatographic (GC) method on an Agilent 5890 system coupled to an Agilent 5973N mass spectrometer (MS) (Palo Alto, CA). Chromatographic separation was performed using a capillary column (HP-5, 30 m × 0.32 mm, 0.25 µm phase film thickness) from Agilent Technologies. The initial temperature was 45°C for 3 min and then increased by 8°C/min to a final temperature of 220°C for 20.5 min. The injector temperature was set to 200°C. Nitrogen was used as the carrier gas at a flow rate of 38.5 mL/min.

MS was operated in the electron ionization mode at 70 eV. The mass range was scanned at 40~350 m/z and for 0.60 seconds per scan for the full-scan mode. Temperatures for the trap, manifold, and transfer line were set to 250, 50, and 280°C, respectively. All data for quantification were collected in the selected ion monitoring mode at 83 and 85 m/z for chloroform, 127 and 129 m/z for dibromochloromethane, and 173 m/z for bromoform.

Statistical Analysis

Three replicates were conducted, and each sample was assayed in duplicate. Data collected from the experiments were analyzed by an analysis of variance (ANOVA) and Duncan's multiple range test using the SAS 8.2 program [32]. Significant differences between tested parameters were determined based on a 95% confidence level ($P<0.05$).

RESULTS AND DISCUSSION

Effects of Different Ratios of $NaClO_2$ for High Concentrations of ClO_2

Table 1 shows that 10% $NaClO_2$ and 20% NaCl generated 4749 ppm of total chlorine and 99.8% pure ClO_2, respectively. The pH was 2.18, and ORP was 1440 mV (Table 1). When the purity of ClO_2 varied from 99.5% to 99.8%, the pH increased from 2.36 to 2.18. Under an acidic condition, Cl_2 easily disassociated into Cl^-, and ClO_2 mainly disassociated into ClO_2^- and ClO_3^-, with a small portion disassociating into Cl^- [21, 33, 34]. A great quantity of Cl^- resulted from excessive NaCl in the raw materials which contained NaCl and sodium hypochlorite. At the anode side, NaCl was converted into $NaClO_2$ and then into $NaClO_3$. $NaClO_3$ was affected by the reducing reaction from the cathode side, producing Cl_2, Cl^- and H^+. The Cl^-, and H^+ then formed into very small amounts of HCl. These reaction cycles generated NaOH and ClO_2, producing Cl_2.

Table 1: Constituents of chlorine dioxide aqueous solution on different constituent ratios of electrolytes.

Electrolyte constituent ratio	20% NaCl			
	7% $NaClO_2$	8% $NaClO_2$	9% $NaClO_2$	10% $NaClO_2$
Constituents of chlorine dioxide aqueous solution				
ClO_2 (ppm)	3193.76^d	3612.31^c	4232.49^b	4738.36^a
ClO_2^- (ppm)	16.02^a	ND	ND	ND
ClO_3^- (ppm)	ND	ND	ND	ND
Cl_2 (ppm)	ND	9.75^b	10.64^a	10.64^a
Total chlorine* (ppm)	3209.78^d	3622.06^c	4243.12^b	4749.00^a
ClO_2 purity** (%)	99.50^a	99.73^a	99.75^a	99.78^a
pH (unit)	2.36^a	2.34^a	2.38^a	2.18^b
Oxidation reduction potential (millivolt)	1.32^c	1.38^b	1.44^a	1.44^a

a-d Means in the same column followed by different superscripts are significantly different at $P < 0.05$ (Duncan's multiple range test).
*Total chlorine (ppm) = $ClO_2 + ClO_2^- + ClO_3^- + Cl_2$.
**ClO_2 purity = $[ClO_2/(ClO_2 + ClO_2^- + ClO_3^- + Cl_2)] \times 100\%$.

Effect of ClO_2 Purity on Tilapia Fillet Disinfection

Tilapia fillets were soaked in 400 ppm of 99% pure ClO_2 for 5, 15, and 25 min. Results indicated that total plate counts on tilapia fillets decreased from 5.72 log CFU/g to 3.23, 2.1, and 1.09 log CFU/g, respectively (Table 2). Although ClO_2 solutions contained 45%, 50%, and 60% of freely available chlorine, the bactericidal effect was not so obviously effective. One of the explanations could be that the Cl_2 is not as effective as ClO_2, because active oxygen molecules diminish the number of electrons on biological cell membranes and cause damage to biological enzymes on biological membranes therefore amino acid and nucleic bodies are hindered from generating proteins in biological cells [33, 35, 36]. Another reason could be that ClO_2 not only reacts with electrons on biological cell membranes but also reacts with Cl_2 to achieve disassociation and oxidation under an acidic condition and then forms ClO_2^- and ClO_3^-, and Cl^- byproducts [21, 33]. The less Cl_2 there was, the higher the disinfection effect was.

Table 2: Total bacterial counts of tilapia fillets for various purities in 400 ppm chlorine dioxide solutions.

Treatment time (min)	ClO_2 purity (%)						
	45%	50%	60%	70%	80%	90%	99%
	Total bacteria count log (CFU/g)*						
5	5.60^{a}	5.51^{a}	4.53^{b}	4.38^{c}	4.13^{d}	3.96^{e}	3.23^{f}
15	5.44^{a}	5.43^{a}	4.3^{b}	4.09^{c}	3.98^{d}	3.29^{e}	2.1^{f}
25	5.33^{a}	4.3^{b}	3.13^{c}	2.91^{b}	2.74^{e}	2.54^{f}	1.09^{g}

$^{a-g}$Means in the same column followed by different superscripts are significantly different at $P < 0.05$ (Duncan's multiple range test).
*Original microbial load: 5.72 log (CFU/g).

Effect of Various Treatments

Three different concentrations (50, 100, and 200 ppm) of 99% ClO_2 solutions were used for soaking or shaking disinfection treatment on tilapia fillets for 5, 15, and 25 min. Results are shown in Table 3. After the fish fillets were shaken in the solutions for 5 min, total plate counts were 3.49, 2.41, and 1.29 log CFU/g, respectively, and all were nondetectable after 15 and 25 min, compared to the control groups at 5.84~5.78 log CFU/g (control).

Table 3: Total bacterial counts for tilapia fillets disinfected by soaking or shaking treatments at different chlorine dioxide concentrations.

Washing method	Concentration (ppm)	Cleaning time (min)		
		5	15	25
		Total bacteria count log (CFU/g)		
Soaking	Control	5.84^{ax}	5.80^{ax}	5.78^{ax}
	50	5.73^{abx}	5.62^{abxy}	5.53^{aby}
	100	5.43^{bx}	5.01^{aby}	4.63^{bz}
	200	4.73^{cx}	4.23^{by}	3.64^{cz}
Shaking	Control	5.83^{ax}	5.74^{ax}	5.85^{ax}
	50	3.49^{dx}	ND	ND
	100	2.41^{ex}	ND	ND
	200	1.29^{fx}	ND	ND

$^{a-f}$Means in the same column followed by different superscripts are significantly different at $P < 0.05$ (Duncan's multiple range test).
$^{x-z}$Means in the same row followed by different superscripts are significantly different at $P < 0.05$ (Duncan's multiple range test).

Similar results for coliforms and E. coli are shown in Tables 4 and 5. The control groups of coliform (control) were 5.23, 5.29, and 5.25 CFU/g. When tilapia fillets were treated with 50, 100, and 200 ppm of high-purity ClO_2 solutions with the shaking method for 5 min, the coliform counts, respectively, decreased to 1.78, 1.07 log CFU/g, and nondetectable (Table 4). Escherichia coli also showed a > 4 log reduction after 5 min and was nondetectable after 15 and 25 min of shaking (Table 5). Both the soaking and shaking methods eliminated microbial populations; however, the results show that the soaking method was not as effective as the shaking method. Microorganisms attached to fish skin may more easily be washed out by shaking with mechanical forces [37]. Aloisio and Francisco [38] claimed that ClO_2 being bound to water molecules by static attraction forces under a steady state hindered the bactericidal effect.

Table 4: Coliform reduction of tilapia fillets disinfected by soaking or shaking treatments at different chlorine dioxide concentrations.

Wash method	Concentration (ppm)	Cleaning time (min)		
		5	15	25
		Coliforms log (CFU/g)		
Soaking	Control	5.23^{ax}	5.33^{ax}	5.21^{ax}
	50	4.80^{bx}	3.22^{bz}	4.06^{by}
	100	4.17^{cx}	2.33^{cy}	ND
	200	3.23^{dx}	1.23^{dy}	ND
Shaking	Control	5.23^{ax}	5.29^{ax}	5.25^{ax}
	50	1.78^{ex}	ND	ND
	100	1.07^{fx}	ND	ND
	200	ND	ND	ND

$^{a-f}$Means in the same column followed by different superscripts are significantly different at $P < 0.05$ (Duncan's multiple range test).
$^{x-z}$Means in the same row followed by different superscripts are significantly different at $P < 0.05$ (Duncan's multiple range test).

Table 5: Escherichia coli reduction of tilapia fillets disinfected by soaking or shaking treatments at different chlorine dioxide concentrations.

Wash method	Concentration (ppm)	Cleaning time (min) 5	15	25
		E. coli log (CFU/g)		
Soaking	Control	5.23^{ax}	5.33^{ax}	5.22^{ax}
	50	3.08^{bx}	2.21^{by}	1.1^{cz}
	100	2.25^{cx}	1.14^{cy}	ND
	200	1.13^{dx}	ND	ND
Shaking	Control	5.56^{ax}	5.51^{ax}	5.58^{ax}
	50	1.1^{dx}	ND	ND
	100	ND	ND	ND
	200	ND	ND	ND

a–d Means in the same column followed by different superscripts are significantly different at $P < 0.05$ (Duncan's multiple range test).
x–z Means in the same row followed by different superscripts are significantly different at $P < 0.05$ (Duncan's multiple range test).

Detection of THMs

THM residuals are a problem for the safety of chlorine-treated food materials [39, 40]. After tilapia fillets were washed by soaking or shaking in the ClO_2 solution with the highest concentration (200 ppm) for 25 min, the waste solutions were analyzed for THMs using GC/MS. THMs include chloroform, dichloromethane, and methyl chloride. The results show that no THMs were detected in a used ClO_2 solution after soaking (GC-MS detection limit was 0.12 ppb), as shown in Table 6. These results conform to Taiwan's environmental protection regulations and act governing food sanitation. Furthermore, a LC-MS analysis also showed that when using a 200 ppm bactericide solution for 25 min at 25°C, residual of ClO_2 solution was no detected in solution (the instrument detection limit was 0.1 ppb).

Table 6: Total trihalomethane (THM) residuals by GC/MS.

Item	Total trihalomethane[a] (ppb)
Methyl chloride	ND
Dichloromethane	ND
Chloroform	ND

[a] The instrument detection limit was 0.12 ppb.

CONCLUSIONS

The results demonstrated the feasibility of stably producing ClO_2 using electrochemical technology. The maximum concentration and purity of ClO_2 were obtained when using a mixture that blended 20% NaCl and 7%~10% $NaClO_2$ together as the electrolytes. The concentration and purity of ClO_2 were 3200~4700 ppm and 99.5~99.7%, respectively. Disinfection results indicate that a 200 ppm ClO_2 solution reduced the total bacterial, coliform, and E. coli counts on tilapia fillets by 3.0~4.0 log CFU/g (P<0.05). The soaking wash treatment was more effective than the shaking method. A GC-MS analysis also showed that when using a 200 ppm bactericide solution for 25 min, residual THMs of the ClO_2 solution were nondetectable. Bactericidal treatment with a ClO_2 solution for tilapia fillets also conforms to Taiwan's environmental protection regulations and act governing food sanitation. The ClO_2 solution is indeed a safer method for treating seafood, and our novel electrolysis equipment can produce highly purified, low-cost ClO_2 to disinfect with water, for immediate use for agricultural product and seafood treatment.

Conflict of Interests

The authors declare that there is no conflict of interests regarding the publication of this paper.

ACKNOWLEDGMENT

This work was financially supported by the Council of Agriculture of Taiwan under Grant no. 97AS-1.2.1-ST-A3.

REFERENCES

1. Q. Ma, B. Li, C. Wang, Y. Ji, S. Wang, and W. Cao, "Efficiency of electrolyzed oxidizing water for inactivation of salmonella spp. and inoculated shell eggs," International Journal of Food Engineering, vol. 5, no. 3, article 3, 2009.
2. Z. Liu, Y. Dong, and W. Jiang, "Decontamination efficiency of slightly acidic electrolyzed water on fresh-cut cucumbers," International Journal of Food Engineering, vol. 7, no. 5, article 13, 2011.
3. L. M. Friedrich, R. Goodrich-Schneider, M. E. Parish, and M. D. Danyluk, "Mitigation of Alicyclobacillusspp. spores on food contact surfaces with aqueous chlorine dioxide and hypochlorite," Food Microbiology, vol. 26, no. 8, pp. 936–941, 2009.
4. R. Vaid, R. H. Linton, and M. T. Morgan, "Comparison of inactivation of Listeria monocytogenes within a biofilm matrix using chlorine dioxide gas, aqueous chlorine dioxide and sodium hypochlorite treatments," Food Microbiology, vol. 27, no. 8, pp. 979–984, 2010.
5. C. D. Cruz and G. C. Fletcher, "Assessing manufacturers' recommended concentrations of commercial sanitizers on inactivation of Listeria monocytogenes," Food Control, vol. 26, no. 1, pp. 194–199, 2012.
6. H. Gallard and U. Von Gunten, "Chlorination of natural organic matter: kinetics of chlorination and of THM formation," Water Research, vol. 36, no. 1, pp. 65–74, 2002.
7. S. Sorlini and C. Collivignarelli, "Trihalomethane formation during chemical oxidation with chlorine, chlorine dioxide and ozone of ten Italian natural waters," Desalination, vol. 176, no. 1–3, pp. 103–111, 2005.
8. M. K. Ramseier, A. Peter, J. Traber, and U. von Gunten, "Formation of assimilable organic carbon during oxidation of natural waters with ozone, chlorine dioxide,

chlorine, permanganate, and ferrate," Water Research, vol. 45, no. 5, pp. 2002–2010, 2011.

9. S. D. Richardson, "Drinking water disinfection byproducts," in The Encyclopedia of Environmental Analysis and Remediation, R. A. Meyers, Ed., pp. 1398–1421, John Wiley & Sons, New York, NY, USA, 1998.

10. Ch. Rav-Acha, A. Serri, E. G. Choshen, and B. Limoni, "Disinfection of drinking water rich in bromide with chlorine and chlorine dioxide, while minimizing the formation of undesirable by-products," Water Science and Technology, vol. 17, no. 4-5, pp. 611–621, 1985.

11. J. M. Kim, T.-S. Huang, M. R. Marshall, and C.-I. Wei, "Chlorine dioxide treatment of seafoods to reduce bacterial loads," Journal of Food Science, vol. 64, no. 6, pp. 1089–1093, 1999. ·

12. W.-F. Lin, T.-S. Huang, J. A. Cornell, C.-M. Lin, and C.-I. Wei, "Bactericidal activity of aqueous chlorine and chlorine dioxide solutions in a fish model system," Journal of Food Science, vol. 61, no. 5, pp. 1030–1034, 1996.

13. S. Zhang and J. M. Farber, "The effects of various disinfectants against Listeria monocytogenes on fresh-cut vegetables," Food Microbiology, vol. 13, no. 4, pp. 311–321, 1996.

14. I. H. Chen, T. S. Huang, J. Kim et al., "The bactericidal effect of chlorine dioxide against E. coli O157 : H7, Listeria monocytogenes, and Salmonella spp. inoculated on strawberries, cucumbers, and cantaloupes," inProceedings of the Institute of Food Technology Annual Meeting, Abstract 78F-22, Dallas, Tex, USA, 2000.

15. Y. Han, T. L. Selby, K. K. Schultze, P. E. Nelson, and R. H. Linton, "Decontamination of strawberries using batch and continuous chlorine dioxide gas treatments," Journal of Food Protection, vol. 67, no. 11, pp. 2450–2455, 2004.

16. S.-Y. Lee, M. Costello, and D.-H. Kang, "Efficacy of chlorine dioxide gas as a sanitizer of lettuce leaves,"Journal of Food Protection, vol. 67, no. 7, pp. 1371–1376, 2004.

17. C. C. Chung, T. C. Huang, C. H. Yu, F. Y. Shen, and H. H. Chen, "Bactericidal effects of fresh-cut vegetables and fruits after subsequent washing with chlorine dioxide," International Proceedings of Chemical, Biological & Environmental Engineering, vol. 9, article 21, pp. 107–112, 2011.

18. W. J. Masschelein, "Experience with chlorine dioxide in Brussels: generation of chlorine dioxide," Journal American Water Works Association, vol. 76, no. 1, pp. 70–76, 1984.

19. P. Westerhoff, P. Chao, and H. Mash, "Reactivity of natural organic matter with aqueous chlorine and bromine," Water Research, vol. 38, no. 6, pp. 1502–1513, 2004. ·

20. R. W. Jordon, A. J. Kosinski, and R. J. Baker, "Improved method generates more chlorine dioxide," Water & Sewage Works, vol. 127, no. 10, pp. 44–46, 1980.

21. E. M. Aieta and J. D. Berg, "A review of chlorine dioxide in drinking water treatment," Journal American Water Works Association, vol. 78, no. 6, pp. 62–72, 1986.

22. G. Gordon and B. Bubnis, "Ozone and chlorine dioxide: similar chemistry and measurement issues,"Ozone Science and Engineering, vol. 21, no. 5, pp. 447–464, 1999.

23. B. R. Deshwal and H.-K. Lee, "Kinetics and mechanism of chloride based chlorine dioxide generation process from acidic sodium chlorate," Journal of Hazardous Materials, vol. 108, no. 3, pp. 173–182, 2004. ·

24. G. Petrucci and M. Rosellini, "Chlorine dioxide in seawater for fouling control and post-disinfection in potable waterworks," Desalination, vol. 182, no. 1-3, pp. 283–291, 2005.

25. H. Bergmann and S. Koparal, "The formation of chlorine dioxide in the electrochemical treatment of drinking water for disinfection," Electrochimica Acta, vol. 50, no. 25-26, pp. 5218–5228, 2005.

26. Y. Qian, Y. Chen, Y. Jiang, and L. Zhang, "A clean production process of sodium chlorite from sodium chlorate," Journal of Cleaner Production, vol. 15, no. 10, pp. 920–926, 2007.

27. H. H. Chen, T. C. Huang, and R. J. Tung, "Chlorine dioxide electrolysis equipment," Inventors: National Pingtung University of Science and Technology, assignee, Taiwan Patent, no. 200722557, December 2005.

28. L. Wang, J. L. Huang, and H. B. Li, "Determination of ClO_2, Cl_2, ClO2 − and ClO3 − in water by using continuous iodimetry," Journal of Harbin University of Civil Engineering and Architecture, vol. 30, no. 4, pp. 70–75, 1997.

29. CNS, "Method of Test for Food Microbiology-Test of Standard Plate Count," General No 10890. Classified No N6186, 1991.

30. S. Suwansonthichai and S. Rengpipat, "Enumeration of coliforms and Escherichia coli in frozen black tiger shrimp Penaeus monodon by conventional and rapid methods," International Journal of Food Microbiology, vol. 81, no. 2, pp. 113–121, 2003.

31. M. A. Stack, G. Fitzgerald, S. O'Connell, and K. J. James, "Measurement of trihalomethanes in potable and recreational waters using solid phase micro extraction with gas chromatography-mass spectrometry,"Chemosphere, vol. 41, no. 11, pp. 1821–1826, 2000.

32. S. A. S. Institute, "User's Guide: Statistics," Ver. 6.4, SAS Institute, Cary, NC, USA, 1990.

33. H. Junli, W. Li, R. Nanqi, M. Fang, and J. Juli, "Disinfection effect of chlorine dioxide on bacteria in water," Water Research, vol. 31, no. 3, pp. 607–613, 1997. ·

34. G. L. Amy, J. Debroux, S. Sinha, P. Brandhuber, and J. Chao, "Occurrence of disinfection by-products (DBPs) precursors in source waters and DBPs in finished Waters," in Proceedings of the 4th International Workshop on Drinking Waters Quality Management and Treatment Technology, pp. 59–70, Taipei, Taiwan, 1998.

35. N. Singh, R. K. Singh, A. K. Bhunia, and R. L. Stroshine, "Efficacy of chlorine dioxide, ozone, and thyme essential oil or a sequential washing in killing Escherichia

coli O157:H7 on lettuce and baby carrots,"LWT—Food Science and Technology, vol. 35, no. 8, pp. 720–729, 2002.

36. L. R. Beuchat, C. A. Pettigrew, M. E. Tremblay, B. J. Roselle, and A. J. Scouten, "Lethality of chlorine, chlorine dioxide, and a commercial fruit and vegetable sanitizer to vegetative cells and spores of Bacillus cereus and spores of Bacillus thuringiensis," Journal of Industrial Microbiology and Biotechnology, vol. 32, no. 7, pp. 301–308, 2005.

37. C.-S. Lin, C. Wu, J.-Y. Yeh, and F. K. Saalia, "The evaluation of electrolysed water as an agent for reducing micro-organisms on vegetables," International Journal of Food Science and Technology, vol. 40, no. 5, pp. 495–500, 2005.

38. S. Aloisio and J. S. Francisco, "A density functional study of H_2O-OClO, $(H_2O)_2$-OClO and H_2O-ClOO complexes," Chemical Physics, vol. 254, no. 1, pp. 1–9, 2000. ·

39. G. P. Li and H. L. Xia, "Review on the application of chlorine dioxide in food sterilization and disinfection," Food Science and Technology, vol. 31, no. 9, pp. 21–25, 2006.

40. S. Q. Hu and R. Y. Jin, "Research on sterilization of chlorine dioxide gas and its application perspective,"China Safety Science Journal, vol. 17, no. 3, pp. 153–155, 2007.

CHAPTER 17

SHELF LIFE OF AIR AND MODIFIED ATMOSPHERE-PACKAGED FRESH TILAPIA (OREOCHROMIS NILOTICUS) FILLETS STORED UNDER CHILLED AND SUPERCHILLED CONDITIONS

Odoli Cyprian, Hélène L. Lauzon, Ragnar Jóhannsson, Kolbrún Sveinsdóttir, Sigurjón Arason, Emilía Martinsdóttir

PRACTICAL APPLICATIONS

The rapidly developing retail market of tilapia fillets and demand for fresh products emphasize the need for prolonged shelf life of fresh fillets. In this study, different storage conditions for fresh fillets are assessed to establish optimal storage conditions. The study shows that air packaged in a high-barrier film bag combined with low-temperature storage maintains superior fillet quality longer and prolonged the shelf life of fillets stored at −1°C by 5–7 days as compared with storage at +1°C.

INTRODUCTION

With increased awareness, the consumers are progressively demanding high food quality and have corresponding expectations that quality will be maintained at a high level during the period between production and consumption (Fisheries and Aquaculture Department 2002). However, the shelf life of fresh fishery products is usually limited by microbial activities that are influenced most importantly by storage temperature (Huss 1995; Simpson et al. 2003). Typical shelf life under icing storage conditions ranges from 6 to 20 days (Martinsdottir et al. 2001; Cyprian et al. 2008) depending on species, harvest location, and season, and short shelf life can result in heavy economic loss (Reddy et al. 1995; Sivertsvik et al. 2002). This has made packaging an integral part of the food industry, as in addition to preservation function, it has several other important roles to play in delivering safe, wholesome, and attractive foods to the market (Kilcast and Subramaniam 2000).

Modified atmosphere packaging (MAP) of fishery products has been shown to inhibit the normal spoilage flora and increase shelf life significantly at refrigerated temperatures (Sivertsvik et al. 2002). The application of this technology to foods has become increasingly more available in recent years (Sveinsdottir et al. 2010), as food manufacturers have attempted to meet consumer demands. Studies have shown MAP to perform an effective synergy with superchilling in prolonging the shelf life of fresh cod (*Gadus morhua*) loins (Wang et al. 2008; Lauzon et al. 2009) and Atlantic salmon (*Salmo salar*) fillets (Sivertsvik et al. 2003; Duun and Rustad 2008). Duun and Rustad (2008) reported superchilled salmon fillets stored at −2°C in combination with MAP to have maintained good quality with negligible microbial growth for more than 24 days based on both sensory and microbial analyses, whereas ice-chilled reference fillets maintained good quality up to 17 days.

Citation: Odoli Cyprian, Hélène L. Lauzon, Ragnar Jóhannsson, Kolbrún Sveinsdóttir, Sigurjón Arason, Emilía Martinsdóttir, Shelf life of air and modified atmosphere-packaged fresh tilapia (Oreochromis niloticus) fillets stored under chilled and superchilled conditions, DOI: 10.1002/fsn3.18.

Nevertheless packaging technologists should be aware of a major concern limiting the development of MAP for this product group, namely *Clostridium botulinum. Clostridium botulinum* (anaerobic) has often been reported in fresh seafood products that have been packaged under regimes designed to limit the availability of oxygen (Aberoumand 2010). Specifically, *C. botulinum* type E is of most concern in this type of packaging as it is a naturally occurring aquatic organism and grows at refrigeration temperature as low as 2.9°C (Conner et al. 1989). To ensure safety and success of MAP, packaged products should be kept at refrigeration temperature below 2.9°C.

MA packaging of tilapia (*Oreochromis niloticus*) fillets in 75% CO_2/25% N_2 stored at 4°C was previously evaluated and promising results reported (Reddy et al. 1995). This study aimed at investigating the combined effect of MA packaging (50% CO_2/50% N_2) and superchilling on the sensory, microbiological, and chemical changes during storage of fresh tilapia fillets.

MATERIALS AND METHODS

Experimental Design

Tilapia (*O. niloticus*) used for the study was farmed in recirculation aquaculture system (RAS) at Straumfraedihus Keldnaholti in Reykjavik, Iceland. A total of 340 fish weighing between 450 and 700 g were harvested, bled, and transported iced to the laboratory within 3 h. On arrival, fish was hand filleted and deskinned, and 24 fillets (control) were immediately used for sensory, microbiological, and chemical analyses. The rest was kept chilled at 0°C while packaging. Four to five deskinned fillets weighing 250–290 g (altogether) were placed with skin side up on a preweighed foam tray (expanded polystyrene, Linstar E 39-34, LINPAC, Plastprent, Iceland) with a built-in absorption mat and packaged with a high-barrier film bag (250 × 400 × 0.120 mm coextruded PA/PE, Plastprent hf., Iceland). Four groups of 40 packs each were prepared (160 packs altogether) based on atmospheric and storage conditions.

In the first two groups (T1 and T2), each bag was sealed containing air using a vacuum packaging machine (HENKOVAC Heavy duty 2000, Hertogenbosch, the Netherlands) equipped with a built-in vacuum pump and gas flush. The third and fourth groups (T3 and T4) were evacuated and packaged under MA (50% CO_2/50% N_2). CO_2 and N_2 were mixed using a gas mixer MAP Mix 9000 (PBI-Dansensor, Ringsted, Denmark). After packaging, T1 and T3 packages were stored at 1°C, whereas, T2 and T4 were stored at −1°C throughout the study. The description of groups T1, T2, T3, and T4 is given in Table 1.

Table 1. Definition of treatment groups studied and sampling scheme

Treatment codes	Atmosphere at packaging	Storage temperature	Number of packs (*n*)	Sampling days
Control	Not packaged	Iced		0
T1	Air	1°C	40	2, 6, 9, 13, 16, 20
T2	Air	-1°C	40	2, 6, 9, 13, 16, 20
T3	50% CO_2/50% N_2	1°C	40	2, 6, 9, 13, 16, 20, 23
T4	50% CO_2/50% N_2	-1°C	40	2, 6, 9, 13, 16, 20, 23

On every sampling day, five packs per treatment were taken (20 altogether). The five packs were analyzed for headspace gas composition, and thereafter, two packs were used for microbial and chemical analysis, with the remaining three packs used for exudate measurements and sensory evaluation (raw and cooked fillets).

Gas Analysis

The headspace gas composition in air and MA packaged was determined upon packaging and on every sampling in five packs for each treatment, using the gas analyzer (CheckMate 9900 Analyzer, PBI-Dansensor, Ringsted, Denmark). The concentrations of sampled packages with respect to treatments were averaged ($n = 5$).

Sensory evaluation of raw and cooked fillets

Evaluation of raw fillets was done using a Quality Index Method (QIM) scheme developed for deskinned tilapia fillets (Table 2) according to the methodology described earlier by Martinsdottir et al. (**2001**) and Hyldig et al. (**2007**). The QIM scheme was based on significant, well-defined characteristics of appearance, odor, and texture attributes changing through storage time. All observations of tilapia fillets were conducted under standardized conditions following the general guidance for the design of test room and testing conditions described in ISO 8589 (**1988**). On every sampling day, three fillets were taken randomly from three different packs for each group and placed on a clean table. Each fillet was blind coded with a number consisting of three digits that did not indicate storage conditions. Eight to twelve trained panelists, all employees of Matis ohf with good use of vocabulary and knowledge of the fish species, individually evaluated changes in color, mucus, texture, and odor. The selection and training of the panel members on tilapia attributes were carried out during three training sessions (preliminary study) according to international standards ISO 8586 (ISO **1993**), which include detection and recognition of tastes and odors, use of scales, and the development of descriptors. A score from 0 to 3 demerit points was given for every attribute evaluated (Table 2).

Table 2. A modified Quality Index Method scheme developed from preliminary scheme for deskinned tilapia fillets (O. Cyprian et al. unpubl. data)

Quality parameter	Description	Demerit point
1. a Lateral/longitudinal stripes at the middle of the loin.		
Skin side		
Colora	Dark red, red brown	0
	Red brown, lighter color	1
	Light brown	2
Flesh		
Color, loin	Light, beige, trace of red or bluish	0
	A little darker color, a little brownish or grayish	1
	Grayish, brownish, yellowish	2
Color, flap	Bluish, transparent	0
	Light, milky color	1
	Grayish or brownish	2
Mucus	Fresh shiny texture, thin clear mucus	0
	Trace of mucus, a little thick	1
	Milky or greenish mucus	2
Texture	Firm	0
	Rather soft	1
	Soft	2
Odor	Fresh, neutral	0
	Seaweed, marine, grass	1
	Sour milk, silage	2
	Acetic, putrid	3
Quality index (0–13)		

Sensory evaluation of cooked tilapia fillets was performed in parallel to the QIM evaluation. Fillet loins from the three packs for each group were cut into pieces of about 4–5 cm long and 3–4 cm wide. The pieces were placed in aluminum boxes coded with three-digit random numbers and cooked in a preheated electric oven Convostar (Convotherm GmbH, Eglfing, Germany) with circulation air and steam at 95–100°C for 6 min. Eight to twelve panelists trained in recognition of sensory characteristics of the samples and to describe the intensity of each attribute using an unstructured scale from 0% to 100% (Stone and Sidel **1985**) were served with cooked samples. Each panelist evaluated duplicates of samples in a random order for each group based on sensory vocabulary of cooked tilapia (Table 3). A computerized system (FIZZ, Version 2.0, 1994–2000, Biosystèmes, France) was used for data recording.

Table 3. Sensory vocabulary for cooked tilapia

Short name	Sensory attribute	Description of attribute
Odor		
O-Boilpot	Boiled potatoes	Whole newly boiled potatoes
O-Boilmilk	Boiled milk	Hot milk, fruity odor
O-Earthy	Earth	Fresh earth
O-Musty	Musty	Moldy
O-Rancid	Rancid	Rancidity
Appearance		

A-Color (T)	Light/dark color	Light, white or dark, yellowish color
A-Color (U)	Dark brown/light gray	Dark brownish or light gray color
A-BlackThr	Black threads	Black threads in flesh
Flavor		
F-Arctic charr	Arctic charr	Arctic charr, new trout
F-Sweet	Sweet	Typical sweet flavor of fresh fish
F-Metallic	Metallic	Metallic flavor
F-Earthy	Earthy	Fresh earth
F-Musty	Musty	Moldy flavor
F-Sour	Sour	Sour taste, spoilage sour
F-Pungent	Pungent	Pungent flavor, bitter
F-Rancid	Rancid	Rancidity, cod liver oil, reminds of paint or solvent/thinner
F-Spoilage	Spoilage	Spoilage, queasy sweet flavor
Texture		
T-Flakes	Flakiness	Fish portion slides into flakes when pressed with a fork
T-Soft	Firm/soft	How firm or soft the fish is during the first bite
T-Juicy	Dry/juicy	When chewed, dry: pulls liquid from mouth; juicy: gives liquid
T-Fibers	Fiber	Roughness of muscle fiber
T-Mushy	Mushy	Mushiness when chewed (mushy texture)
T-Sticky	Sticky	Glues together teeth when biting the fish

Microbiological Analysis

Fillets were sampled aseptically and analyzed in duplicate for each pack for two packages per treatment. Fillets were minced in a sterile blender, assessing two pooled fillets for each replicate sample ($n = 2$). Minced flesh (20 g) was mixed with 180 g of chilled Maximum Recovery Diluent (MRD, Oxoid, U.K.) in a stomacher for 1 min. Successive 10-fold dilutions were done as required. Total viable counts (TVC) were obtained by spread plating of aliquots on iron agar (IA) modified from Gram et al. (**1987**), containing 1% NaCl (no overlay), followed by aerobic incubation at 17°C for 5 days. Counts of H_2S-producing bacteria were also evaluated by counting black colonies on IA. Presumptive *Pseudomonas* counts (22°C, 3 days) were obtained using the modified cephaloridine–fucidin–cetrimide (CFC) agar as described by Stanbridge and Board (**1994**). *Pseudomonas* agar base (Oxoid) with CFC Selective Agar Supplement (Oxoid) was used. *Pseudomonas* spp. form pink colonies on this medium.

pH measurement

Measurements of pH were done using 5 g of minced fish mixed with 5 mL of deionized water. The pH meter Radiometer PHM 80 was calibrated using the buffer solutions of pH 7.00 ± 0.01 and 4.01 ± 0.01 (25°C) (Radiometer Analytical A/S, Bagsvaerd, Denmark). All measurements were done in duplicate per package (for two packs per group) and results presented as an average.

Assessment of exudate in the packages

Exudate accumulation in the packages during storage was measured gravimetrically using three packs per treatments ($n = 3$). The mass of the exudate (g) was divided by the initial mass of the fish product (g) and reported as a percentage of drip loss.

Total Volatile Basic Nitrogen and Trimethyl-amine

Duplicates of fillet mince for each treatment that remained after microbiological analyses were used for total volatile basic nitrogen (TVB-N) and trimethylamine (TMA) analysis. The method of Malle and Tao (**1987**) was used. TVB-N was measured by steam distillation (Struer TVN distillatory, STRUERS, Copenhagen, Denmark) and titration, after extracting the fish muscle with 7.5% aqueous trichloroacetic acid (TCA) solution. The distilled TVB-N was collected in boric acid solution and then titrated with sulfuric acid solution. TMA was measured in TCA extract by adding 20 mL of 35% formaldehyde, an alkaline binding mono- and diamine, TMA being the only volatile and measurable amine.

Data Analysis

The mean values of measured parameters were plotted separately against storage time using Microsoft Excel (2007). Multivariate comparison of different sensory attributes and samples was performed with principal component analysis (PCA) on mean sensory attribute values using full cross-validation in the statistical program Unscrambler® (Version 8.0 CAMO, Trondheim, Norway). Analysis of variance (ANOVA) of the results was performed in the statistical program NCSS 2000 (NCSS, UT). The program (ANOVA) calculates multiple comparisons using Duncan's test to determine if sample groups are different. Significance of difference was defined at the 5% level.

RESULTS AND DISCUSSION

Headspace Gas Composition

The gas product ratio in the packages was 5:1. At packaging, air packaged (T1 and T2) had approximately 21% oxygen and 0% carbon dioxide as shown in Figure 1a. Oxygen levels in all the groups showed similar trends of decline with storage time as reported elsewhere (Reddy et al. **1995**; Hovda et al. **2007**). The largest decline occurred in air-packaged samples stored at 1°C (T1) which corresponded with the occurrence of high bacterial numbers (Fig. 4) suggesting that the oxygen was being consumed during microbial metabolism. At −1°C, oxygen consumption by bacteria was delayed and occurred to a lesser extent, being first observed after 16 days of storage coinciding with a bacterial load of about log 6 colony-forming units (CFU) g^{-1}. Similar trends in the opposed direction were observed with CO_2 composition in the headspace of air-packaged fillets where CO_2 increased as a secondary product of microbial metabolism (Hovda et al. **2007**).

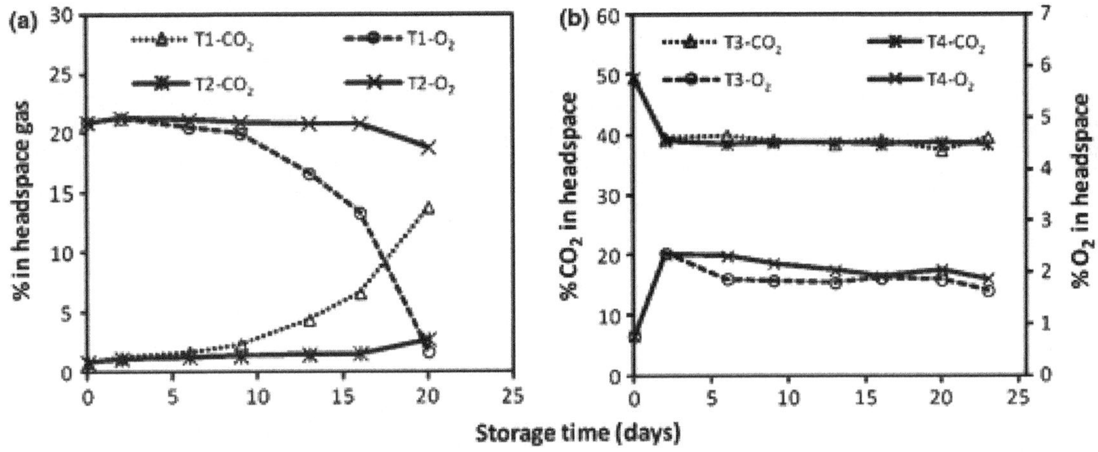

Figure 1. Gas composition changes in headspace of air (a) and modified atmosphere (MA) (b) packaged tilapia fillets during storage at 1°C and −1°C (average values for packs,$n = 5$).

Initially, the headspace of MA-packaged fillets (T3 and T4) had a low oxygen level (<1%), which was observed to be higher on subsequent sampling on day 2 (>2%) and thereafter gradually declined (Fig. 1b). This may be explained by the presence of oxygen in the muscle being released to the headspace upon dissolution of carbon dioxide in the muscle water phase after packaging and gas equilibrium was reached. In packages stored at 1°C, the oxygen drop was more pronounced than at −1°C due to a faster microbial proliferation. However, O_2 did not decrease below the initial 1% as less microbial growth was reported in MA-packaged groups. On the other hand, CO_2 reduced from 50% at packaging to around 40% during early storage as it dissolved into fillets due to its high solubility in water and fat at low temperature (Soccol and Oetterer **2003**).

Sensory Evaluation Using QIM (raw fillets)

The QIM scheme is constructed in that fish evaluated shortly after catch should be scored low and subsequently increases with storage time reaching close to maximum score at the end of shelf life (Martinsdottir et al. **2001**). The first change observed in air- and MA-packaged tilapia fillets was deterioration in muscle appearance on day 2 of storage at both 1°C and −1°C. At the beginning of storage trial, tilapia fillets were dark red in color on the skin side, but the backbone side had a light appearance with trace of red color or hue (Table 2 and Fig. 2a and b). According to the appearance parameters, color faded soon after packaging under MA. On day 2, it was characterized by red brown and light brown on the skin and backbone sides, but the color progressed to light brown and grayish brown, respectively, toward the end of storage time. As from day 2, MA treatments recorded significantly higher scores ($P < 0.05$) for the appearance parameters compared with air packaged except for T1 toward end of storage time (day 16 and 20). Skin discoloration of some MA-treated fish samples has previously been reported (Stansby and Griffiths **1935**; Lauzon et al. **2002**) and is due to denaturation of muscle and pigment proteins. The discoloration process was slower in air packaged as fillets retained the dark red color characteristic longer, especially with storage at −1°C which maintained a stable color until the end of storage time at day 20. This could be explained by the maintenance of the oxymyoglobin pigment in the muscle in presence of oxygen, being more soluble under superchilled conditions (Sivertsvik et al. **2003**; Ando et al. **2004**; Ando et al. **2005**).

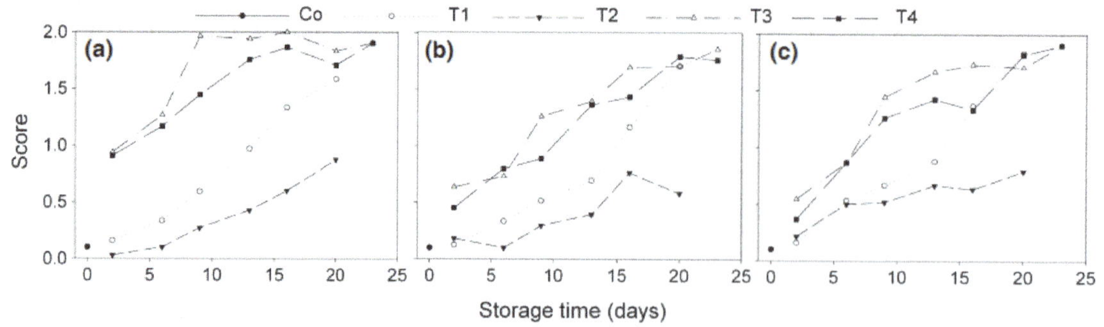

Figure 2. Average scores (three fillets from three packs, $n = 3$) of the appearance parameters as assessed by Quality Index Method (QIM) scheme for deskinned tilapia fillets during storage at 1°C and −1°C. (a) Flesh color from skin side; (b) flesh color from loin side; and (c) color of belly flap.

Sensory Evaluation of Cooked Tilapia Fillets

Figure 3 shows how sample groups were described by sensory attributes of cooked tilapia fillets with storage time. Sensory quality attributes detected at the beginning of storage time (sweet, metallic, and arctic charr flavor) considered freshness attributes were positioned left along the first principal component (PC1), whereas attributes detected at the end of storage time (rancid, spoilage, musty, and pungent flavor) were positioned to the right and considered spoilage attributes (Sveinsdottir et al. **2002**; Cyprian et al. **2008**). The samples varied mainly with regard to differences in flavor and odor attributes along PC1, explaining 65% of the variation between the sample groups. Samples also varied with regard to differences in appearance and texture attributes along the second principal component (PC2), explaining 10% of the variation. The main difference occurred with storage time, as the sample groups are located to the left side at the beginning of storage but on the right side toward end of storage time (Fig. 3a).

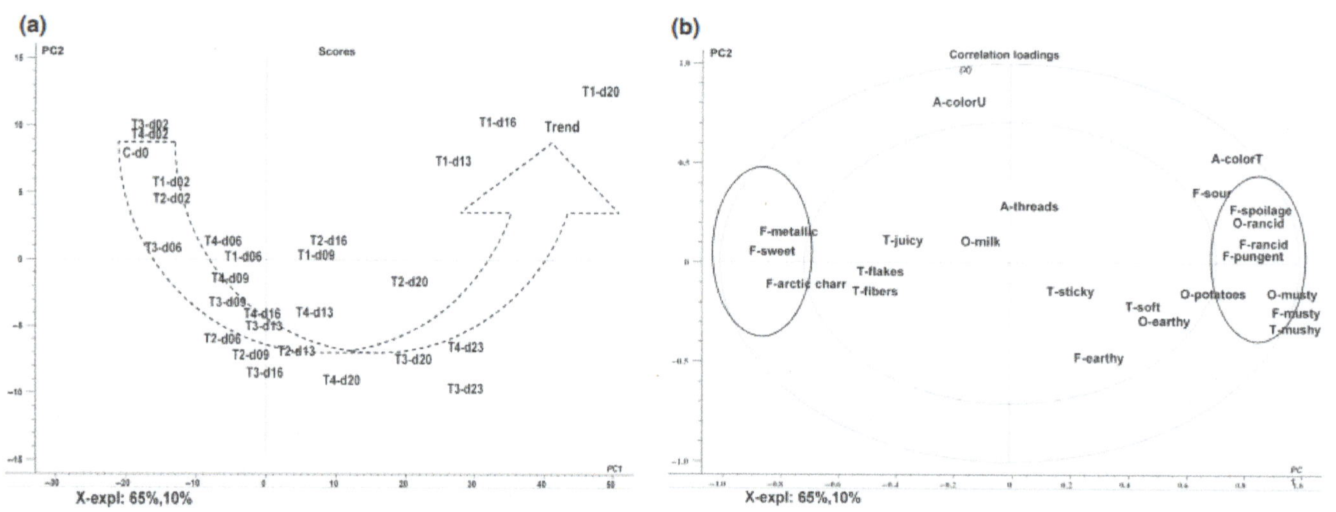

Figure 3. Scores (a) and correlation loadings (b) describing sensory quality of cooked tilapia fillets as evaluated by a trained sensory panel. Principal component 1 (PC1) (65%) versus PC2 (10%). d, storage days; T1, T2, T3, and T4, treatment groups; F, flavor; O, odor; A, appearance; T, texture.

As time progressed, the first noticeable difference was on day 9 when rancid flavor was noticed in T1, as well as other negative parameters from day 13 (Table 4). End of shelf life is usually determined when spoilage-related sensory attributes become evident and most panelists detect them. The average score of above 20 (on the scale 0–100) for these spoilage attributes has been applied by various authors as an indication that fish sample is approaching the end of shelf life (Sveinsdottir et al. **2002**; Magnússon et al. **2006**; Bonilla et al. **2007**; Cyprian et al. **2008**). Air-packaged fillets stored at 1°C (T1) showed first spoilage characteristics on day 13. On day 16 and day 20, groups T2, T3, and T4 also showed signs of spoilage. However, the scores for spoilage attributes were below 20 on day 23 for groups T3 and T4. On the other hand, a part of the panel could not taste T1 samples due to strong spoilage odor indicating that the group (T1) was no longer fit for human consumption on that day. Similar observation was noted by Sveinsdottir et al. (**2002**) who reported salmon to be unfit for consumption when part of the panel did not taste samples due to spoilage odor. T2 and T3 recorded spoilage attributes scores close to 20 on day 20 and 23, respectively, implying the groups may have been approaching the end of shelf life.

Table 4. Mean sensory scores (scale 0–100) of odor, flavor, and appearance/texture attributes for cooked tilapia fillets

Storage time (days)	Group	Odor		Appearance/texture		Flavor			
		Musty	Rancid	Color T	Juicy	Arctic charr	Sweet	Rancid	Spoilage
1. Data within the same column with respect to storage time (days) with different letters are significantly ($P < 0.05$ in bold) different.									
0	Co	23	3	28	63	33	33	4	3
2	T1	28	2	30	60	37	36	2	2
	T2	25	2	34	63	37	35	2	2
	T3	24	3	34	62	34	35	2	2
	T4	23	2	29	64	36	35	2	2
	P-value	0.905	0.699	0.347	0.775	0.737	0.999	0.997	0.990
6	T1	26	9	31	59	34	30	7	7
	T2	26	4	28	60	33	33	6	4
	T3	20	3	27	52	34	33	4	3
	T4	25	6	26	57	32	31	7	5
	P-value	0.252	0.080	0.636	0.095	0.937	0.765	0.483	0.580
9	T1	27	8	34	59	32	25	12^{a}	6
	T2	24	5	32	60	38	30	6^{b}	3
	T3	23	6	26	57	36	29	4^{b}	6
	T4	25	5	30	55	38	27	4^{b}	1
	P-value	0.586	0.554	0.320	0.512	0.075	0.479	**0.020**	0.130
13	T1	31^{a}	15^{a}	45^{a}	53	30	22	24^{a}	20^{a}
	T2	30^{a}	4^{b}	31^{b}	61	32	24	5^{b}	9^{b}
	T3	25^{b}	5^{b}	31^{b}	56	24	25	7^{b}	5^{b}
	T4	24^{b}	6^{b}	34^{b}	54	20	23	12^{b}	8^{b}
	P-value	**0.004**	**<0.001**	**<0.001**	0.122	0.254	0.804	**<0.001**	**0.002**
16	T1	29	20^{a}	44^{a}	55	27^{a}	22^{a}	22^{a}	26^{a}
	T2	24	12^{a}	30^{b}	60	34^{b}	23^{a}	13^{b}	12^{b}
	T3	25	8^{b}	25^{b}	57	36^{b}	30^{b}	10^{b}	8^{b}
	T4	27	7^{b}	27^{b}	54	34^{b}	29^{b}	10^{b}	7^{b}
	P-value	0.328	**0.014**	**<0.001**	0.450	**0.002**	**0.006**	**0.024**	**<0.001**
20	T1	32	25^{a}	47^{a}	55^{a}	28	14^{a}	25^{a}	36^{a}
	T2	26	14^{b}	32^{b}	64^{b}	34	24^{b}	18	18^{b}
	T3	28	14^{b}	30^{b}	55^{a}	33	22^{b}	18	19^{b}
	T4	28	10^{b}	28^{b}	58	32	25^{b}	12^{b}	12^{b}
	P-value	0.368	**0.004**	**<0.001**	**0.042**	0.075	**0.006**	**0.049**	**<0.001**
23	T3	29	13	28^{a}	58	26	21	18	14
	T4	34	14	37^{b}	62	28	20	19	18
	P-value	0.377	0.893	**0.038**	0.646	0.588	0.432	0.472	0.256

Microbiological Analyses

Huss (**1995**) reported *Pseudomonas* spp. to be the specific spoilage bacteria of iced stored tropical freshwater fish. Figure 4A and B shows the growth curves for TVC and pseudomonads in air- and MA-packaged tilapia fillets stored at 1°C and at −1°C. TVC maximum limit of log 8 CFU g^{-1} has previously been used in tilapia as the limit for human consumption (Reddy et al. **1995**; Waliszewski and Avalos **2001**). In this study, TVC and *Pseudomonas* spp. had reached levels greater than the above-reported consumption limit for air-packaged fillets during storage at 1°C (T1) on day 16 (Fig. 4a and b) indicating that they had surpassed the consumption limit. No further bacterial growth was observed after day 16. This observation could also be attributed to change in headspace gas composition (Fig. 1a) whereby the exponential aerobic growth resulted to high oxygen consumption and an increase in CO_2 as a secondary product of metabolism whose accumulation may have restricted further growth (Adams and Moss **2008**). Pseudomonads were prevailing among the spoilage flora of tilapia fillets. A count of log 7.3 CFU g^{-1} was reached at onset of sensory spoilage in T1 samples (day 13). This is close to the reported pseudomonad spoilage level of log 7 CFU g^{-1} in refrigerated cod products (Gospavic et al. **2011**). In air-packaged fillets stored at −1°C, TVC and *Pseudomonas* spp. counts were close to log 7 CFU g^{-1} on day 20, at the onset of spoilage for T2. The low count in air-packaged fillets during storage at −1°C compared with 1°C is due to the effective delay of bacterial growth during storage at superchilling conditions (Huss **1995**; Church **1998**). MA groups recorded extended lag phase during early storage days and aerobic counts of less than log 4 CFU g^{-1}were seen up to day 23. This indicates that aerobic counts were too low to cause microbial deterioration in MA-packaged fillets (Fig. 4), in agreement with the sensory data reported. The delayed growth and low counts of aerobic microorganisms in MA-packaged fillets can be attributed mainly to CO_2 inhibition (Sivertsvik et al. **2002**) and the presence of low level O_2(Reddy and Armstrong **1992**) but also to low storage temperature (Huss **1995**) used in this study. Throughout the storage time, low counts (<log 3.0 CFU g^{-1}) of H_2S-producing bacteria, a recognized spoilage bacterial group including several *Shewanella spp.* (Vogel et al. **2005**), were enumerated in all the groups (data not provided), which was attributed to the poor growth of these bacteria at low flesh pH (Table 5) as previous reported by Sivertsvik et al. (**2003**) and Gram and Huss (**1996**) as well as probable competition with pseudomonads (Gram and Melchiorsen **1992**).

Table 5. pH changes in air- and modified atmosphere (MA)-packaged tilapia fillets during storage at 1°C and −1°C (average values of duplicates per pack for two packs and SD shown,$n = 2$). Data within the same column with different letters are significantly ($P < 0.05$ in bold) different

Group	Day							
	0	2	6	9	13	16	20	23
1. a Data within the same column in respect to treatment (group) with different letters (superscript) are significantly (P < 0.05 in bold) different.								
Control	6.5±0	–	–	–	–	–	–	–
T1		6.5±0	6.6±0	6.6±0.1	6.5±0.1	6.6±0.3	6.7±0	–
T2		6.6±0.1	6.7±0.1	6.4±0	6.6±0.1	6.4±0	6.5±0.1	–
T3		6.4±0a	6.5±0.1	6.4±0	6.5±0.1	6.4±0.1	6.5±0.2	6.5±0
T4		6.5±0	6.5±0.1	6.5±0	6.4±0	6.4±0.2	6.5±0.1	6.5±0.1
P-value		**0.049**	0.111	0.138	0.242	0.501	0.150	0.422

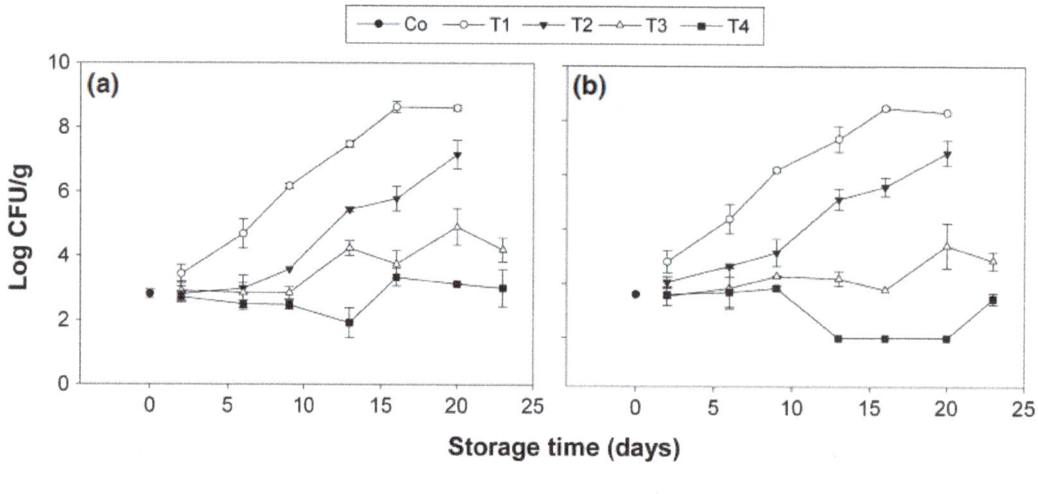

Figure 4. Changes in total viable counts on iron agar (a) and counts of presumptive pseudomonads (b) in air- and modified atmosphere (MA)-packaged tilapia fillets during storage at 1°C and −1°C (duplicates per pack for two packs, $n = 2$).

pH Changes and Drip Loss

The pH values measured were similar and slightly higher in air- than MA-packaged tilapia fillets (Table 5), as could be expected due to the acidic effect of dissolved CO_2 (Sivertsvik et al. **2002**). This partly explains the bacteriostatic effects of CO_2 in MAP (Huss **1995**). Overall, storage time and atmosphere interaction did not affect significantly pH values of fillets ($P > 0.05$). Similarly Reddy et al. (**1995**) did not find significant differences in pH between packing atmospheres (air and 75% CO_2/25% N_2) during storage of tilapia fillets at abused and refrigerated temperature, but their higher temperature (4–16°C) is expected to have caused a lesser CO_2 quantity to dissolve. The slight difference in pH observed contributed significantly to drip loss reported among the groups (Fig. 5). Postmortem pH according to Huss (**1988**) is the most significant factor influencing the texture of the meat and the degree of "gaping," that is, the rapture of the connective tissue. One of the reasons for this is that even minor changes in pH drastically affect the properties of the connective tissue directly affecting the water-holding capacity of the fish muscle and in turn the drip loss.

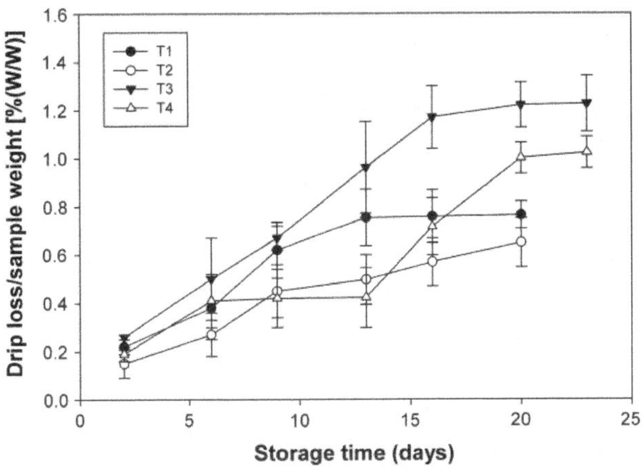

Figure 5. Drip loss in air- and modified atmosphere (MA)-packaged tilapia fillets during storage at 1°C and −1°C. Values (percentage of initial sample weight) are given as mean±SEM or SD (average values for packs, $n = 3$).

Higher drip was observed in MA-packaged groups with T3 recording higher values that were significantly different from T2 ($P < 0.05$) on day 13, 16, and 20 for T2 and T1 (Fig. 5). With respect to packaging atmosphere, higher drip was recorded with storage at refrigeration (1°C) than superchilling (−1°C). This agrees with Sivertsvik et al. (**2003**) who reported that superchilled storage leads to less exudation probably due to high water-holding capacity at low temperature. According to FAO (**1996**), drip loss is defined as the nutrients from and value of fish, which is potentially available for human consumption, but fails to be consumed or sold as products. Drip during transportation of product from source to final consumer or storage may result to weight below legally allowed tolerance even if it is wholesome and fit for consumption. It is therefore evidenced that air-packaged tilapia fillets stored at −1°C (T2) showed better quality related to analyzed physical properties.

TVB-N and TMA

TVB-N, TMA, and other volatile amines are commonly used as indicator for fish deterioration in fresh and lightly preserved seafood (Kilcast and Subramaniam **2000**). Volatile bases result from degradation of proteins and nonprotein nitrogenous compounds. The study depicted slow TVB-N accumulation in all sample groups (<25 mg N/100 g) during storage at 1°C and −1°C (data not shown). Nonetheless at the end of storage time (day 20), T1 was different (21 mg N/100 g) to other groups but within concentration of 30 mg/100 g, above which fish is considered unfit for consumption. This is because changes in some of volatile nitrogen bases are influenced most importantly by microorganisms which were >log 8 CFU g^{-1} in T1 on day 20. Ababuocha et al. (**1996**) showed sardines accumulated TVB-N faster when stored at ambient temperature than on ice probably because of increased growth and activity of spoilage and mesophilic bacteria.

The amount of TMA in tilapia fillets during storage time was very low (<1 mg/100 g) and to some extent not detected in most groups. This is probably because TMAO is virtually absent from freshwater species and terrestrial organisms (Hebard et al. **1982**). Conversely, some studies have reported TMA in tilapia fish (Reddy et al. **1995**; Waliszewski and Avalos **2001**). According to Waliszewski and Avalos (**2001**), the presence of TMA is attributed to marine fish meal fed on tilapia.

Shelf Life

Descriptive analysis can be useful in shelf life studies, as it provides information about maximum shelf life (Bonilla et al. **2007**; Cyprian et al. **2008**). The results from descriptive analysis and microbiological analysis indicated that air-packaged tilapia fillets had a shelf life of 13–15 days when stored at 1°C (T1) in contrast to about 20 days when stored at −1°C (T2). On the other hand, MA-packaged fillets on day 23 were within consumption limits as evaluated with descriptive analysis (score <20 for spoilage attributes) and microbiologically suitable for consumption. Nevertheless, the MA-packaged groups were unwanted around day 6 based on appearance of raw fillets caused by changes in color as well as mucus and texture. This is because color of the fresh meat is an important indicator of tilapia quality and a major factor in influencing retail purchase decisions. Reddy et al. (**1995**) reported MA-packaged (75% CO_2/25 N_2) tilapia fillets stored at 4°C to have a shelf life of >25 days based on microbial evaluation, which is longer than the findings from this study despite storage at low temperature. This could be attributed mainly to emphasis placed on appearance in the current study.

CONCLUSIONS

MAO performed an effective synergy with superchilling in extending microbiological quality of fillets as the increase in viable cell counts was suppressed compared with counterparts refrigerated and air packaged. Conversely, MA-packaging negatively influenced color, drip loss, and texture of products which limited its application in extending the shelf life of tilapia fillets. Based on sensory evaluation of cooked tilapia and microbial counts, the maximum storage time was 13–15 days for air-packaged fillets stored at 1°C (T1) and 21 days for those superchilled (T2).

Superchilled, air-packaged samples (T2) retained freshness color characteristic longer, recorded less drip loss, and suppressed microbial growth compared with refrigerated air-packaged samples (T1). Therefore, primarily based on sensory evaluation, but also on physical properties and microbiological and chemical data, it can be concluded that optimal condition for storage of fresh tilapia fillets was air packaged, the suitable temperature being −1°C, and the storage time 20 days. It was observed in this study that both TVB-N and TMA did not provide a useful index of shelf life to reflect spoilage in tilapia fillets during storage.

ACKNOWLEDGMENTS

First and foremost is to acknowledge the United Nations University Fisheries Training Programme (UNU-FTP) and Technological Development and Innovation Fund of Iceland (Rannis.is) for financing this project. We also thank Adalheidur Olafsdottir and Hannes Magnusson for the guidance during preparation for sensory analysis and microbiological analysis, and all the panelists who performed sensory evaluation and the staff of the microbial and chemical laboratory at MATIS.

REFERENCES

1. Ababouch, L. H., L. Souibri, K. Rhaliby, and O. Ouahdi. 1996. Quality changes in sardines (*Sardina pilchardus*) stored in ice and at ambient temperature. *J. Food Microbiol.* **13**:123–132.
2. Aberoumand, A. 2010. Occurrence of *Clostridium botulinum* in fish and fish products in retail trade. *World J. Fish Mar. Sci.* **2**:246–250.
3. Adams, R. M., and O. M. Moss. 2008. Food microbiology. 3rd ed. RSC Publishers, University of Surrey, Guildford.
4. Ando, M., H. Nakamura, R. Haramada, and A. Yamane. 2004. Effects of superchilling storage on maintenance of freshness of Kuruma prawn. *Food Sci. Technol. Res.* **10**:25–31.
5. Ando, M., T. Takenaga, S. Hamase, and A. Yamane. 2005. Effects of superchilling storage on maintenance of quality and freshness of Swordtip squid (*Loligo edulis*). *Food Sci. Technol. Res.* **11**:355–361.
6. Bonilla, C. A., K. Sveinsdottir, and E. Martinsdottir. 2007. Development of Quality Index Method (QIM) scheme for fresh cod (*Gadus morhua*) fillets and application in shelf life study. *Food Control* **18**:352–358.
7. Church, N. 1998. MAP fish and crustaceans-sensory enhancement. *Food Sci. Technol. Today* **12**:73–83.
8. Conner, D. E., V. N. Scott, D. T. Bernard, and D. A. Kautter. 1989. Potential *Clostridium botulinum* hazards associated with extended shelf-life refrigerated foods. *J. Food Safety* **10**:131–153.
9. Cyprian, O., K. Sveinsdottir, H. Magnusson, and E. Martinsdottir. 2008. Application of Quality Index Method (QIM) scheme and effects of short-time temperature abuse in shelf life study of fresh water Arctic charr. *J. Aquat. Food Product Technol.* **17**:303–321.
10. Duun, S. A., and T. Rustad. 2008. Quality of superchilled vacuum packed Atlantic salmon (*Salmo salar*) fillets stored at −1.4 and −3.6°C. *J. Food Chem.* **106**:122–131.
11. FAO. 1996. Pp. 31–33 *in* Storage quality of three pelagic species. Food and Agriculture Organization Fisheries Technical Report Paper 574.
12. Fisheries and Aquaculture Department. 2002. FAO 2002 (Internet). Available athttp://www.fao.org/fishery/culturedspecies/Oreochromis_niloticus (accessed 20 October 2010).
13. Gospavic, R., H. L. Lauzon, V. Popov, E. Martinsdottir, M. N. Haque, and E. Reynisson. 2011. Microbial growth models for shelf life prediction in an Icelandic cod supply chain. *WIT Trans. Ecol. Environ.* **152**:69–79.
14. Gram, L., and H. Huss. 1996. Microbiological spoilage of fish and fish products. *J. Food Microbiol.***33**:121–137.
15. Gram, L., and J. Melchiorsen. 1992. Interaction between fish spoilage bacteria *Pseudomonas* sp. and*Shewanella putrefaciens* in fish extracts and on fish tissue. *J. Food Microbiol.* **4**:65–72.
16. Gram, L., W. C. Neergaard, and H. H. Huss. 1987. Detection of specific spoilage bacteria from fish stored at low (0°C) and high (20°C) temperatures. *J. Appl. Bacteriol.* **80**:65–72.
17. Hebard, C. E., J. G. Flick, and E. R. Martin. 1982. Occurrence and significance of trimethylamine oxide and its derivatives in fish and shellfish. Pp. 149–304 *in* R. E. Martin, G. J. Flick, C. E. Hebard, and D. R. Ward, eds. Chemistry and biochemistry of marine food products. AVI Publishing Company, Westport, CT.
18. Hovda, M. B., B. T. Lunestad, M. Sivertsvik, and J. T. Rosnes. 2007. Characterization of the bacterial flora of modified atmosphere packaged farmed Atlantic cod by PCR-DGGE of conserved 16S rRNA gene regions. *J. Food Microbiol.* **117**:68–75.
19. Huss, H. H. 1988. Fresh fish – quality and quality changes training manual. United Nations, FAO/DANIDA,Rome.
20. Huss, H. H. 1995. Quality and quality changes in fresh fish. FAO Fisheries Technical Paper, Rome.
21. Hyldig, G., A. Bremer, E. Martinsdottir, and R. Schelvis. 2007. Quality index method. Pp. 529–549 *in* L. M. L. Nollet, ed. Handbook of meat, poultry and seafood quality. Blackwell Publishing, Oxford, U.K.
22. Kilcast, D., and P. Subramaniam, eds. 2000. The stability and shelf life of food. Woodhead Publishing Limited, Cambridge, U.K.
23. Lauzon, H. L., G. Stefansson, G. Jonsson, and S. Kolbrun. 2002. Sensory quality and shelf life of MAP redfish (*Sebastes marinus*) fillets as affected by previous MA bulk storage of whole fish. *J. Aquatic Food Prod. Technol.* **11**:225–273.
24. Lauzon, H. L., H. Magnússon, K. Sveinsdóttir, M. Gudjónsdóttir, and E. Martinsdóttir. 2009. The effect of brining, modified atmosphere and superchilling on the shelf life of cod (*Gadus morhua*) loins. *J. Food Sci.***74**:M258–M267.
25. Magnússon, H., K. Sveinsdóttir, H. L. Lauzon, Á. Thorkelsdóttir, and E. Martinsdóttir. 2006. Keeping quality of desalted cod fillets in consumer packs. *J. Food Sci.* **71**:70–76.

26. Malle, P., and S. H. Tao. 1987. Rapid quantitative determination of trimethylamine using steam distillation.*J. Food Prot.* **50**:756–760.

27. Martinsdottir, E., K. Sveinsdottir, J. B. Luten, R. Schelvis-Smit, and G. Hyldig. 2001. P. 58 *in* Sensory evaluation of fish freshness. Reference manual for the fish sector. QIM-Eurofish, Ijmuiden.

28. Reddy, N. R., and D. J. Armstrong. 1992. Shelf life extension and safety concerns about fresh fish products packed under MA. *J. Food Safety* **12**:87–118.

29. Reddy, N. R., M. Villanueva, and D. A. Kautter. 1995. Shelf life of modified-atmosphere-packaged fresh tilapia fillets stored under refrigeration and temperature-abuse conditions. *J. Food Prot.* **58**:908–914.

30. Simpson, R., S. Almonacid, C. Acevedo, and C. Cortes. 2003. Mathematical model to predict effect of temperature abuse in MAP systems applied to Pacific Hake (*Merluccius australis*). *J. Food Eng.* **26**:413–434.

31. Sivertsvik, M., W. K. Jeksrud, and T. J. Rosnes. 2002. A review of modified atmosphere packaging of fish and fishery products – significance of microbial growth, activities and safety. *J. Food Sci.* **37**:107–127.

32. Sivertsvik, M., J. T. Rosnes, and G. H. Kleiberg. 2003. Effect of modified atmosphere packaging and superchilled storage on the microbial and sensory quality of Atlantic salmon (*Salmo salar*) fillets. *J. Food Sci.* **68**:1467–1472.

33. Soccol, M. C. H., and M. Oetterer. 2003. Use of modified atmosphere in seafood preservation. *J. Braz. Arch. Biol Technol.* **46**:569–580.

34. Stanbridge, L. H., and R. G. Board. 1994. A modification of the *Pseudomonas* selective medium, CFC, that allows differentiation between meat pseudomonas and Enterobacteriaceae. *Lett. Appl. Microbiol.* **18**:327–328.

35. Stansby, M. E., and F. P. Griffiths. 1935. Handling fresh fish. *J. Ind. Eng. Chem.* **27**:1452–1458.

36. Stone, H., and J. L. Sidel. 1985. Sensory evaluation practices. Academic Press Inc., Boca Raton, FL.

37. Sveinsdottir, K., G. Hyldig, E. Martinsdottir, B. Jorgensen, and K. Kristbergsson. 2002. Application of Quality Index Methods (QIM) scheme in shelf life study of farmed Atlantic salmon (*Salmo salar*). *J. Food Sci.***6**:1570–1579.

38. Sveinsdottir, K., E. Martinsdottir, G. Hyldig, and S. Sigurgisladottir. 2010. Sensory characteristics of different cod products. *J. Sensory Stud.* **25**:294–314.

39. The International Organization for Standardization [ISO]. 1988. Pp. 1–9 *in* Sensory analysis – general guidance for the design of test rooms. Standard Nr 8589. The International Organization for Standardization, Geneva.

40. The International Organization for Standardization [ISO]. 1993. Pp. 1–10 *in* Sensory analysis – general guidance for the selection, training and monitoring of assessors. Part I: Selected assessors. Part II: Experts. Standard Nr 8586–1. The International Organization for Standardization, Geneva.

41. Vogel, B. F., K. Venkateswaran, M. Satomi, and L. Gram. 2005. Identification of *Shewanella baltica* as the most important H2S-producing species during iced storage of Danish marine fish. *J. App. Environ. Microbiol.* **71**:6689–6697.

42. Waliszewski, K. N., and M. Avalos. 2001. Shelf life extension of tilapia in crushed ice with additives. *J. Food Proc. Pres.* **25**:299–307.

43. Wang, T., K. Sveinsdottir, H. Magnusson, and E. Martinsdottir. 2008. Combined application of modified atmosphere packaging and superchilled storage to extend shelf life of fresh cod (*Gadus morhua*) loins. *J. Food Sci.* **73**:S11–S19.

CHAPTER 18 INFLUENCE OF GOOD MANUFACTURING PRACTICES ON THE SHELF LIFE OF REFRIGERATED FILLETS OF TILAPIA (OREOCHROMIS NILOTICUS) PACKED IN MODIFIED ATMOSPHERE AND GAMMA-IRRADIATED

Maria Lúcia Guerra Monteiro, Eliane Teixeira Mársico, Sérgio Borges Mano, Claudia Emília Teixeira, Anna Carolina Vilhena da Cruz Silva Canto, Helio de Carvalho Vital, and Carlos Adam Conte-Júnior

INTRODUCTION

Tilapia (*Oreochromis niloticus*) is one of the most important species in aquaculture because of its genetic, reproductive, and marketing characteristics. Although these fish can be raised under controlled conditions, their flesh is a complex biochemical structure that is easily metabolized by microorganisms and is rapidly affected by autolytic enzymes, due to its high water content, near-neutral pH, and predominance of unsaturated fat (Mársico et al. [2006]). Therefore, investigations of alternative methods that could enhance the safety of this important food by preserving its physical, chemical, microbiological, and sensory attributes over longer storage times without significantly altering its nutritional content are desirable.

Modified atmosphere packaging (MAP) has the potential to retard microbial growth, extending the shelf life of foodstuffs (Mano et al. [2000]). Irradiation is highly effective in reducing the initial microbial load, and often prolongs the shelf life of food without significantly changing its sensory characteristics and nutritional value (Vital and Freire [2008]).

Consumers are becoming increasingly concerned with issues related to the nutritional, microbiological, sensory, physical, and chemical quality of fresh foods. However, despite the significant effects of MAP and irradiation on increasing the shelf life of products, the two methods do not reverse food spoilage and do not produce persistent effects. This makes it necessary to adopt good manufacturing practices (GMP) in all stages of processing in order to ensure effective treatment and hygienic quality of food (Lopes et al. [2004]).

Although MAP and irradiation have been extensively studied in order to develop proper procedures for extending the shelf life of tilapia, to our knowledge, no studies have examined the influence of manufacturing practices on the technologies used to increase the shelf life of this fish species.

The goal of this study was to evaluate the effects of GMP on the shelf life of refrigerated fillets of Nile tilapia (*O. niloticus*) treated with irradiation and packed in a modified atmosphere.

Citation: Maria Lúcia Guerra Monteiro, Eliane Teixeira Mársico, Sérgio Borges Mano, Claudia Emília Teixeira, Anna Carolina Vilhena da Cruz Silva Canto, Helio de Carvalho Vital, and Carlos Adam Conte-Júnior, Influence of good manufacturing practices on the shelf life of refrigerated fillets of tilapia (Oreochromis niloticus) packed in modified atmosphere and gamma-irradiated, doi: 10.1002/fsn3.41.

MATERIALS AND METHODS

Fish Samples

In the first step, 120 (25 kg) fresh tilapia fillets were obtained from a commercial fish processing firm in Rio de Janeiro, during its routine functioning, under normal sanitary conditions established by the firm. In the second step, 200 (45 kg) fresh tilapia fillets were obtained from a pilot plant in Rio de Janeiro, which was designed to minimize the operational deficiencies that usually occur in commercial fish processing. The filleting process in the pilot plant was performed according to GMP established by the International Standards Organization – ISO (ISO – International Standards Organization [2005]). The mean weight was 619 (±52) and 622 (±49) g for whole fish and 128 (±9) and 132 (±11) g for filleted fish that were commercially produced and produced in the pilot plant, respectively. The mean weights of the two sets of samples did not differ significantly. Both sets of samples were obtained in the same period (August) in order to avoid any effects of seasonal variations such as ambient temperature, humidity, and precipitation.

The samples were divided into eight lots according to treatment and origin. The commercially produced samples were labeled: T1 (control sample), T2 (MAP – 40/60 co_2/N_2), T3 (gamma radiation – 1.5 kGy), and T4 (MAP and gamma radiation, using the same conditions as treatments 2 and 3). The samples produced in the pilot plant were: T5 (control sample), T6 (MAP – 40/60 CO_2/N_2), T7 (gamma radiation – 1.5 kGy), and T8 (MAP and gamma radiation, using the same conditions as treatments 6 and 7). The samples were transported in isothermal containers with ice (0 ± 1°C) to the laboratory, and the physical, chemical, microbiological, and sensory analyses were performed periodically.

MAP and Gamma Radiation Process

The MAP was performed under laboratory conditions. Each fillet was packaged individually in a Cryovac BB4L bag, in a FAMABRAS (Model TEC MAQ AP 450; Itaquaquecetuba, São Paulo, Brazil) packaging machine (Conte Junior et al. [2010]). The irradiation used a cesium-137 source, in the Centro Tecnológico do Exército (CTEx), located in Guaratiba, Rio de Janeiro. The samples, packed in plastic bags, were placed in the drawers of the irradiator (80 L capacity). Dry ice was placed in the drawers to prevent any excessive temperature increase during the irradiation process. The source activity of the radiation equipment was 45 KCi, where 0.9 kGy/h was used as the maximum dose rate. A computer program was used to calculate the effective doses and irradiation time, taking into account the density, size, and date of the samples. The samples for each treatment were irradiated for 1 h and 40 min.

Physical and Chemical Analyses

The total volatile bases (mg TVB-N/100 g of sample) were determined according to the procedure of Conway, as stated in the manual of Siang and Kim ([1992]).

Microbiological Analysis

Microbiological methods were used to analyze the mesophilic and psychrophilic bacteria content, as recommended by the American Public Health Association (APHA – American Public Health Association[2001]).

Consumer Sensory Testing

Sensory evaluation was performed by the acceptance test with a 9-point hedonic scale, scored from 9 (strongly like) to 1 (strongly dislike), for a group of 30 untrained panelists (Stone and Sidel [1998]). The attributes evaluated were color and general appearance. A score of 5 (neither liked nor disliked) was considered as the indifference region of the affective relationship of the judge to the product. Scores from 6 to 9 were considered as the acceptance region, and scores from 1 to 4 as the rejection region (Pereira et al.[2010]).

Statistical Analysis

The results of the physical and chemical tests were analyzed using a second-degree polynomial regression. The linear regression with the Baranyi equation was applied to the parameters for bacterial development (latency phase and doubling time) of the curves obtained for the bacterial counts (Baranyi and Roberts [1994]). In the sensory evaluation, the means and standard deviation associated with the judgments of the different attributes and treatments were calculated. An analysis of variance and Tukey test, at the 5% significance level, were then applied to these results, using the GraphPad Program InStat 3.0 (Motulsky [1998]).

RESULTS AND DISCUSSION

Microbiological Analysis

Brazilian legislation does not specify the maximum allowed limits for mesophilic and psychrotrophic aerobic heterotrophic bacteria counts in fish products. In this experiment, we established 7.0 log CFU/g as the maximum limit, in accordance with data provided by the International Commission on Microbiological Specifications for Foods ([1988]), because the fillets were rejected by sensory inspection when the bacterial count of the samples exceeded this value.

The mesophilic and psychrophilic bacteria counts (Fig. (Fig.1)1) in the commercially produced fillets were performed after a shorter period of time (hours) because of the rapid growth of the bacterial population in these samples. After 3 h of storage, the bacterial growth in group T1 was 4.2 log CFU/g for mesophilic bacteria, without psychrotrophic growth. Mesophilic (5.5 log CFU/g) and psychrophilic (4.0 log CFU/g) bacterial growth started on the second day. After this storage period, the bacteria progressively multiplied, with a doubling time of 3.6 and 2.9 h, respectively. Before the third and fourth days of storage, the mesophilic and psychrophilic counts were 7.0 log CFU/g, respectively; that is, both counts reached the established limit.

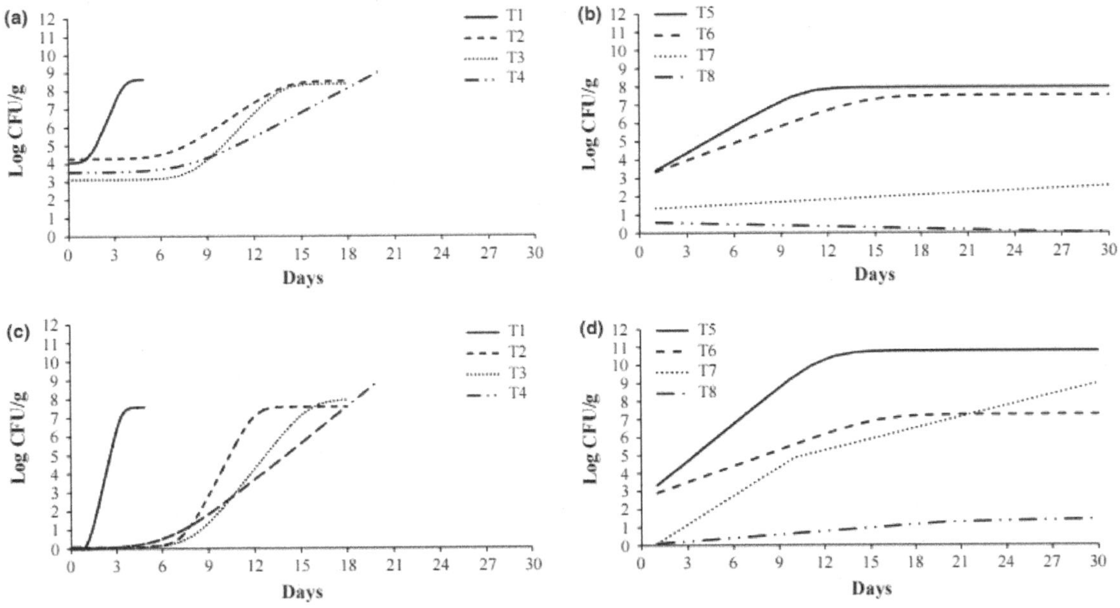

Figure 1: Linear regression of the mesophilic aerobic heterotrophic bacterial counts (a) and (b), and the psychrotrophic aerobic heterotrophic bacterial counts (c) and (d) of the tilapia fillets following treatments T1 (control sample), T2 (modified-atmosphere packaging – 40/60 CO_2/N_2), T3 (gamma radiation – 1.5 kGy), T4 (modified-atmosphere packaging and gamma radiation, using the same conditions as treatments 2 and 3), samples from the fishery cooperative; and T5 (control sample), T6 (modified-atmosphere packaging – 40/60 CO_2/N_2), T7 (gamma radiation – 1.5 kGy) and T8 (modified-atmosphere packaging and gamma radiation, using the same conditions as treatments 6 and 7), samples from the pilot plant.

The initial count of mesophilic and psychrophilic bacteria in the control samples produced in the pilot plant (T5) was 3.6 log CFU/g, and exceeded the limit (7.0 log CFU/g) only on the 12th and eighth days of storage, respectively. As seen in Figure Figure1,1, the initial count of the commercially produced samples was higher than in the samples produced in the pilot plant. Furthermore, the doubling time of mesophilic (3.6 h) and psychrophilic (2.9 h) bacteria of the commercially produced samples was lower than the doubling time of mesophilic (14.4 h) and psychrophilic (10.6 h) bacteria in the samples produced in the pilot plant. Therefore, the commercially produced samples showed significantly greater bacterial growth. This can be explained by the processing conditions in the plants where the samples were obtained. Under appropriate processing conditions, the shelf life of tilapia fillets can be increased by up to 4 days under refrigeration. This is 2.5 times longer than the storage period observed for the commercially produced samples.

The control samples (T1 and T5) showed higher bacterial counts than the samples from the treatments (2, 3, 4, 6, 7, and 8) during the entire storage period. This illustrates the beneficial effect of irradiation and MAP on extending the shelf life of tilapia fillets. However, the sanitary conditions in the processing plants where the samples were obtained may also have influenced the effectiveness of these treatments (Fig.1).

The initial counts of the mesophilic and psychrophilic bacteria in the commercially produced samples with MAP (T2) were 4.2 and 2.9 log CFU/g, while the samples produced in the pilot plant showed 3.4 and 2.5 log CFU/g, respectively. Under controlled sanitary conditions, the bacterial counts were lower than in the commercially produced samples. MAP increased the doubling time of mesophilic and psychrophilic bacteria of the samples from both sites (pilot plant and commercially produced). During the storage period, the bacterial counts obtained in the commercially produced samples with MAP and the samples from the pilot plant (T2 and T6) remained stable, and were lower than the bacterial counts of the respective control samples (T1 and T5). For MAP, the bacterial doubling times were 12.5 h (mesophilic) and 13.7 h (psychrophilic) for the commercially produced samples (T2). These values were lower than the samples produced in the pilot plant (T6), which showed a 1-day doubling time for mesophilic and psychrophilic bacteria. Therefore, during the storage period, the samples produced in the pilot plant showed slower bacterial growth than in the commercially produced samples. This illustrates the importance of GMP, even when conservation methods are used. The mesophilic count of the commercially produced samples exceeded the limit on the 12th day, and the psychrophilic count exceeded the limit on the 15th day. The samples produced in the pilot plant reached the limit after the 15th and 17th days, but the mesophilic and psychrophilic count remained at around 7.2 and 7.5 log CFU/g, respectively, until the end of the experiment (30 days of storage under refrigeration). The MAP had a positive effect on reducing bacterial growth (Mano et al. [2000]), which tripled the shelf life of the samples from both the commercial fish processing firm and the pilot plant. Soccol et al. ([2005]) found that MAP (40% CO_2 and 60% N_2) increased the shelf life of tilapia fillets. However, these authors observed lower bacterial counts (nearly 4.0 log CFU/g on the 20th day of refrigeration storage) compared to our study. This could be explained by the higher CO_2 concentration and, consequently, the greater bacteriostatic effect of CO_2 used by Soccol et al. ([2005]). The MAP effect was mainly observed in the samples produced in the pilot plant, which had a longer storage time than the commercially produced samples. The MAP apparently reduced the bacterial doubling time, especially in the samples produced in the pilot plant, which showed a lower initial count of microorganisms.

With respect to the irradiation method, the mesophilic counts were 2.2 and 1.8 log CFU/g in the commercially produced samples (T3) and the samples produced in the pilot plant (T7), respectively. In other words, the gamma radiation reduced the mesophilic bacteria count by 50%. The commercially produced samples exceeded the limit on the 12th day. The samples produced in the pilot plant did not reach 7.0 log CFU/g during the period of the experiment, because these samples were obtained from a plant with controlled sanitation conditions and showed lower initial counts of microorganisms (Fig. 1). The psychrophilic counts for treatments 3 and 7 reached the limit on the 15th and 23th storage days, respectively. The commercially produced samples showed an 8 h doubling time for mesophilic bacteria, and 11.3 h for psychrophilic bacteria. The bacterial doubling times (mesophilic and psychrophilic) of the samples produced in the pilot plant were 170.4 and 35.0 h, respectively. Irradiation damages the bacterial nucleic acids, reduces the initial count of microorganisms (Vital and Freire [2008]; Monteiro et al. [2010]), and increases the bacterial doubling time. Cozzo-Siqueira et al. ([2003]) found that 1.0, 2.2, and 5.0 kGy decreased the microbe content during storage when compared with nonirradiated samples. In agreement with our findings, these authors observed that the irradiated samples did not reach 7.0 log CFU/g until the 30th day of refrigerated storage.

The gamma radiation tripled the shelf life of the commercially produced tilapia fillets and the fillets produced in the pilot plant, suggesting the importance of GMP, even when conservation methods are used, because controlled sanitation conditions augmented the irradiation effect. According to Gunes et al. ([2012]), lower irradiation doses would likely be sufficient for commercial products, as the contamination levels would be lower than 5–6 log.

The mesophilic and psychrophilic bacteria counts for the samples treated with both MAP and irradiation (T4 and T8) were lower than the bacterial counts for the other treatments during the entire storage period.

The mesophilic and psychrophilic bacteria counts for the commercially produced samples (T4) exceeded the limit around the 16th storage day, and both showed a 15-h bacterial doubling time. The samples produced in the pilot plant (T8) did not reach the limit by the end of the experiment (30 days), and both counts indicated 15 days of bacterial doubling time (Fig. 1). Both the MAP and irradiated samples produced in the pilot plant and those commercially produced (T4 and T8) showed a longer bacterial doubling time than for the samples submitted to other treatments. This probably resulted from the combined effect of gamma radiation, which reduced the initial count of microrganisms, and MAP, which decreased the rate of bacterial growth. The dissolved CO_2 lowers the pH, impeding the development, adaptation, and population growth of irradiation-resistant microorganisms (Monteiro et al. [2012]). On the other hand, the mesophilic and psychrophilic bacterial doubling times of the commercially produced samples were lower than the bacterial doubling time of the samples produced in the pilot plant.

The MAP combined with gamma radiation increased the shelf life of tilapia fillets from the fishery cooperative and pilot plant by four and three times, respectively. Thus, the combined treatment was more effective than the individual treatments, possibly due to a combined effect of the irradiation and MAP. Gunes et al. ([2012]) observed that the use of MAP (50% CO_2 and 50% N_2) combined with irradiation (2 and 4 kGy) did not affect the radiation-induced inactivation of the pathogens, and inhibited the growth of the surviving pathogens and other microorganisms, increasing the shelf life of meatballs during refrigerated storage.

Independently of the treatment, the samples from the pilot plant were higher in quality than the commercially produced samples during the entire storage period (Figs.1,2), because samples produced in the pilot plant deteriorated more slowly due to the lower initial microbial load. This difference in microbial loads suggests that the pilot plant had better sanitation conditions.

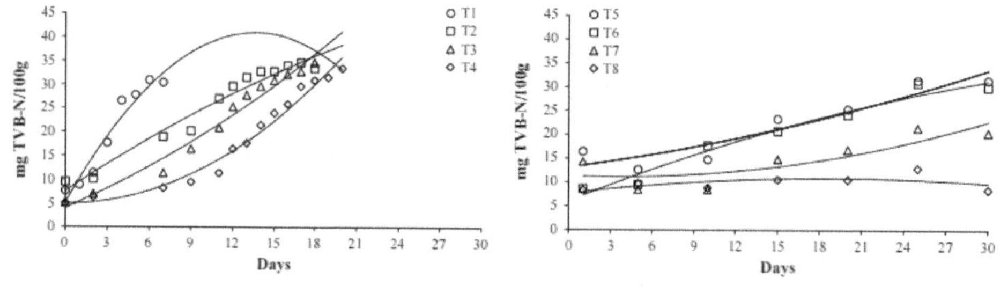

Figure 2: Second-degree polynomial regression of total volatile bases (TVB-N) of tilapia fillets following treatments T1 (control sample), T2 (modified-atmosphere packaging – 40/60 co_2/N_2), T3 (gamma radiation – 1.5 kGy), T4 (modified-atmosphere packaging and gamma radiation, using the same conditions as treatments 2 and 3), samples from the fishery cooperative; and T5 (control sample), T6 (modified-atmosphere packaging – 40/60 CO_2/N_2), T7 (gamma radiation – 1.5 kGy) and T8 (modified-atmosphere packaging and gamma radiation, using the same conditions as treatments 6 and 7), samples from the pilot plant.

The filleting process was considered the critical control point in an evaluation of the hazard analysis, and also the critical control point in the processing of tilapia fillets. To ensure the longest possible shelf life, it is essential to use preventive measures such as personnel training (Carbonera et al. [2006]). In all the processing steps, the adoption of GMP is essential, because conservation methods such as MAP and irradiation cannot reverse the deterioration process (Satin [2002]).

Physical and Chemical Analyses

The TVB-N results consistently indicated an increasing deterioration of the fillets with storage time in all treatments, demonstrating a progressive reduction in the quality of the samples during the storage period (Fig.2). However, the samples treated with irradiation, MAP, or a combination of both (2, 3, 4, 6, 7, and 8) deteriorated more slowly and had longer shelf lives in comparison with the control samples, demonstrating the positive effects of MAP, gamma radiation, and the two conservation methods combined.

Brazilian legislation (Ministry of Agriculture, Livestock and Supply [1997]) does not define the allowed limits for the total volatile base (TVB) in freshwater fish. Therefore, it is necessary to correlate the physical, chemical, bacteriological, and sensory results to provide a basis for suggesting official limits for this food. In this study, the limit of 19 mg of TVB-N/100 g was adopted, as recommended by Soccol et al. ([2005]).

The TVB of the commercially produced control samples (T1) was 17.64 mg of TVB-N/100 g on the third storage day. The control samples produced in the pilot plant (T5) contained 13.86 mg of TVB-N/100 g on the eighth day under refrigeration. In this period, the sensory evaluation correlated with the other results of this experiment, and the commercially produced tilapia fillets showed undesirable sensory characteristics for consumption. The samples produced in the pilot plant gave better results than the commercially produced samples, because the samples were obtained under controlled sanitary conditions and had low initial counts of microorganisms.

The commercially produced samples with MAP, irradiation, and both conservation methods combined (T2, T3, and T4) showed 20.16, 20.79, and 21.42 mg TVB-N/100 g, exceeding the limit established in this study on the ninth, 11th, and 14th storage days, respectively. In this period, the bacteria count of the tilapia fillets exceeded the limit (7.0 log CFU/g), and the panel judges described a product with poor quality. In contrast, the samples produced in the pilot plant and submitted to the same treatments showed 20.70 mg of TVB-N/100 g (T6) and 21.42 mg of TVB-N/100 g (T7) on the 15th and 25th storage days, respectively. On the 30th storage day, samples produced in the pilot plant (T8) showed 8.44 mg of TVB-N/100 g. The bacterial counts for these samples did not reach the limit (7.0 log CFU/g), and the panel judges described no inappropriate sensory characteristics.

Cozzo-Siqueira et al. ([2003]) reported that doses of irradiation up to 5 kGy did not affect proteins (amino acids) and lipids (fatty acids) in tilapia muscle during 30 days of storage under refrigeration. However, MAP and irradiation can cause biochemical changes in tilapia fillets. The TVB showed a poor correlation with the other results of this study, and therefore was an inappropriate quality evaluation parameter for the refrigerated, modified atmosphere-packaged and irradiated fillets.

Consumer Sensory Testing

Tables 1 and 2 show the consumer sensory scores of the commercially produced samples (T1, T2, T3, and T4) and the samples produced in the pilot plant (T5, T6, T7, and T8). The control samples (T1 and T5) showed mean scores that were intermediate between "strongly dislike" and "neither like or dislike," on the fifth and 20th storage days for general appearance and color, respectively. This indicated that the judges perceived an indifferent sample quality, followed by a tendency to reject the product (Pereira et al. [2010]). The commercially produced control samples (T1) deteriorated rapidly, which was reflected in the rapid development of undesirable sensory characteristics. The control samples produced in the pilot plant (T5) were judged acceptable for a period four times longer than the commercially produced control samples (T1).

Table 1: Consumer sensory scores of tilapia (*Oreochromis niloticus*) fillets commercially produced

	Attribute							
	Appearance				Color			
Days	T1	T2	T3	T4	T1	T2	T3	T4
0	5.90^a	6.20^a	7.00^a	6.23^a	5.96^a	5.53^a	6.76^a	6.26^a
2	6.06^a	5.75^a	6.46^a	5.90^a	5.73^a	5.70^a	6.30^a	5.96^a
5	2.63^b	5.70^a	6.33^a	6.20^a	2.66^b	5.50^a	6.23^a	5.86^a
8	ANP	6.00^a	6.10^a	5.00^a	ANP	4.76^a	6.06^a	5.26^a
12	ANP	4.73^a	5.40^a	5.33^a	ANP	4.33^a	5.13^a	5.20^a
15	ANP	3.33^b	2.83^b	5.10^a	ANP	3.00^b	2.33^b	5.06^a
19	ANP	ANP	ANP	4.16^b	ANP	ANP	ANP	4.13^b

T1 (control sample), T2 (MAP – 40/60 CO_2/N_2), T3 (gamma radiation – 1.5 kGy), T4 (MAP and gamma radiation, using the same conditions of the treatment 2 and 3), samples produced in the pilot plant.
ANP: Analysis not performed, because the others parameters evaluated showed undesirable sensory characteristics of consumption.
Different letters in the same line indicate significant differences ($P < 0.05$).

Table 2: Consumer sensory scores of tilapia (*Oreochromis niloticus*) fillets produced in the pilot plant

	Attribute							
	Appearance				Color			
Days	T5	T6	T7	T8	T5	T6	T7	T8
1	6.76^a	4.26^b	5.73^a	3.96^b	6.53^a	4.30^b	5.66^a	3.83^b
10	6.63^a	3.20^c	4.90^b	3.26^c	6.73^a	3.60^c	4.96^b	3.40^c
20	5.63^a	3.33^b	5.20^a	4.36^{ab}	5.53^a	3.43^c	5.00^{ab}	4.00^{bc}

T5 (control sample), T6 (MAP – 40/60 CO_2/N_2), T7 (gamma radiation – 1.5 kGy), T8 (MAP and gamma radiation, using the same conditions of the treatment 2 and 3), samples produced in the pilot plant.
Different letters in the same line indicate significant differences ($P < 0.05$).

On the first day, the treated samples (2, 3, and 4) received higher scores than the control samples (T1), but there was no significant difference ($P > 0.05$). The irradiated samples produced in the pilot plant (T7) received lower scores than the commercially produced control samples (T5), and again, the scores were not significantly different. Therefore, initially, irradiation did not affect the sensory quality of the tilapia fillets. In agreement with our findings, Ózden et al. ([2007]) reported that acceptability scores for texture and odor of sea bass were not affected by irradiation doses up to 5 kGy, but these scores decreased with longer periods of refrigerated storage. On the other hand, the modified atmosphere-packaged samples (T6) and the samples treated with both conservation methods (T8) differed significantly ($P < 0.05$) from the control samples (T5). This can be explained by the absence of oxygen and presence of CO_2, which causes myoglobin deoxygenation and consequently discoloration of the tilapia fillets (Soccol and Oetterer [2003]).

Modified atmosphere packaging slows bacterial multiplication (Mano et al. [2000]) and gamma radiation reduces the initial count of microrganisms (Vital and Freire [2008]). This delays the appearance of undesirable sensory changes and extends the shelf life of products. Therefore, the modified atmosphere-packaged and irradiated samples, and those treated with both conservation methods showed desirable quality characteristics for longer periods than the control samples. However, during the sensory analysis, the modified atmosphere-packaged samples showed "drip," flaccid texture, and pale color, all critical attributes for the judges' evaluation. Meats packaged in a modified atmosphere with CO_2 shows increased exudation, because the CO_2 dissolved on the surface lowers the pH. Consequently, the water-retention capacity of the proteins decreases, which causes the increase in "drip," flaccid texture, and pale color in the package (Soccol and Oetterer [2003]). In addition, tilapia is considered a thin fish and has a large amount of water in its composition, which tends to produce exudation and affected the other attributes evaluated in this experiment (general appearance and color).

Modified atmosphere packaging combined with irradiation (T4) did not significantly influence ($P > 0.05$) the general appearance and color of the fillets by the 15th storage day, and maintained desirable sensory characteristics for a longer period. This suggests that the irradiation minimized the negative effects of MAP (dripping, flaccid texture, and pale color) (Monteiro et al. [2012]).

CONCLUSIONS

The control samples produced in the pilot plant maintained their quality over a longer storage period than the commercially produced control samples, demonstrating that GMP during the filleting process can increase the shelf life of refrigerated tilapia fillets. Furthermore, the conservation methods used (MAP, irradiation, and both methods combined) gave better results for the samples produced in the pilot plant than for the commercially produced samples during the entire storage period. Therefore, the adoption of GMP before, during, and after the filleting process is essential to obtain a product with a longer shelf life, even when using conservation methods that are effective, but have no persistent effect and do not reverse the deterioration process of foods.

ACKNOWLEDGMENTS

The authors are grateful for the financial support provided by Fundação Carlos Chagas Filho de Amparo à Pesquisa do Estado do Rio de Janeiro, process number E-26/111.196/2011. M. L. G. Monteiro and C. E. Teixeira were supported by the Conselho Nacional de Desenvolvimento Científico e Tecnológico and Coordenação de Aperfeiçoamento de Pessoal de Nível Superior, process numbers 551079/2011-8 and 23038.007363/2011-13, respectively. The Cooperative and the university, located in Rio de Janeiro, are gratefully acknowledged for providing samples. Janet W. Reid revised the English text.

Conflict of Interest

None declared.

REFERENCES

1. APHA – American Public Health Association. Compendium of methods for the microbiological examination of foods. 4th ed. Washington, DC: Editora APHA; 2001. p. 676.
2. Baranyi J, Roberts TA. A dynamic approach to predicting bacterial growth in food. Int. J. Food Microbiol. 1994;23:277–294.
3. Carbonera N, Mirapalheta T, Vivian VB, Espírito Santo MLP. 2006. Análise de Perigos e Pontos Críticos de Controle no Processamento de Filés Congelados de Tilápia (Oreochromis niloticus). Available athttp://200.169.53.89/download/CD%20 congressos/2006/CRICTE%202006/trabalhos/657672-ega-01-09-62134.pdf (accessed 15 July 2012)
4. Conte Junior CA, Peixoto BTM, Lopes MM, Franco RM, Freitas MQ, Fernández M, et al. Effect of modified atmosphere packaging on the growth/survival of *Yersinia enterocolitica* and natural flora on fresh poultry sausage. In: Méndez Vilas A, editor. Current research, technology and education topics in applied microbiology and microbial biotechnology. Badajoz: Formatex; 2010. pp. 1217–1223. (Org.).
5. Cozzo-Siqueira A, Oetterer M, Gallo CR. Effects of irradiation and refrigeration on the nutrients and shelf-life of tilapia (*Oreochromis niloticus*. J. Aquat. Food Prod. Technol. 2003;12:85–101.
6. Gunes G, Yilmaz N, Ozturk A. Effects of irradiation dose and O_2 and co_2 concentrations in packages on foodborne pathogenic bacteria and quality of ready-to-cook seasoned ground beef product (meatball) during refrigerated storage. Sci. World J. 2012;2012:274219. doi: 10.1100/2012/274219. Available at http://www.ncbi.nlm.nih.gov/pmc/articles/PMC3329864/ (accessed 15 March 2013)
7. ICMSF – International Commission on Microbiological Specifications for Foods. Microorganisms in foods: 4. Application of hazard analysis critical point (HACCP) system to ensure microbiological safety and quality. London: Blackwell; 1988.
8. ISO – International Standards Organization. Geneva, Switzerland: International Standards Organization; 2005. ISO/TS 22004: Food safety management systems – Guidance on the application of ISO 22000.
9. Lopes MM, Mársico ET, Sobreiro LG, Silva LP, Conte-Junior CA, Pardi HS, et al. Efeito da embalagem em atmosfera modificada sobre a conservação de

sardinhas (*Sardinella brasiliensis*. Rev. Port. Ciênc. Vet. 2004;99:207–210.

10. Mano SB, Ordoñez JA, Fernando GDG. Growth/survival of natural flora and *Aeromonas hydrophila*on refrigerated uncooked pork and turkey packaged in modified atmospheres. Food Microbiol.2000;17:657–669.
11. Mársico ET, Oliveira CM, Ferreira PV, Antunes L, Sobreiro LG. Avaliação da qualidade de sushis e sashimis comercializados em shoppings centers. Hig. Alimentar. 2006;20:63–65.
12. Ministry of Agriculture, Livestock and Supply. Pecuária e Abastecimento, 6: Brasília: Ministério da Agricultura; 1997. Regulamento Técnico de Identidade e Qualidade de Peixe Fresco (Inteiro e Eviscerado) approved by Portaria n° 185 de 13 de maio de 1997.
13. Monteiro MLG, Mársico ET, Vital HC. Avaliação físico-química dos efeitos da irradiação e da evisceração na conservação de atum (*Thunnus atlanticus*) refrigerado. Rev. Port. Ciênc. Vet.2010;109:45–48.
14. Monteiro MLG, Mársico ET, Teixeira CE, Mano SB, Conte-Júnior CA, Vital HC. Validade comercial de filés de Tilápia do Nilo (*Oreochromis niloticus*) resfriados embalados em atmosfera modificada e irradiados. Ciênc. Rural. 2012;42:737–743.
15. Motulsky HJ. InStat 3.0 Program. InStat guide to choosing and interpreting statistical tests. San Diego, CA: GraphPad Software, Inc; 1998. Available at http://www.graphpad.com (accessed 21 May 2012)
16. Özden O, İnugur M, Erkan N. Effect of different dose gamma radiation and refrigeration on the chemical and sensory properties and microbiological status of aquacultured sea bass (*Dicentrarchus labrax*. Radiat. Phys. Chem. 2007;76:169–178.
17. Pereira C, Schmidt CAP, Ishihara RY, Koelln FTDS, Scussel VM. Québec City, Canada: Canadian Society for Bioengineering (CSBE/SCGAB); 2010. XVIIth World Congress of the International Commission of Agricultural and Biosystems Engineering (CIGR), 13–17 June. Available athttp://www.csbe-scgab.ca/docs/meetings/2010/CSBE101536.pdf (accessed 13 March 2013)
18. Satin M. Use of irradiation for microbial decontamination of meat: situation and perspectives. Meat Sci. 2002;62:277–283.
19. Siang NC, Kim LL. Determination of trimethylamine oxide, trimethylamine and total volatile basic nitrogen by Conway's micro-diffusion method. In: Miwa K, Ji LS, editors. Laboratory manual on analytical methods and procedures for fish and fisheries products. Kuala Lumpur: South East Asia Fisheries Development Center; 1992. pp. 3.1–3.6.
20. Soccol MCH, Oetterer M. Use of modified atmosphere in seafood preservation. Braz. Arch. Biol. Technol. 2003;46:569–580.
21. Soccol MCH, Oetterer M, Gallo CR, Spoto MHF, Biato DO. Effects of modified atmosphere and vacuum on the shelf life of tilapia (*Oreochromis niloticus*) fillets. Braz. J. Food Technol. 2005;8:7–15.
22. Stone H, Sidel JL. Quantitative descriptive analysis: developments, applications, and the future. Food Technol. 1998;5:48–52.
23. Vital HC, Freire M., Jr . A irradiação de alimentos. In: Rosenthal A, editor. Tecnologia de Alimentos e Inovação: Tendências e Perspectivas. Brasília, DF: Embrapa Informação Tecnológica; 2008. pp. 149–164. (Org.).

INDEX